Unifying Themes in Complex Systems

Volume IIIA

Overview

Springer Complexity

Springer Complexity is a publication program, cutting across all traditional disciplines of sciences as well as engineering, economics, medicine, psychology and computer sciences, which is aimed at researchers, students and practitioners working in the field of complex systems. Complex Systems are systems that comprise many interacting parts with the ability to generate a new quality of macroscopic collective behavior through self-organization, e.g., the spontaneous formation of temporal, spatial or functional structures. This recognition, that the collective behavior of the whole system cannot be simply inferred from the understanding of the behavior of the individual components, has led to various new concepts and sophisticated tools of complexity. The main concepts and tools – with sometimes overlapping contents and methodologies – are the theories of self-organization, complex systems, synergetics, dynamical systems, turbulence, catastrophes, instabilities, nonlinearity, stochastic processes, chaos, neural networks, cellular automata, adaptive systems, and genetic algorithms.

The topics treated within Springer Complexity are as diverse as lasers or fluids in physics, machine cutting phenomena of workpieces or electric circuits with feedback in engineering, growth of crystals or pattern formation in chemistry, morphogenesis in biology, brain function in neurology, behavior of stock exchange rates in economics, or the formation of public opinion in sociology. All these seemingly quite different kinds of structure formation have a number of important features and underlying structures in common. These deep structural similarities can be exploited to transfer analytical methods and understanding from one field to another. The Springer Complexity program therefore seeks to foster cross-fertilization between the disciplines and a dialogue between theoreticians and experimentalists for a deeper understanding of the general structure and behavior of complex systems.

The program consists of individual books, books series such as "Springer Series in Synergetics", "Institute of Nonlinear Science", "Physics of Neural Networks", and "Understanding Complex Systems", as well as various journals.

New England Complex Systems Institute

President
Yaneer Bar-Yam
New England Complex Systems Institute
24 Mt. Auburn St.
Cambridge, MA 02138, USA

For over 10 years, The New England Complex Systems Institute (NECSI) has been instrumental in the development of complex systems science and its applications. NECSI conducts research, education, know-ledge dissemination, and community development around the world for the promotion of the study of complex systems and its application for the betterment of society.

NECSI was founded by faculty of New England area academic institutions in 1996 to further international research and understanding of complex systems. Complex systems is a growing field of science that aims to understand how parts of a system give rise to the system's collective behaviors, and how it interacts with its environment. These questions can be studied in general, and they are also relevant to all traditional fields of science.

Social systems formed (in part) out of people, the brain formed out of neurons, molecules formed out of atoms, and the weather formed from air flows are all examples of complex systems. The field of complex systems intersects all traditional disciplines of physical, biological and social sciences, as well as engineering, management, and medicine. Advanced education in complex systems attracts professionals, as complex systems science provides practical approaches to health care, social networks, ethnic violence, marketing, military conflict, education, systems engineering, international development and terrorism.

The study of complex systems is about understanding indirect effects. Problems we find difficult to solve have causes and effects that are not obviously related. Pushing on a complex system "here" often has effects "over there" because the parts are interdependent. This has become more and more apparent in our efforts to solve societal problems or avoid ecological disasters caused by our own actions. The field of complex systems provides a number of sophisticated tools, some of them conceptual helping us think about these systems, some of them analytical for studying these systems in greater depth, and some of them computer based for describing, modeling or simulating them.

NECSI research develops basic concepts and formal approaches as well as their applications to real world problems. Contributions of NECSI researchers include studies of networks, agent-based modeling, multiscale analysis and complexity, chaos and predictability, evolution, ecology, biodiversity, altruism, systems biology, cellular response, health care, systems engineering, negotiation, military conflict, ethnic violence, and international development.

NECSI uses many modes of education to further the investigation of complex systems. Throughout the year, classes, seminars, conferences and other programs assist students and professionals alike in their understanding of complex systems. Courses have been taught all over the world: Australia, Canada, China, Colombia, France, Italy, Japan, Korea, Portugal, Russia and many states of the U.S. NECSI also sponsors postdoctoral fellows, provides research resources, and hosts the International Conference on Complex Systems, discussion groups and web resources.

New England Complex Systems Institute
Book Series

Series Editor

Dan Braha
New England Complex Systems Institute
24 Mt. Auburn St.
Cambridge, MA 02138, USA

New England Complex Systems Institute Book Series

The world around is full of the wonderful interplay of relationships and emergent behaviors. The beautiful and mysterious way that atoms form biological and social systems inspires us to new efforts in science. As our society becomes more concerned with how people are connected to each other than how they work independently, so science has become interested in the nature of relationships and relatedness. Through relationships elements act together to become systems, and systems achieve function and purpose. The study of complex systems is remarkable in the closeness of basic ideas and practical implications. Advances in our understanding of complex systems give new opportunities for insight in science and improvement of society. This is manifest in the relevance to engineering, medicine, management and education. We devote this book series to the communication of recent advances and reviews of revolutionary ideas and their application to practical concerns.

Unifying Themes in Complex Systems

Overview

Volume IIIA

*Proceedings from the Third International
Conference on Complex Systems*

Edited by Ali Minai and Yaneer Bar-Yam

Ali A. Minai

Univeristy of Cincinnati
Department of Electrical and
Computer Engineering, and Computer Science
P.O. Box 210030, Rhodes Hall 814
Cincinnati, OH 45221-0030
USA
Email: Ali.Minai@uc.edu

Yaneer Bar-Yam

New England Complex Systems Institute
24 Mt. Auburn St.
Cambridge, MA 02138-3068
USA
Email : yaneer@necsi.org

This volume is part of the
New England Complex Systems Institute Series on Complexity

ISBN-3-540-35870-6 Springer Berlin Heidelberg New York

Library of Congress Control Number: 2006928841

Springer is a part of Springer Science+Business Media
springer.com
© NECSI Cambridge, Massachusetts 2006
Printed in the USA

CONTENTS

INTRODUCTION

The mysteries of highly complex systems that have puzzled scientists for years are finally beginning to unravel thanks to new analytical and simulation methods. Better understanding of concepts like complexity, emergence, evolution, adaptation and self-organization have shown that seemingly unrelated disciplines have more in common than we thought. These fundamental insights require interdisciplinary collaboration that usually does not occur between academic departments. This was the vision behind the first International Conference on Complex Systems in 1997; not just to present research, but to introduce new perspectives and foster collaborations that would yield research in the future.

As more and more scientists began to realize the importance of exploring the unifying principles that govern all complex systems, the Third ICCS attracted a diverse group of participants representing a wide variety of disciplines. Topics ranged from economics to ecology, particle physics to psychology, and business to biology. Through pedagogical, breakout and poster sessions, conference attendees shared discoveries that were significant both to their particular field of interest, as well as the general study of complex systems. These volumes contain the proceedings from that conference.

Even with the third ICCS, the science of complex systems is still in its infancy. In order for complex systems science to fulfill its potential to provide a unifying framework for various disciplines, it is essential to establish a standard set of conventions to facilitate communication. This is another valuable function of the conference; it allowed an opportunity to develop a common foundation and language for the study of complex systems.

These efforts have produced a variety of new analytic and simulation techniques that have proven invaluable in the study of physical, biological and social systems. New methods of statistical analysis led to better understanding of polymer formation and complex fluid dynamics; further development of these methods has deepened our understanding of patterns and networks. The application of simulation techniques such as agent-based models, cellular automata, and Monte Carlo simulations to complex systems has increased our ability to understand or even predict behavior of systems that once seemed completely unpredictable.

The concepts and tools of complex systems are of interest not only to scientists, but to corporate managers, doctors, political scientists and policy makers. The same rules that govern neural networks apply to social or corporate networks, and professionals have started to realize how valuable these concepts are to their individual fields. The ICCS conferences have provided the opportunity for professionals to learn the basics of complex systems and share their real-world experience in applying these concepts.

Third International Conference on Complex Systems: Organization and Program

Organization:

Host:

New England Complex Systems Institute

Partial financial support:

National Science Foundation
Perseus Books
Harvard University Press

Conference Chair:

*   Yaneer Bar-Yam - NECSI

Executive Committee:

*   Larry Rudolph - MIT
†   Ali Minai - University of Cincinnati

Organizing Committee:

Philip W. Anderson - Princeton University
Kenneth J. Arrow - Stanford University
* Michel Baranger - MIT
Per Bak - Niels Bohr Institute
Charles H. Bennett - IBM
William A. Brock - University of Wisconsin
* Charles R. Cantor - Boston University
Noam A. Chomsky - MIT
Leon Cooper - Brown University
Daniel Dennett - Tufts University
* Irving Epstein - Brandeis University
Michael S. Gazzaniga - Dartmouth College
* William Gelbart - Harvard University
Murray Gell-Mann - CalTech/Santa Fe Institute
Pierre-Gilles de Gennes - ESPCI
Stephen Grossberg - Boston University
Michael Hammer - Hammer & Co
John Holland - University of Michigan
John Hopfield - Princeton University
* Jerome Kagan - Harvard University
Stuart A. Kauffman - Santa Fe Institute
Chris Langton - Santa Fe Institute
Roger Lewin - Harvard University
Richard C. Lewontin - Harvard University
Albert J. Libchaber - Rockefeller University
* Seth Lloyd - MIT
Andrew W. Lo - MIT
Daniel W. McShea - Duke University
Marvin Minsky - MIT
Harold J. Morowitz - George Mason University
Alan Perelson - Los Alamos National Lab
Claudio Rebbi - Boston University
Herbert A. Simon - Carnegie-Mellon University
* Temple F. Smith - Boston University
H. Eugene Stanley - Boston University
* John Sterman - MIT
* James H. Stock - Harvard University
Gerald J. Sussman - MIT
Edward O. Wilson - Harvard University
Shuguang Zhang - MIT

Session Chairs:

   Dan Stein - University of Arizona
   Jeffrey Robbins - Addison-Wesley
* Yaneer Bar-Yam - NECSI
   Steve Lansing - University of Arizona
   David Litster - MIT
* Irving Epstein - Brandeis University
   Richard Bagley - Digital Equipment Corporation
   Yasha Kresh - Drexel University
   Tim Keitt - SUNY Stony Brook
* Les Kaufman - Boston University
   Mark Bedau - Reed College
† Dan Braha - Ben-Gurion University
   Dan Frey - MIT
   Sean Rice - Yale University
   Max Garzon - University of Memphis
   Bob Savit - University of Michigan
* Larry Rudolph - MIT
   Jerry Chandler - George Mason University
   Richard Cohen - MIT
   Kosta Tsipis - MIT
   Walter Willinger - AT&T Bell Laboratories
† Helen Harte - NECSI
   Farrell Jorgensen - Kaiser Permanente
   Joel MacAuslan - Speech Technology and Applied Research
   Anjeli Sastry - MIT
   Walter Freeman - UC Berkeley
† Ali Minai - University of Cincinnati
   Michael Jacobson - University of Georgia
   William Fulkerson - Deere & Company
* Tom Petzinger - Wall Street Journal

* NECSI Co-faculty
† NECSI Affiliate

Subject areas: Unifying themes in complex systems

The themes are:

EMERGENCE, STRUCTURE AND FUNCTION: substructure, the relationship of component to collective behavior, the relationship of internal structure to external influence.

INFORMATICS: structuring, storing, accessing, and distributing information describing complex systems.

COMPLEXITY: characterizing the amount of information necessary to describe complex systems, and the dynamics of this information.

DYNAMICS: time series analysis and prediction, chaos, temporal correlations, the time scale of dynamic processes.

SELF-ORGANIZATION: pattern formation, evolution, development and adaptation.

The system categories are:

FUNDAMENTALS, PHYSICAL & CHEMICAL SYSTEMS: spatio-temporal patterns and chaos, fractals, dynamic scaling, non-equilibrium processes, hydrodynamics, glasses, non-linear chemical dynamics, complex fluids, molecular self-organization, information and computation in physical systems.

BIO-MOLECULAR & CELLULAR SYSTEMS: protein and DNA folding, bio-molecular informatics, membranes, cellular response and communication, genetic regulation, gene-cytoplasm interactions, development, cellular differentiation, primitive multicellular organisms, the immune system.

PHYSIOLOGICAL SYSTEMS: nervous system, neuro-muscular control, neural network models of brain, cognition, psychofunction, pattern recognition, man-machine interactions.

ORGANISMS AND POPULATIONS: population biology, ecosystems, ecology.

HUMAN SOCIAL AND ECONOMIC SYSTEMS: corporate and social structures, markets, the global economy, the Internet.

ENGINEERED SYSTEMS: product and product manufacturing, nano-technology, modified and hybrid biological organisms, computer based interactive systems, agents, artificial life, artificial intelligence, and robots.

Program:

Sunday, May 21, 2000

PEDAGOGICAL SESSION - **David Meyer** - Session Chair

    George Cowan - Complexity: Past and Future

    Michel Baranger - Physics and the Complexity Revolution

    Atlee Jackson - Unifying Principles

    Ronnie Mainieri - Dynamical Systems

    Mitchell Feigenbaum - Chaos

    Robert Berwick - Language

    Don Ingber - Biomedicine

RECEPTION SESSION

    Edward Lorenz - Climate

Monday, May 22, 2000

    Yaneer Bar-Yam - Welcome

EMERGENCE - **Michel Baranger** - Session Chair

    Stuart Kauffman - Emergence [Herbert A. Simon Award Lecture]

    Eugene Stanley - Correlations and Dynamics

    Simon Levin - The Ecology and Evolution of Commmunties

    Dave Clark - Emergent Dynamics of the Internet

DESCRIPTION AND MODELING - **Jack Cohen** - Session Chair

    Greg Chaitin - Fundamentals of Mathematics

    Per Bak - Self-organization

    Kathleen Carley - Agents in Societies

HIGH DENSITY PARALLEL SESSIONS

    Thread A

POSTER SESSION

Ilija Dukovski - Invaded Cluster Monte Carlo Algorithm for Critical Points with Continuous Symmetry Breaking

J. Bhattacharya, R. Kariyappa, E. Pereda, & P. P. Kanjilal - Application of Nonlinear Analysis to Intensity Oscillations of the Chromospheric Bright Points

Tuesday, May 23, 2000

SELF-ORGANIZATION - David Campbell - Session Chair

> **George Whitesides** - Complex Chemical Systems
>
> **Irv Epstein** - From Nonlinear Chemistry to Biology
>
> **Chris Adami** - Artificial Life
>
> **Duncan Watts** - Small World Networks

AFTERNOON PARALLEL BREAKOUT SESSIONS

> **John Symons** - Philosophy of Brain/Mind
>
> > **John Bickle**
> > **Bill Bechtel**
> > **Alfredo Pereira**
> > **Joao Teixera**
> > **Alex Rueger**
>
> **Irv Epstein** - Evolution
>
> > **Jack Cohen** - Evolution is Complex ...
> >
> > **Homayoun Bagheri-Chaichian** - Evolvability of Multi-Enzyme Systems
> >
> > **J. Peter Gogarten, W. Ford Doolittle, & Lorraine Olendzenski** - Does HGT Shape Microbial Taxonomy?
> >
> > **Yukihiko Toquenaga** - Critical States of Fitness Landscapes
> >
> > **John Pepper** - Positive assortment among cooperators through environmental feedback
>
> **David Meyer & Ronnie Mainieri** - Spatio-Temporal Patterns
>
> > **Alfred Hubler** - Adaptation to the Edge of Chaos of Self-Adjusting Dynamical Systems
> >
> > **Andreas Rechtseiner & Andrew M. Fraser** - Hidden States for Modeling Interactions Between Disparate Spatiotemporal Scales

S. J. Nasuto, J. L. Krichmar, R. Scorcioni, & G. A. Ascoli - Algorithmic Statistical Analysis of Electrophysiological Data for the Investigation of Structure-Activity Relationship in Single Neurons

Peter Cariani - Emergence of a Perceptual Gestalt: Neural Correlates of the Pitch of the "Missing Fundamental"

Jeff Stock - Ecology

Madhur Anand - Ecological Communities: More than the Sum of their Parts

Vasyl Gafiychuk & I. A. Lubashevsky - Synergetic Self-Regulation in Complex Hierarchical Systems

Guy Hoelzer - The Self-Organization of Population Structure in Biological Systems

Hiroki Sayama, Les Kaufman & Yaneer Bar-Yam - The Role of Spontaneous Pattern Formation in the Creation and Maintenance of Biological Diversity

Michael Hauhs, Holger Lange, & Alois Kastner-Maresch - Computer-Aided Managing of Ecosystems: The Case of Forestry

Michael Jacobson - Business & Management

Bob Wiebe

Linda Testa

Pierpaolo Andriani - Complexity, Knowledge Creation, and Distributed Intelligence in Industrial Clusters

Harold E. Klein - Representation of the Strategic Organization Environment as a Complex System

Luis Mateus Rocha - A Complex Systems Approach to Knowledge Management

Phillip Auerswald - The Complexity of Production and Inter-Industry Difference in the Persistence of Profits Above the Norm

Ted Fuller & Paul Moran - Thinking for Organizational Learning Complexity as a Social Science Methodology in Understanding the Impact of Exogenous Systemic Change on Small Business

Wednesday, May 24, 2000

COMPLEX ENGINEERED SYSTEMS - Dan Braha - Session Chair

Nam Suh - Complexity and Design Engineering

Steven Eppinger - Product Development Complexity

Bill Mitchell - Complexity in Architecture

Michael Caramanis - Scale Decomposition of Production

AFTERNOON PARALLEL SESSIONS

Thread A

Education

Lynn Andrea Stein - Changing Educational Concepts in Computer Science

Uri Wilensky - Using networked handheld devices to enable participatory simulations:of complex systems

Robert Tinker - Emergence in Precollege Education

Karen Vanderven & Carlos Antonio Torre - Towards Transforming Education: Applications of Complexity Theory

Len R. Troncale - Stealth Systems Science at All Universities: Integrated Science General Education.

Art

Igor Yevin - Complexity Theory of Art and Traditional Study of Art

Jack Ox - A Complex System for the Visualization of Music, Including the Journey from 2D to Virtual Reality

Thread B

Particle Physics Special Talk

Claudio Rebbi - Multiple Scales in Particle Physics

Socio-Economic Systems Special Talk

Lionel Sacks - A complexity analysis of Integrity in Telephony Networks

Applications

Fred M. Discenzo - Intelligent Devices Enable Enhanced Modeling and Control of Complex Real-Time Systems

Funding

Mariann Jelinek - NSF

James J. Anderson - NIH

Eric Hamilton & Anthony E. Kelly - NSF

Thread C

Pattern Formation

Mark Kimura, Yuri Mansury, Thomas S. Deisboeck, & Jose Lobo - A Model of Spatial Agglomeration

Friday, May 26, 2000

MEDICAL COMPLEXITY - **Thomas Deisboeck** / **Yasha Kresh** - Session Chairs

> **Clay Easterly** - The Virtual Human
>
> **Alan Perelson** - Theory of the Immune System
>
> **Jim Collins** - Dynamics in Multiscale Biology
>
> **Timothy Buchman** - Multiorgan Failure
>
> **Ary Goldberger** - Fractal Mechanisms and Complex Dynamics in Health, Aging and Disease
>
> **Stephen Small** - Medical Errors
>
> **Mark Smith** - Medical Management

Publications:

Proceedings:

Conference proceedings (this volume)
Video proceedings are available to be ordered through the New England Complex Systems Institute.

Journal articles:

Individual conference articles were published through the refereed on-line journal InterJournal and are available on-line (http://interjournal.org/) as manuscripts numbered 217-271.

Other products:

An active email discussion group has resulted from the conference. Access and archives are available through links from http://necsi.org/.

Web pages:

http://necsi.org/
Home page of the New England Complex Systems Institute with links to the conference pages.
http://necsi.org/html/iccs2.html
Second International Conference on Complex Systems (this volume).
http://necsi.org/events/iccs/iccs2program.html

Conference program.
http://necsi.org/html/iccs.html
First International Conference.
http://necsi.org/html/iccs3.html
Third International Conference.
http://interjournal.org/
InterJournal: refereed papers from the conference are published here.

Chapter 1

Pedagogical Sessions

David Meyer
Session Chair
George Cowan
Complexity: Past and Future
Michel Baranger
Physics and the Complexity Revolution
Atlee Jackson
Unifying Principles
Ronnie Mainieri
Dynamical Systems
Dan Braha, Ali Minai, Helen Harte, David Meyer, Yaneer Bar-Yam, and Daniel Miller
Panel Discussion
Don Ingber
Biomedicine

Introduction

Dave Meyer: I think it's worth making a few comments about why we have a pedagogical session at this meeting. Although people have been thinking about complex systems ever since they started wondering how the world worked, they didn't really think of it as a subject until fairly recently.

And that means that there are a lot of translation problems. There are people thinking about different kinds of complex systems. They don't all speak the same language. They don't all know the same ideas and techniques. And one of the things that the pedagogical session should do is try to explain some of the systems that people think about and some of the common ideas.

So we have five talks today that will try to do some of that. In the spirit of this being a pedagogical session, I should encourage sort of a homework assignment for everybody. And that is, you know, although not every speaker is going to be speaking about a subject which is in your particular area, or field of interest, I think it would be useful for you and for everybody else that you talk to, to try to make some connection between what you work on and what the speakers are saying.

So by this I mean, if you hear a talk about physics and you're working in biology, or you're working in management, or whatever aspect of complex systems you're interested in, try to understand how the ideas that the person is speaking about fit in with what it is that you're thinking about, and try to do it—I mean, this is a long-term homework assignment. Try to do it in a way which is a little bit more than analogical. I mean, try to make it as precise as you can. Because that's one of the things that I always get worried about in complex systems discussions is that there is a lot of talking, but maybe not as much precision of ideas as there might be.

That being said, I think the speakers today are going to be very precise, partly because almost all of them come from physics, with one exception. And that shouldn't scare people off who aren't physicists, and I think Michel, when he speaks, is going to explain why there are so many physicists here, actually.

George Cowan
Complexity: Past and Future

Dave Meyer: So with that, let me introduce the first speaker. The first speaker is George Cowan. He has his doctorate of science in physical chemistry, which is kind of physics, I guess, from the Carnegie Institute of Technology.

He is currently emeritus senior fellow at Los Alamos National Labs, and a consultant there, and also a research member of the Santa Fe Institute. He was one of the founders of the Santa Fe Institute, and the president of it, from '84 to '91.

He has won several important awards, the EO Lawrence Award from the U.S. Atomic Energy Commission, 1965, presumably for work he can't tell us about, and the Enrico Fermi award in '91, from the U.S. Department of Energy.

In '97, he was elected to the American Academy of Arts and Sciences. So he's going to be speaking today about complexity past and future. So that's an appropriate way to start the morning, I would think.

George Cowan: On this occasion I'll recall some early history of the Santa Fe Institute including the decision to study complex adaptive systems and the use of the term "complexity" to describe our major theme. I want also to mention some topics in complexity that show particular promise for the future.

The first general discussion of a new institution occurred at a meeting of the Los Alamos Senior Fellows some time in 1983. On that occasion I suggested expanding our interests in "hard" science to include some parts of the "soft" sciences. My immediate concern rose from my membership on the White House Science Council where I was constantly reminded of the mismatch between the political and policy-making culture of the executive office and the culture of physical science. I increasingly felt that scientists should be deeply concerned about this mismatch.

The problem is largely due to the complexity of politically important issues, which means that scientific and sociopolitical considerations are almost always hopelessly entangled and consideration of the scientific issues by themselves can seem irrelevant to executive and legislative leaders. Let me illustrate this point by listing some of the questions we discussed in the early eighties:

1. What are the best options for defense against nuclear attack?
2. What, if anything, should be done about a new disease called AIDS?
3. Where should we turn for energy sources that would decrease our dependence on imported oil, a controversial option being fast breeder reactors?
4. Should space probes carry humans or be entirely robotic?

Each of these issues involved questions that went beyond physical science and engineering considerations. They concern complex systems and they interact with people. The most troublesome property of such systems is that you can't predict their behaviors by considering the separate properties of each of their parts. Like most systems in our daily world, their dynamics are nonlinear. Panels of scientific experts that are asked to examine such systems are usually given charters that are insufficiently broad.

But many good scientists are reluctant to go beyond their major professional expertise and, understandably, would hesitate to accept a very broad charter. If they are selected to serve on panels purely for their expertise and are asked to advise on topics similar to those that I have mentioned, it's understandable that policy-makers sometimes choose to give little weight to their advice.

The proximate question was whether it was possible to make any useful concessions toward executive office needs without departing too far from the culture of hard science but adding nevertheless, to the breadth and relevance of scientific advice. The more basic problem had to do with an appropriate agenda for modern science. When I talked with the group at Los Alamos, I asked whether it was possible to improve the scientific review process. I then suggested that we discuss various possible charters for a new research center that would bring physical science, mathematics, and numerical simulation of nonlinear processes into closer contact with questions in molecular biology and social and political science.

The members of the group included Herb Anderson, Nick Metropolis, Peter Carruthers, Louis Rosen, Stirling Colgate, and Mark Bitensky. They all expressed interest in pursuing this discussion and it became the main topic of our regular meetings. The group rather quickly expanded to include Darragh Nagle, Dick Slansky, David Campbell, and George Bell and visiting fellows Gian-Carlo Rota, David Pines, and Tony Turkevich. During this period, Murray Gell-Mann called and said that he had been considering ideas for such a center for a long time. He was immediately and happily made a member.

Our discussions produced a variety of suggestions. They included founding an undergraduate college that would develop a hybrid curriculum in hard and soft science, a research center that would major in cognition and some aspects of artificial intelligence, and a computational center that would apply the exploding technology of computer science to various nonlinear dynamical systems. Murray proposed establishing a graduate university without disciplinary departments with an endowment of 3 units, a unit being one hundred million dollars, and described a new modus operandi for such an institution.

We agreed that there should be a new institution but no general consensus emerged during our first several weeks of discussion concerning its central theme. Then Herb Anderson proposed that we sponsor a select meeting of certain scientists who might be attracted to a new departure and ask them to talk about questions that they regarded as important and interesting. Murray persuaded his friends at the Carnegie Corporation to underwrite such a meeting and, as the list of proposed participants grew, I approached the MacArthur Foundation, Murray abstaining due to conflict of interest, and received a similar grant to support a second founding meeting.

At Murray's and David Pine's suggestion, we chose "Emerging Syntheses" as the subject of the meetings, the intent being to explore questions that involved the interfaces between two or more of the conventional departmental disciplines in academia. Participants at the founding meetings were encouraged to choose subjects that related to such syntheses. We favored discussion of topics that went beyond analytically solvable problems.

At our founding meetings, repeated references were made to the complexity of the various topics discussed which meant, of course, that the systems of interest were composed of many parts. I was encouraged by the fact that a majority seemed attracted to the idea of working on important but inelegant problems of this type. They liked the notion of providing advocacy for research on questions that were frequently considered outside the realm of serious discussion by physical scientists. A number of the participants confessed that they had long been attracted to such problems, despite their "softness."

I began rereading the work of von Bertalanffy and others on general systems analysis, Wiener on cybernetics, Haken on synergetics, Prigogine on systems far from equilibrium, and eventually came to Warren Weaver's 1948 paper in the American Scientist on ordered and disordered complexity. I was impressed by his far-sighted prediction that work on ordered complexity would become a major scientific enterprise in the foreseeable future. I agreed with this view. The broad theme I came to prefer was "complexity." Most of the problems of interest to us concerned the dynamics of complex adaptive systems. This theme was generally adopted although the choice was not unanimous. Murray preferred and continues to prefer "simplicity and complexity." Others simply refer to complex adaptive systems or CAS. We are, of course, all reductionists but obviously vary in degree of dedication.

At the time, we felt very daring and vulnerable. In retrospect, the time was right and the theme of complexity resonated strongly. Advocacy by the Institute seemed to help. Complex system studies are now much more widely pursued by physical scientists than was the case in 1985. The number of research scientists who use numerical simulation when analytic solutions are impossible and who can tolerate considerable uncertainty has multiplied. Collaboration between people who come from the physical and the social sciences has increased. Clearly, interest in complexity is more than a passing fad. Weaver's prediction has, by now, been echoed by a number of great scientists, including Stephen Hawking's, and is being validated by meetings such as this. I think that the entire intellectual agenda and society in general will benefit from these developments.

I want now to go to the second part of my talk and mention some of my own current interests in complexity. From here on I claim no expertise. When I retired from administration nine years ago, I also abandoned research on the subjects I knew best and chose to become a student in a field I knew very little about. The subject that fascinated me was the human brain, the most complex adaptive system in our every day world and the least understood. The Institute agreed to serve as host for meetings of a number of outstanding people in developmental and clinical psychology, in animal behavioral studies, and in aspects of neuroscience. We have had a large number of such informal meetings at the Institute over the past eight years.

By now, this combination of interests has produced several new fields of study with various hybrid names. Despite the length of the name, I am attracted to "developmental cognitive neuroscience" with emphasis on the earliest years of human life. Development of the brain and mind in the fetus and then in the neonate through infancy and toddler years is an enormously complicated process but should be more accessible to study than the development of human behavior over a lifetime of experience and over evolutionary epochs. I believe that it is also a period when much of the ability of an individual to benefit from teaching and experience is determined. This potential for learning appears to be closely related to the development of a more or less widely extended and interconnected neuronal architecture. A considerable body of data exists indicating that the level is strongly coupled to the nature of the infant's and toddler's environment. Learning, the realization of that potential, can continue indefinitely but is more difficult if the potential is low. Some current research indicates the occurrence of a second period of neuronal rewiring around puberty. If the Bell curve is ever to be shifted strongly to the right, it may be critical to raise the average potential for learning during these critical years in addition to nurturing the full realization of an improved potential throughout life. The rapid development of new imaging modalities to examine neural activity and neurophysiological structure has stimulated the establishment of many research laboratories that investigate the brain. The field is becoming increasingly rich in descriptive and quantitative data, particularly with respect to location of cognitive, emotional, and sensorimotor evoked responses although most are so broadly distributed that it can often be inappropriate to speak of centers of function.

Extended studies of dynamics or temporal functions in the brain are less common. Better understanding of this complex adaptive system will emerge from a growing emphasis on longitudinal studies, focused on the strong coupling between neurophysiological and cognitive development and the nature of the external environment. I think that the work of Bill Greenough at Urbana on the brains of young rats beautifully exemplifies this kind of research but can only suggest what might emerge from future studies with humans.

Related studies on human infants are not yet anywhere nearly as detailed as animal studies. They require further development of benign, non-invasive modalities to examine, for instance, the development of the cerebrovascular system in the various sensory domains during the early years and its relationship to the richness of sensory stimuli originating in the external environment. I believe that applications of these modalities, based on fMRI or electric field measurements or MEG or optical imaging or other techniques, will multiply in the relatively near future and, in addition to important clinical applications, will greatly affect what we know about developmental processes in the human brain. Longitudinal studies, particularly during the first few years of life, will produce exciting movies of development rather than snapshots. They can also have an impact on social policy if these studies demonstrate relationships between neurophysiological development, early mental development, and various kinds of environmental enrichment.

I want to finish my remarks by speculating a little about some truly difficult problems. I refer to improving our understanding of the complex of influences that determines the behavior of living systems and, an even more difficult problem, to deepening our understanding of the behavior of human beings. The term "behavioral science" was invented by James Grieg Miller in the early 'fifties when he was at the University of Chicago. He spent his professional life working on a general theory of the behavior of living systems, from cells to organisms to human organizations. He authored a huge volume,

"Living Systems," summarizing this work but, apparently, it had no great impact.

I'd like to be better informed about recent efforts to produce general theories of behavior. A great deal has been done that should lead to improvements on past efforts. Much has been learned in the last two decades concerning evolutionary behavior, cell behavior, animal behavior, genetically directed behavior, culturally directed behavior, organizationally directed behavior, etc. They all have common features. Miller made a list of nineteen features, more recently twenty, that he believes are shared by all living systems and that he considers a prerequisite for life. The powerful computers and algorithms now available might test the relative importance of these features against a large body of new data and improve on his pioneering efforts.

Perhaps it's too early and a definitive general theory is still far down the road. But the remarkable progress that is being made in our understanding of the elements that contribute to behavior is very encouraging. All of these elements, ranging from the genetic and neurophysiological aspects to the cultural and including the dynamics and variability of learning processes, are being intensively examined by increasing numbers of scholars with an increasing arsenal of techniques and instruments. We should begin to examine the possibilities for synthesis.

All discussions about our prospects for a desirable future world come down, in the end, to the behavioral patterns and life styles of individuals and aggregates of individuals as yet unborn. It seems evident that they will have to be much better adapted to the modern condition and to accelerating rates of change than are the patterns of present day individuals and societies. Clearly the rate of change in the world around us has greatly outpaced the rate of change in social attitudes and behavior. It is my hope, admittedly a highly optimistic one, that our improving understanding of human behavior will help inform and accelerate changes in individual and social behavioral patterns in ways that make them more compatible with a sustainable and desirable world. Behaviors do change, perhaps more rapidly in modern times than in the past. Only a few generations ago, people picnicked with their children at public hangings.

One promising initial approach to a better understanding of the plasticity of behavioral patterns is to nurture the current explosion of interdisciplinary studies of mental and behavioral development and to emphasize opportunities, now too frequently wasted, for improving the enduring potential for learning during the early years of human life.

Michel Baranger
Physics and the Complexity Revolution

Dave Meyer: Now we have Professor Michel Baranger from MIT. He got his Ph.D. in Physics from Cornell, where he worked with Feynman and

Bethe. He was a professor at Carnegie Mellon and then at MIT. He has been emeritus there since '97. His work started in quantum electrodynamics, and then he worked on quantum many-body theory for years. In the 'Eighties he got interested in classical and quantum chaos, and I think most recently he's been working on non-equilibrium thermodynamics, and its applications to complex systems.

Today he is going to talk about physics and the complexity revolution, and he is going to explain why there are physicists here.

Michel Baranger: Good morning. The 21st century, which you are starting, is really starting with a big bang. For the person in the street, the bang is about the big technical revolution, which looks like it may well dwarf the Industrial Revolution of the 18th and 19th century.

And for us scientifically minded people, the bang is also about complexity, which is a totally different way of looking at things. It seems to change the whole focus of basic thinking even to biology and medicine. Being a physicist, I wanted to ask the question, what role does physics, the simplest and oldest of all sciences, have to play in this 21st century? Since I'm a theoretical physicist, as you will see, I will put my emphasis on theoretical physics.

So this is the way I see it. I think physics also needs to undergo a radical change, from its traditional focus on simplicity to a new focus on complexity.

Now, for us theoretical physicists, the revolution started a few decades ago with chaos. I'm going to talk about chaos, and I'm going to talk about its connection to complexity. Chaos is a purely mathematical concept. It is an undeniable mathematical fact. And it is well known that theoretical physicists are all applied mathematicians, that theoretical physics is built with mathematics.

Therefore, I would like to ask the following question. Why is it that among all the scientific disciplines, both pure applied and social science types of disciplines, why is it that among all of these people the physicists were the last ones to become interested in chaos?

We heard twice that there are lots of physicists here. Well, that is a good thing. But the physicists who are here are the exception. And we all know that we are the exception. If we are physicists in a university somewhere, we know that we have a lot of trouble getting our colleagues interested. We have a lot of trouble getting tenure. For me, that aspect is much too late, because I am retired. But why is that?

In addition to people who are here, there are other exceptions. I want to mention especially the people who built the large particle accelerators. These people knew about chaos. In fact, they discovered a lot of it. So I do know there are some big exceptions. But by and large, if you go to an average university physics department, the people there do not know about chaos, and they don't want to know about it.

Twice in my teaching career at MIT, I tried to introduce some notions of chaos, very simple notions, in an undergraduate course for physics majors. I did that twice in two undergraduate courses for physics majors. And at MIT, they switch us around from course to course very quickly. They say that after two or three years you get stale, and so you have to go and teach something different.

So twice the person who took my succession in teaching the course said, "I'm sorry, but I don't know anything about that stuff. I cannot teach it," and took it out of the curriculum.

And if you look at the most popular physics textbook in classical mechanics, and if you're a physicist you all know what book that is, there is no chaos in it. And he also apologizes meekly in his introduction saying, "Oh well, that subject, it would need a whole book. I can't put it in this book," which is a big cop out.

So why is that? Physicists are smart people. There must be a good reason. The reason is very simple. Physicists did not have the time to learn about chaos because they were fascinated by something else. And what they were fascinated by was 20th century physics: relativity, quantum mechanics, and the millions of consequences over the fascinating stuff that has come out of those two subjects throughout the 20th century.

But I will say more. I will say physicists not only are unfamiliar with chaos, but they find it somewhat distasteful. This is exactly that point. Why do physicists find chaos and complex systems distasteful? To go into that, I'm afraid I have to go several centuries back in time.

I go back to Newton and Leibniz who back in the 17th century more or less independently invented calculus. And in doing so, they provided the scientific world with the most powerful mathematical tool ever, except for the invention of numbers themselves.

One way to appreciate the reach of calculus is to look at it in geometrical terms. Before calculus, geometry could deal with straight lines, circles, ellipses, parabolas and hyperbolas with a bit of trouble, and a couple of other things, and that was all. After calculus, geometry could deal with any curve whatsoever, just as easily as with straight lines and circles and parabolas, as long as it was smooth.

I'm going to give you a course in calculus which is going to last about three minutes, and which has all of calculus in it. Now don't laugh. This is your complete course in calculus.

Imagine that I have a function, $y = f(x)$. Okay, what does that mean? That means that there is a variable, x, and if you give me the numerical value of x, I have a mechanism—either a formula, a computer program, an experiment that I do in my laboratory, or all other kinds of things—such that if you give me the numerical value of x, I can determine the numerical value of y. The mathematicians call that a function, and that's the way they write this. y is a function of x.

And you can draw a graph with x on the horizontal axis, and y on the vertical axis, and for every pair of values x from which you calculate a y, from every pair of values you can put a point on the graph. And if you do that for all values of x, then you get a graph of your function. And I'm going to show you a graph of a function which is a smooth function.

So here is my function, which I made smooth. I'm going to approximate that function by pieces of straight line that I'm going to draw. A piece of straight line there, a piece of straight line there, a piece of straight line there. And you have to put the axis in. So this is a smooth function, and I have replaced it by three pieces of straight line.

It's not a very good approximation, so I'm going to do it again, but this time I'm going to make the pieces of straight line smaller. So you may have trouble seeing them because they are smaller. And you will all agree that this is a better approximation than before. And now, if I had made the pieces of straight line even smaller than that, the approximation would be even better. And if we keep making the pieces of straight line smaller and smaller by a process which the mathematicians call "the passage to the limit," then the approximation becomes excellent. And that's calculus. This is a complete course. We are finished. (Laughter)

Everything else follows from what I just said. What calculus says is that every smooth function can be approximated by pieces of straight line—very, very small ones. Since we know everything about the geometry of straight lines, we know everything about every single smooth function whatsoever.

And you see that it is essential that the curve is smooth. If that curve had not been smooth, I could not have approximated it to any desired degree of accuracy by tiny little pieces of straight line. Smoothness is the key to the whole thing, and there are functions that are well known to mathematicians which are not smooth, but calculus does not apply to them. But if you restrict yourself to smooth functions, everything is okay, and calculus is in the saddle.

Now, for at least 200 years, theoretical science fed on this calculus idea. The mathematicians invented concepts like continuity and analicity to describe smoothness more precisely. And the discovery of calculus led to an explosion of further discoveries. The branch of mathematics so constituted is called "analysis" and it is by far the most powerful, the most useful, and the most used, branch of all the branches of mathematics. It is used by engineers, physicists, and chemists—all theoretical people. All people who are not prevented by lack of time, because they have things to do in the lab, are applied mathematicians. (Laughter)

And, of course, when you take your freshman course at MIT, the first thing they teach you is calculus because you are going to need it, no matter what tests you are going to do at MIT in later years. So, the mathematicians came up with things like derivatives, and integrals, and differential equations, and series expansions, and integral presentations or special functions and all of these things which are the tools of calculus, and in which we solved lots and lots and lots of problems in science and in engineering.

After many decades of unbroken success with analysis, theories became imbued with the notion that analysis was the way of the world. That all problems would eventually yield to it, given enough effort and enough computing power. This idea was not expressed explicitly. It was unconscious, and that was the trouble. But this idea pervades most of the science of the 20th century. It is reflected in the choices made in the university curriculum, and by the textbooks. The two great physical theories of the 20th century, relativity and quantum mechanics, are both solidly based in analysis.

This idea of the invincibility of analysis was extremely powerful and as I said, it was unconscious. People forgot that there were assumptions at the beginning. They made the assumption that the curve is smooth. The conditional truth became absolute truth. And I know, because I was one of them. I was there myself. None of us theoretical physicists ever discussed the possibility that there might be practical problems where calculus did not apply.

When we heard about unsmooth curves we said, and I quote, "Those are pathological phenomena of interest only to mathematicians." (Laughter) You see that phrase hundreds of times in the 20th century papers, published work.

If you go to the bottom of this belief, you find the following. Everything can be reduced to little pieces of straight lines, everything. Therefore, everything can be known and understood, if we analyze it on a fine enough scale, of course. It may take lots of work, and it may not be worth doing. But in principle, we have absolute power.

The enormous success of calculus is, in large part, responsible for the decidedly reductionist attitude of most 20th century science. The belief in absolute control arising from detailed knowledge. And yes, the mathematicians were telling us all along that smooth curves were the exceptions, not the rule, but we did not listen.

Chaos. Chaos is the anti-calculus revolution. Chaos is the rediscovery that calculus does not have infinite power. In its widest possible meaning, chaos is the collection of those mathematical results which have nothing to do with calculus. (Laughter)

And of course, this is why it is distasteful to 20th century physicists. You've got your reason right there. There are a few physicists who agree about this. In terms of applications, chaos theory solves a wide variety of scientific and engineering problems, which do not respond to calculus. That is not saying that the calculus from now would be considered outdated, and that we must focus all our attention on chaos.

No, no, no. Calculus retains all of its power, but its power is limited. Calculus is only part of the truth. It is one member of the couple. Chaos is the wife chosen by the mathematicians to complement calculus. When calculus had chaos chosen for him, we did not like her. (Laughter) But there will be no divorce. They are married forever, and calculus will become very fond of her eventually. (Laughter)

Of all the dimensions that we use to describe physical objects, two of the most important ones are space and time. So let's talk about chaos in space. A geometrical object which is chaotic in space is called a fractal. There are many possible definitions of a fractal. A very loose definition for a fractal goes like this. A fractal is a geometric figure which does not become simpler when you analyze it into smaller and smaller parts. In other words, a fractal is something which never becomes smooth when you look at it under a microscope.

A simple example of fractals have been known to mathematicians for a long time. The simplest one is called the Cantor Set, for instance. Then there are slightly more complicated ones which have two dimensions. One which I take out of Yaneer's book is called the "Sierpinski Triangle". And those are all very simple fractals which can be defined in a couple of lines. And in fact, this one is defined in the figure caption.

Then there is the famous Mandelbrot set which is so popular in all the picture books which you have all seen. It's slightly more complicated, but still the formula for it just takes a couple of lines. And out of this very, very simple formula comes this amazing complexity of the Mandelbrot set.

It goes without saying that you cannot print a fractal on a piece of paper. You can do something that approximates a fractal, you can make it better, and better, and better, but you would have to work for an infinite time in order to put the complete fractal on the paper. In the case of the Mandelbrot set, even such approximations to the nth fractal are so beautiful that they ended up in all the museums of the world. So that is chaos in space. Of course, the fractals are not just made by mathematicians. Nature has lots of fractals. In fact, we live among fractals. A beautiful cloudy sky is a fractal, I should say an approximation to a fractal. But it does carry over many scales.

A range of mountains is a fractal. In the picture book on fractals you will see many examples of desert vegetation, or forest vegetation, or imprints made by little crabs on the sand beach. All of these are fractals. Of course, our bodies are fractals. So there are fractals everywhere. Isn't it funny that the word fractal was only invented in 1974? That says something about the power of calculus. And in fact, when the word was invented by Leibniz, people said, "Oh, this kind of stuff. This is not science, this is just fooling around."

Let's go to chaos in time. For chaos in time, you have to have the notion of phase space. Phase space is, if you have a certain dynamical system, it has a certain number of variables. And then in phase space you have one dimension for each one of those variables.

So for instance, if the variables are x, Y and Z, then phase space will be a three-dimensional space with an x-axis, a y-axis and a z-axis. But if you have 1,000 variables in your dynamical system, then phase space will have 1,000 dimensions because there is one dimension for each variable. Every point in phase space defines the value of all the variables of the system, and therefore it differentiates the state of the system.

The signature of chaos is sensitivity to initial conditions. And this says that if you take two points in phase space that are very close to each other, and then you let them evolve over time, the two trajectories are going to separate exponentially. This is sensitivity to initial condition. So suppose that I start here, and I have this trajectory. But if I take a different point in phase space, very close to the previous one, the trajectory will maybe do something like that, the distance between the two trajectories increases exponentially. That is the signature of chaos in time. There are lots of books about it, and lots of things you can do with it.

What's the connection between the space aspect of chaos, which is fractals, and the time aspect of chaos, which is sensitivity to initial conditions? Well, in phase space I can take a smooth region like this. So I consider all the points in phase space inside this region here. And now I let my region evolve in time, what will happen to it? Little by little it will get more and more complicated. It will stretch and spread and send tendrils, and bend, and get more and more complicated. And if you let that process go on for an infinite time, you'll get a fractal. So a chaotic, dynamical system is a system which over time changes the smooth picture into a fractal. And conversely, any fractal can be considered as having been drawn by a chaotic engine working for an infinite time. So that's the connection between chaos, sensitivity to initial conditions, the chaos of dynamical systems, and fractals. It's the same thing, but one of them is looked for in space, and for the other you look in time.

That's all for chaos. Chaos destroys reductionism's dream, which is a dream that we have absolute power if we know enough about the details. But chaos and calculus need each other, and they will be fruitful and multiply.

I want to point out one more thing about chaos. As a general rule, the opposite of chaos in mechanics is called regularity. Chaos is a general rule. Regularity is the exception. If you look at the textbooks, the textbooks treat only problems which are regular. The textbooks in classical mechanics for physics students, as I said, they do not mention chaos. They give 400 pages of examples of classical mechanics. They are all regular examples, and that is how generations of students were misled. And that is one explanation why physicists find chaos distasteful. What I want to do is talk about complex systems and what their connection with chaos is. I'm going to show very quickly a collection of transparencies about properties of complex systems, which I either got out of Yaneer, or Yaneer's book, and I should say Yaneer is in no way responsible if I say anything that's wrong. So complex systems contain many constituents interacting nonlinearly. Chaos also demands nonlinearity. So there is a connection between complexity and chaos right there.

The constituents of a complex system are interdependent. So I give you a little example here. The complex system possesses a structure of spanning several scales. That's very odd, yes?

And the complex systems is capable of emergent behavior. The notion of emergence is extremely important, and we have lots of sessions about that.

And when you put together emergence and structure, you've got self-organization. Now, I want to emphasize the difference between complex systems and chaos, because many people have a tendency to mix them up. I've had many occasions to talk to general audiences, non-scientific audiences, and what usually comes out is, "Well, chaos has to do with nonlinearity. Complex systems have to do with nonlinearity." That's the same thing. You're talking about the same thing with two different words.

Not true at all. Chaos is a very big subject. There are lots of books and lots of papers about it. But complex systems are much bigger. There are not many books and papers about it yet because it's just a starting subject, so not a whole lot is known about it. But if you look at chaos as a tiny little thing you need to learn before you approach complex systems, it's really tiny, tiny, tiny compared to complex systems, so that has to be clear.

Chaos is mathematics. The books and papers on chaos were written by mathematicians. And mathematicians, as you may know, have a way of coming into a subject when it's already very well in progress. They come towards the end to codify it, and to really put it in totally logical shape. But at the beginning, the mathematicians stay away from the subject because it's too fuzzy, it's too woolly.

Well, complex systems are still a subject that most mathematicians stay away from. Not all of them, there are some here, but even fewer than physicists. (Laughter) And the striking difference between complex systems and chaos is this. As I said, complex systems have several scales, and while chaos may reign on scale number n, the scale above it, scale n-1, may be self-organizing, which in a way is the opposite of chaos. So you may have chaos on one scale and self-organization on a different scale, which is just the opposite. So we might say that complexity involves an interplay between chaos and non-chaos.

I will finish this by talking about a notion which is very popular, which is extremely attractive, beautiful, but as I said before, it is still fuzzy and woolly. And so the ?? types say, "I don't want anything to do with that idea." This is the idea that the most interesting complex systems, such as living systems, are at the edge of chaos, which makes sense, in view of what I just said. Presumably, this means this: suppose that in an equation of motions of systems, there exists a parameter, or several parameters which people call control parameters, which can be changed depending on the environment. It could be the temperature, the concentration, the intensity of some external effect like sunlight, or stuff like that.

And we know that chaotic systems are not chaotic all of the time. Under some circumstances they are chaotic, and then if you change the control parameters, they are not. And so there is an edge. There is a point in your control parameters between chaos and non-chaos.

The edge of chaos. That's where things are the most interesting. This is the place where you get the change of phase. This is the place where in physics you get long-range correlations. This is what a physicist would call the critical point, the edge of chaos. And all kinds of interesting things go on at the edge of chaos. The usual laws of statistical mechanics do not apply. You have power laws instead of exponential laws, all kinds of things like that which I don't have time to go into because my time is over.

But much of this conference is actually going to be taking place on the edge of chaos, I believe. Okay, thank you.

UNIDENTIFIED: For a lay person, how do you explain, in simple words, what complexity is? Or is it possible to define complexity in regular words?

Michel Baranger: Go ahead, George. (Laughter)

George Cowan: Well, the dictionary definition is a complex system consists of many parts. You said that it becomes complex when you have a dimensionality of at least three or greater, right?

Michel Baranger: It becomes chaotic.

George Cowan: Chaotic, okay, but not necessarily complex.

Michel Baranger: Oh, no.

George Cowan: I think also by way of emergence as a way to this phase change, and to emergent property as a way to freeze out some of the intrinsic dimensionality of the system by constraints. So that there are many things a system can no longer do, and therefore a pattern emerges.

Michel Baranger: I do not have a one sentence definition of the word "complexity," but those five transparencies that I showed, just told you a lot about complexity. And that's good enough for me. It's not good enough for a certain editor of *Scientific American*, but it's good enough for me. (Laughter)

UNIDENTIFIED: Can we define complexity, in very simple words, as the property of multi-component systems that cannot be accounted for completely in terms of the properties of the components?

Michel Baranger: I would say it's a good start, but it doesn't convey the whole flavor of the thing. (Laughter)

UNIDENTIFIED: I'm a biologist, and you've made a claim for physicists, that most of your colleagues don't know anything about chaos, and don't know anything about complexity. But most biologists don't know about it, and I dare say most sociologists, and most chemists, because most scientists, I think, have their nose to the grindstone. I mean, they mostly don't read between 9 A.M. and 5 P.M., but they don't do much reading after 5 P.M. and before 9 A.M. (Laughter) And because this is outside their area, most professional scientists don't go to somewhere, as we all do, which is outside their particular patch.

Now, as to the definition, if I may take this on, I'm going to be talking about evolutions, particularly about models. And it seems to me, I go for this one. A complex system is one where you can't get all the properties out of consideration of the parts, and their interactions. We ought to put their interactions in as well because you can't even do it then, you get some emergence.

But I think in principle—and you don't like this, as a theoretical physicist, I know—in principle, physics is an invented subject, so it's all contained, and you've got it, and you're there. It's a little toy playground that you've got. (Laughter) But biology and geology and astronomy are given subjects. We are used to the idea that we can't define a human being, or a woman, or a mouse, and we can describe them.

I am going to say that a complex system is one which can be described by two different, non-overlapping descriptions. And if that sounds paradoxical, it isn't at all. I'm going to show you a picture of a sign in Neuchâtel Airport which says "Lost property, objet trouvé." (Laughter)

Michel Baranger: Well, you are right, of course, that not only the physicists don't know about chaos. But after all, as you said also, physics is a simpler science, and therefore physicists should lead the way.

UNIDENTIFIED: Simple-minded. I'm sorry. (Laughter)

Michel Baranger: It's the same thing.

UNIDENTIFIED: You use the term nonlinear a lot. I've been working in this area for some time, and I sometimes think I don't really understand what nonlinear is. Could you explain?

Michel Baranger: Well, if you look at it from the point of view of stimulus and response, if you double the stimulus and that doubles the response, that's linear. If you multiply the stimulus by ten and the response gets multiplied by ten, that's linear.

But if it doesn't work that way, if you double the stimulus and the response gets multiplied by four, say, then that's not linear, but then it could be even worse. Suppose that you then triple the stimulus and the response drops down? That's even more nonlinear than the first example.

Things are nonlinear if when you plot a graph of response versus stimulus, you do not get a straight line. Linear means straight. That's not etymologically correct, but linear means proportional.

There is something called linear thinking, which has to do with smooth lines, which is what you're thinking of. So when you talk about linear thinking, or you accuse somebody of linear thinking, you are not using the same sense for the word "linear" as when you say something is nonlinear.

Atlee Jackson
Unifying Principles

Dave Meyer: Our last speaker this morning is Atlee Jackson. He sent me a little description of what he does, which I liked so much that I have to read it.

He is actively retired from the Physics Department at the University of Illinois at Urbana-Champaign. He got his Ph.D. in physics at Syracuse University. At the University of Illinois, he was the Director for the Center of Complex Systems for some number of years, I'm not sure how many. He held research positions in Sweden, Holland, Italy and Japan, as well as in Illinois, and he spent his career working on nonlinear dynamics of various kinds. So he'll give a very expanded answer to that last question.

He is currently completing a book called *Exploring Nature's Dynamics* which looks like it has some software associated with it, so you can actually play with it. He's going to speak about the evolution of nature's dynamics - unifying principles.

Atlee Jackson: Let me just give you a very brief background.

I started right after I got my Ph.D. I started doing research in nonlinear dynamics in 1960. I've done nonlinear dynamics ever since in a variety of fields, so I'm very much focused in on nonlinear dynamics as such.

I started teaching at the University of Illinois in 1973, and developed several books. If you'd like some broad looks at nonlinear dynamics, these are called *Perspectives on Nonlinear Dynamics*, and they are a two volume set, and they will keep you busy for some time. It kept me busy for a long time in trying to put it together.

So my focus is very much on dynamics, and as such, I've never been in the mainstream of physics. If there is one thing physicists don't know much about, it's dynamics, and I'm very serious about that, and I'll show you why.

In any case, the title of the talk here is concerning what I want to talk about: the idea of the evolution of nature's dynamics, and the issue of unifying principles.

And unifying principles is going to address the question of the fact that we all know that science today has blossomed into a bunch of separate fields of understanding. And the question is, is there any sort of unifying concept we can put together? And I'm going to make a suggestion which is not going to satisfy a lot of people, particularly physicists, in the answer to this.

My mantra is "dynamics, dynamics, dynamics." And the point, first of all, is that life is all about dynamic processes. I don't think that that's a surprise. But I would claim, in fact, nothing has any real meaning to us, unless we move relative to it. That is to say, we come to understand things because of emotions and what we're observing. Soulful situations, anything like this, has to do with dynamics. And I'm going to give you a little illustration to break you in because I just think this is super, because it does several things.

We observe things in nature, and these things are what I would call categorical. That is to say, you only can observe certain features of whatever we're looking at. I know you're happy, I know you're sad, whatever it is. You have certain categories you pin things down in.

You also know that you observe things, and you think that you understand things because you observe something which just looks familiar, and looks familiar because you've observed it before.

I want to show you an example, to show you how much we, in fact, acquire mentally from these experiences, and we don't appreciate the fact that we do acquire them.

This is a little off the beaten track, but I just like it so much I've got to show it to you. This is due to Roger Shepard, who is a psychologist. He is retired from Stanford. And frankly, he changed my mind about psychiatrists when I saw this example, because I learned that I'm wired to recognize things I think I understand but I don't understand.

Obviously you see here two tables, okay? And they are obviously quite different. This comes from a book of his, I think it's *Mind Sights*, and they look very different. But of course, these are not tables, obviously, these are pictures of tables. Nonetheless, we think we understand what we're seeing by just taking a look at it. Now I've watched this closely, because I can show this 100 times, and I still don't believe what I see. Okay?

Now this isn't a long table. This is obviously a square table, more or less. Now just watch. (Laughter) Okay. Now, if that doesn't shake you up a little, you might as well leave, okay? I've got nothing else I can impress you with.

There is a lot you can think about on this thing. What it is, in fact, and I could talk about this a long time. It would be fun to talk about it. So I just want to shake you up a little bit, but you don't really understand everything you see. A lot of it comes from past experiences. You've gotten these experiences which George talked about, the plasticity of the mind picks up experiences we put in our memories, and we take our actions from them.

In any case, then, about eight years ago I was wondering, we know a lot about dynamics, are these going to fall into your lap? What do these dynamic concepts do to contribute to the future of science, and in particular, to this question of the unification of different fields of sciences, and in fact, what is a field of science to begin with, when you get into complex systems in the future?

The question is basically does it represent some fundamental viewpoint within what I call the metamorphosis of science—the change in the basis of science that I think we are approaching.

The thing I want to talk about is dynamics, and the question is about the understanding of dynamics. I'd like to just make the point, the obvious point, that science is an activity that occurs within nature. And scientists are part of nature, and they are subject to exactly the same things as all the rest of nature is subject to, and the dynamic principles that go on.

There is another book by John Moore, *Science is a Way of Knowing*, which is a fine book about the foundations of modern biology. In any case, it's only one of the human modes of understanding, and moreover, it's an evolutionary mode of understanding. It's evolutionary because of the influence of culture, and it's evolutionary because of the influences of technology, to name a few. And so science is not a fixed affair, and it's changing dramatically, I think. This is going to be very exciting to see what turns up in this century.

The first thing is, what is a common scientific mode of understanding that can be applied within all such fields of research? That is to say, can we start getting to the point of trying to elucidate what it is that we would consider to be a scientific activity, even though it's dealing in sociology, or something like this? It's a tough business. But some of these things have to start to be addressed.

And the key word here I'm throwing around is "understanding." Knowledge in the sense of understanding. What does understanding mean? And this is what I want to address particularly, and I want to address it in relationship to dynamics, because I believe that the understanding of dynamics is the key to making the future progress here. And what one has to learn is that there are many modes of understanding, and you have to be flexible enough to appreciate that. So these are the things I want to talk about.

Is there anything basic to nature in the field divisions that occur? Is that something basic, or is that something just human-made? Is there something common to all of these fields? And I will suggest the dynamic commonality in a certain sense.

And then finally, these considerations require the development of entirely new concepts in what is meant by the idea of unity of these different fields. It is not the reductionist idea, it is going to be quite different, and it's going to come from acquiring a sensitivity as to what understanding may mean scientifically. So that's sort of the ball of wax there. I put the ideas in various orders, so stick with me.

John Moore, as I said, wrote this book, *Science as a Way of Knowing*, which I think is a very fine book. I like the title. In his point, for example, in the field of biology, Aristotle got it right. Aristotle is credited with being the founder of putting out the first basic questions of biology. This had to do with the structures of the systems, not the dynamics, the structures.

In the field of dynamics, you unfortunately got it totally wrong. You got it wrong in terms of celestial things, and got it wrong in terms of terrestrial dynamics, both of these. Now, these got quantified by academicians later, and it messed up dynamic studies for 2000 years. But don't blame it on Aristotle. In fact, he still said that these were simply initial ideas and needed to be developed further.

The blame was not on Aristotle; the blame was on the culture that froze him in. And it was that cultural impact that messed up Ptolemy and Copernicus, for example, and to a certain degree, Kepler. So this issue of knowing and understanding—well, I'll talk about that. The present era is in search of understanding dynamic phenomena. I think that that is a universal feature of all of these fields that we're talking about.

There are other things, of course, things like molecular biology - getting the genome information. But information is not understanding. There is a quote from Poincaré which says to the effect that science is built up of facts, like a house is built up of stones. But science is no more a collection of facts, than a house is a pile of stones. Now, if you just think about that a little bit, you realize what makes a pile of stones a house is because they got put in connection with each other, and it's a relationship. So that is a relationship that I want to emphasize.

Let's take, for example, Newton's mode of understanding. We were given some idea of Newton, but the idea which is really most important, and certainly in terms of dynamics, is the idea of a recursive relationship. He introduced this for the first time into the thought of man, and just to put it down, say you have $X(t+1)$ is equal to some function, $X(t)$. That's what he used initially in *Principia*. That's a recursive relationship. It means that if I put a value at the time t, this will tell me what the value plus 1 is. I take that, I put it back in, and this goes on and on. That's dynamics. That's a recursive relationship.

And the one in calculus where you just do the limits we were talking about already. If you have this sort of thing, this dt is simply $t+1$ taken down to the infinitesimal. So that's a calculus expression. Nonetheless, it's a recursive relationship.

Now, all this tells you is how things progressed step by step. What we've learned, and I will show you, is hidden in this relationship. There are a tremendous amount of things we don't know anything about, and were just only able to discover in the end of the last century for the first time.

Moreover, the idea of fundamental laws and the whole Newtonian paradigm that comes out of that should not be blamed on Newton. This should be blamed on the people that claim that this is what Newton said, and I'll give you a quote on that too.

But one thing that is very, very typical of this sort of thing is it deals with closed systems. Systems that you can write down an equation and it tells you the dynamics of that system, without any consideration of an environment and environmental impact.

And this gets carried over, to a certain degree, of course in computers. Computers are extremely important, very useful, but you have to remember they're giving you also algorithmic information. You can only run them one solution at a time, unless you're smart enough to write yourself a multiple search program which has to do with, of course, the human input to this idea.

So out of this has come a culture, particularly from physics, which has been very detrimental. It started off remarkably early. Here is a quote from Einstein, 1918: "The supreme task of the physicist is to arrive at those universal elementary laws from which the cosmos can be built up by pure deduction."

Well, I've shown you mathematically that that is impossible. This is not philosophy. This isn't even mathematics. This is simply not true. This is not only a bad philosophy, but it also is detrimental to the future of science.

In more recent times this was, of course, picked up by Weinberg. He wrote articles about Newton's dreams. The goal of the formation of a few simple principles that can explain why everything is the way it is. This is Newton's dream, and it is our dream. Well, it was Weinberg's dream, it was not

Newton's dream. If you read Newton, you'll find many statements, including ?? of this thing in *Principia* where he's leaving it wide open as to what the future of understanding might be. And of course, he has this famous quote about wandering on the shores and finding a prettier pebble as part of that experience.

He also said, "We are interested in the final principles that we hope we will learn about the study of these elementary particles." So the first lesson is that the ordinary world is not a very good guide to what is important. This is probably the arrogance that has crept into our culture. It's arrogance, and it's because of ignorance as to in fact what it is we do now, even mathematically.

There was a very famous counter to this, with which if you're not familiar, you should look up. This is by Philip Anderson. It was written in 1972, which is remarkably early. And what he was pointing out is that the knowledge that we have in different fields, the elementary entries in x obey the laws of the science y. So that you have this sort of hierarchy of knowledge. This is part of the fields we were talking about before.

The reference to this is in the *Science Magazine*, Vol. 177, 1972. I just extracted a little here, but this hierarchy does not imply that science x is just applied to y. Which is to say, you can't deduce y from x. It isn't said that way, but I will claim that that is, in fact, true. At each stage, entirely new laws, concepts and generalizations are necessary, requiring inspiration and creativity to as great a degree as the previous one.

So I want to put some flesh on these ideas. As sort of a background to all of this, I like these quotes. One is from Boorstin. He's talking about exploring concepts, but this puts it very well here. "The great obstacle of man is the illusion of knowledge. The great significance of voyages was the discovery of ignorance." And that what we have to learn is the discovery of our ignorance, and an appreciation of that, and trying to see what it is we can do to build a meaningful extension of science into these other fields.

Then George Bernard Shaw had said to Major Barbara, "you've learned something, but it always feels at first as if you've lost something." So at first you're going to be, particularly if you're reductionist, sadly disappointed. But nonetheless, you can learn something here.

Newton made the big change in modern science. He introduced the recursive relationship, and he also, with the help of Kepler, established a law of universal gravitation, which became the poster child of ideas of fundamental laws, that there are some basic laws which cover everything.

People have been studying dynamics prior to Newton. How did they understand dynamics? Maybe if we go back, as Churchill said, the farther you look back, the farther you're likely to see forward. If you go back and take a look at the concepts of dynamics prior to Newton, how did they understand dynamics? What was the concept of understanding that they came up with?

And it sounds a little ridiculous to go back, for scientists to go back and try to learn something that may influence them in the future. But I claim that it's not a bad starting point. So, the pre-Newtonian idea of understanding dynamics. The details of this are spelled out. I've got some pre-prints of the paper I'm going to present very shortly on general complexity, which is called something like "The Unbounded Vistas of Science: Evolutionary Limitations." And it goes through these things in some detail.So let's take a look. The first one is Pythagoras, or the Pythagorean school, at least. The one I'm focusing on is simply the idea that if you take vibrating strings of different lengths, you can get harmonious tones, concordant tones, whatever you want to call it, provided that there is a certain simple ratio of the lengths. I think that that is a scientific statement, but it's also an interesting one. It turns out in fact, in a sense, it's one of the most complicated ones.

First of all, it's a connection between a quantitative thing, lengths, to a psychological concept, concordant tones.

Not only that, it's a cultural concept. Because what is harmonious to the western ear may not be harmonious to the Indian ear, or to the Chinese ear.

So you have here a very interesting thing, which one could explore further. You not only get a connection between quantitative and qualitative features, but you also have the idea that in fact this scientific idea may be culturally different.

That is to say there may be aspects in science which were not homogeneous in the world, and that there are principles which you develop because of one's experience within that culture, or as Churchill would say, the plasticity development. So science may be not as homogeneous as we think it is. That's a big step for the first start.

Archimedes was a brilliant person who really didn't much care about applied things. His love was mathematics. He did the first example of the integral under the curve that we just saw develop, which is a beautiful thing. There is a fine book on this. I'll give you a reference in a minute. In any case, this one here has to do with the organizing of Archimedes' principle, the idea of buoyancy. When does a ship float, and when does it sink?

And there are famous anecdotes. In any case, he developed the idea of forces. He was the one, in fact, who gave Galileo the concept of a force. And he had the idea of balancing forces.

This is of course static, so it's not dynamic in the sense of time dependency, but it tells you that if the force that comes up on this thing is equal to the amount of weight of the water displaced, and if it can't displace enough water, then it will sink.

Later on though, this is, I'm not sure if this is true or not, it is accredited with the idea of fluid continuity. This is the idea that if you take water in a pipe, and you constrict the pipe, it will travel faster here than it does here, and in such a way that the amount of mass that is going through is equal. He presumably saw this with river streams, but of course Leonardo was not famous for publishing things that you could find, much less read, unless you read Latin backwards.

Then there is William Harvey. Harvey is never mentioned in these areas, but he was the one that got to understand functional dynamics. Dynamics appears to have a function. That's a new concept. Namely, he understood what the function was, he understood that first of all there was blood circulation, and what the purpose was in terms of nutrients going on.

He also understood the sequential development of embryo development, embryo dynamics, which in fact I didn't know much about until I read some of Needham's work on this stuff. This is very complicated dynamics we're talking about here, but he had an appreciation for the sequential development of an embryo being essentially a purposeful development component.

Tycho. Tycho is a wonderful individual, a curious individual, and very much neglected in appreciation of what he did. Here is where technology came into play. He managed to go to an observatory on the island of a Danish king, and he was able to measure angular separations of a few seconds of light, rather than minutes. And in that process, of course, he took down all of this planetary motion data, which Kepler then made use of later. But what he did also, which is not really quite appreciated, was that he used a well known, simple dynamic process called parallax. If you're driving along in a car and you look out on the horizon, you can tell which objects are nearer to you than further away because they move backwards from you faster than the ones further away.

This is a relative motion thing. So you just hold your head and move back and forth. That's parallax. He had the very good fortune of having both a nova take place in the sky, a brand new bright star take place, which is obviously a dynamic process because it only lasted a year. It came in, and it went away.

He could measure a parallax very accurately. It had no parallax with respect to the other fixed stars. That is to say, it was in that region of the heavens. This was quite contrary to the idea that fixed stars were places that no dynamics took place. So that cancelled out one of Aristotle's ideas, and it brought the heavens into a dynamic context with respect to celestial motions.

Then he had the good fortune of seeing a good comet come in, and he could measure that. Aristotle had decreed that comets are essentially something that takes place in the atmosphere. But that Kepler could find parallax motion with it, he could establish that, in fact, it was a celestial thing. He was still locked up with circular motions, which is a whole new story, so he thought that that was just a circle. But nonetheless he knew that you had the comet, which was in any case, part of the celestial world.

Now the big step, it has to do with different types of dynamic understandings, analytic, recursive and holistic, and this is holistic. I'm claiming that understandings always come about by finding relationships between things that we observe. It's a relationship which is a dynamic relationship, which is to say that it holds even when things change individually. It's a dynamic relationship.

If you're not familiar with Kepler's three relationships, these have to do with planetary motions. And they involve such ideas as the idea that the planet, in fact, travels in an elliptic circle with the sun at one of the full sides of the ellipse.

Now, you don't observe an elliptic circle. You don't observe the full side of an ellipse. These are mental constructs. These are constructs that are done mentally by integrating effectively, putting together a bunch of data, and coming up with a holistic concept of elliptical motion. It's not trivial. It's a mental affair.

He also showed that if you take a line from the center, from the sun out to the planet, that the amount of area that gets swept out in equal times is equal, regardless of where it is. There are no lines from the sun to the planet. This is his construct, and the area is a construct, and looking at these at different times is a construct. These are mental things, holistic types of concepts, and you're getting a relationship, not between observables, but between holistic concepts.

The last one has to do with comparing planets, and it has to do with the period of motion in relationship to the size of the semi-major axis. It's a crazy power law. Kepler spent over 20 years putting these things together from Tycho's information. You're not going to get many grants for that sort of thing. (Laughter) So there is this idea of holistic relationships, which I think is a very important and different way of understanding things.

Galileo was the first to introduce time explicitly into relationships. The time variables had never been introduced, and part of the reason he could do it was once again, technological. One of the reasons nobody else was doing this stuff, there was no such thing as a good clock. You had to get a clock going, which could divide up time in short enough intervals, and then on top of that, he was clever enough to put things on an inclined plane, and so forth, to slow things down. But time came in here.

There is also the idea of partial differential equations, fluid dynamics, which I won't have time to do a discourse upon. In the 19th century, what we found was that there is a breakdown in this Newtonian dynamics. That is to say, there are physical phenomena that occur that you can't capture in those differential equations.

The first one, the most sensational which was neglected was Mendel's relationships which dealt with the beginning of understanding genetic dynamics. The probability ?? of living features, physical features. So the physical features, once again, are qualitative. They're things like peas have a rough or a smooth surface. They are green, or they're yellow. Things of this character. And it's probabilistic. You're not given an answer, you're given the probability that you will have, in the next generation, a certain fraction with certain characteristics. It's much more technical than this, and I don't have time to go into it, but this is where probability comes in.

Maxwell introduced the discussion of electromagnetic fields. The field concept is once again not an observable thing. You can observe the consequences of it, but it is a concept which is outside of the usual observable—the relationship here is a relationship between fields and the propagation of electromagnetic waves.

Boltzmann talked about the dynamics of atoms, and he introduced, for the first time, the concept of a distribution function. Once again, this is a holistic concept. You are talking about the probability that particles have certain velocities and are traveling in certain positions. It's a mental construct. And he, moreover, had intuitively the idea of sensitivity of dynamics, for which he took a tremendous beating from the closed systems society of physicists and mathematicians, which is another sad story.

Darwin, of course, is a fascinating thing. He begat an evolutionary dynamic theory, which is still ongoing. The reason it's ongoing is because technologically we're learning more and more about what, in fact, are part of the elements. In fact, the whole genetic process only got picked up in the beginning of the century. Of course, that's been changed dramatically in the last 20 or 30 years.

There is a nice, interesting book here by Depew and Weber. It's called *Darwinism Evolving*. And it goes through the history, the ideas of the evolution of the Darwinian theory, and how it's still transforming today.

Then I want to introduce you to something which is generally not known, and this is Poincaré. Poincaré used the idea of phase space to its maximum. Let's see what we've got left.

For those who don't know what a phase space is, this is going to have to be terribly brief, I'm afraid. It's another mathematical construct, which is very different from calculus. So you have a position and you have a velocity for a particle. And what you do is you construct a motion, an orbit, for its motion. These are called orbits. These go from plus infinity to minus infinity in time. The first thing about it is that time doesn't appear any place here. You throw out time. You throw out what Galileo brought in. There is no time. One of the fascinating things about this is that you can learn more by discarding some information. And there are a number of examples. So if you have periodic motion for example, you have something that looks like this. That would be a periodic motion.

At a certain time, you'll find the position here, but I didn't tell you what the time was. It's all the history of a particle that started with a certain condition. And between these two, in fact there has to be something fancy that looks something like this. Well, there is one possibility, at least.

There can be various types of very fancy orbits. And you can talk about the fact that this orbit and this orbit are not the same type of orbit. They're not topologically equivalent. The point being that if you put these things on a rubber sheet, you can't stretch one into the other. If you can put the two orbits on the same rubber sheet and stretch one onto the other, then they are topologically equivalent.

So you've got another holistic distinction between things. Things that are periodic are not things like this. These are dynamically quite distinct. So let me introduce something else. So this is topology. I have to do this, because this is something that's totally neglected virtually every place.

We introduce in these phase spaces, an idea of a point-creating map. For example, in some three dimensional phase space, you take a certain section there, and let's say you have a periodic orbit. It starts at P zero, and it comes around and goes back to P zero. That's a periodic orbit.

Now, let's take a point nearby, and let it go back some place else. This is what's called "A first return map of point array." The map itself is a set of points which doesn't care at all whether they're in orbits spinning in between. You throw away everything. You throw most of it away, and you keep just the end part. And you think what the heck did you possibly learn from that?

What you can prove is called the Poincaré-Birkhoff theorem. Birkhoff finally nailed it down in the '20s. But this is really the first example of sensitivity that was ever proved. And it shows you that in the neighborhood of any periodic orbit, there are an infinite number of different periodic orbits—that's orbits with different periods—and in fact unbounded, and there is an uncountable number of non-periodic orbits.

Now, there was only one person who ever picked up on this seriously and understood it. In fact, Poincaré himself did not pick up on this in some respects. But just try to figure out if you're going to say the orbit is periodic, empirically. Empirically our information is always categorical. It's only within some little thing. And we always say, "Well, I'll squeeze it down, and I'll finally tell you what's going on."

Well, this thing says you can't, no matter how you squeeze it. This mathematical thing is telling you that you're still going to find all of these properties squeezed in any zone at all. There is no way you can do that. What it does tell you is that you've got to give up this idea of knowing things forever. You can only know things for certain time scales, at best. In fact, the whole nature is in the process of evolution, and you can only know things for certain time scales. What is so stunning about this is that the very mathematics used for the Newtonian paradigm, is, in fact, the source of destroying this idea of empirical determinism.

The idea that physicists use in many writings is that you can do things in principle. In principle you can take your laws and you can extend them. This tells you that you cannot, even in principle, do that. It's a meaningless statement. It's a philosophical statement of no value. So in principle the arguments are, in fact, fallacious and very widespread.

If you think about it from the development of stars and planets, and life on earth and so forth, the evolution of nature's dynamics has to do with a sequence of processes, and they have very much to do with appreciating the fact that systems come out of an environment.

In fact, the problem of trying to define a system in an environment is a non-trivial problem. When do these things decouple and so forth? The decoupling affair once again, is a question of the time scales you've got to talk about. Because nothing is decoupled ultimately. As Baum said, "There are no things, there are only processes." All structures are ephemeral. They only live for a certain time.

So in understanding types of dynamic relations, what I try to show you is that there is a variety, there is a richness of types of dynamics understanding, which can be applied in different fields, and are just as meaningful. Just as meaningful in those fields as is Newton's equation for a billiard ball, or something like that.

The trick is this idea of physical features. You've got to extract from information. You've got to extract physical features which you feel have, in fact, a relationship to each other. So the idea from information to extracting physical features which have a relationship is a process that one has to develop.

So the evolution of nature's dynamics involves a sequence of coherent dynamic systems, developing in a non-equilibrium fit environment, that is, one that will accept them and not destroy them right away. There is a time scale.

Nature has generated emergent dynamics we observe, and there is no deductive method of understanding this evolutionary process. There is no deductive method for a variety of reasons, one reason being what Poincaré showed us, the other being we don't know what the conditions were. And the environment that exists now is not the environment that things were born in. So there is no deductive method connecting these things. It's a total fallacy. It's folklore.

Science is an activity within our natural environment. That is to say what we are studying is our nature, not nature in general, but our nature. We're trying to uncover that, get an understanding of that. There is a very fine book that has just come out called *Rare Earth*. It is a wonderful, detailed collection of scientific facts. But the idea is a unifying principle, so I'm going to give you what I consider a possibility.

Through the discovery of the compatibility of diverse modes of understanding, I would say the recognition of the fundamental character of understanding of diverse dynamics, emergence in nature, is what science is about. Trying to get this, to accept it as fundamental. Not that you're going to put up with it, but in fact that these people have a lot more difficult system to find these features, and to find the relationships. That's quite true. But nonetheless, there is a scientific mode in which they can try to understand things, and that mode of understanding can be quite diverse and inventive.

I really haven't touched upon the fact that there are a lot of mathematical possibilities here. There are a lot of things out there. This is not an empty field. This is a field full of possibilities and I think it's a very exiting thing for the future. Thanks. (Applause)

Ronnie Mainieri
Dynamical Systems

Ronnie Mainieri: Where does this idea of dynamics come from? Well, if you lived in a time when there was no street lights, you couldn't help but stare at the sky and notice that the stars would remain fixed with respect to each other as time went by and there were a few exceptions. There were these nasty little stars that kept wandering around. So if you looked at them as time went by, they seemed to remain in the same place every day, but then when you looked at them the next day, instead of being exactly here, they would be in some slightly different position.

So people say "Okay, I'm going to do something. I'm going to record where this strange star is one hour before sunrise." And if you do that, what you discover is something like this. Six June it will be at Scorpio. On the 15th it will be on this other constellation Sagittarius, and so on, as the year goes by.

And you start thinking, well, let me see how this thing changes from year, to year, to year. And the Greeks did that. And what they came up with was the discovery that the motion of these wandering stars, which they call planets, is actually predictable. And the way they did this was they imagined that these things moved in circles, and they imagined that these circles rotated at constant speed. So if you do that, you're able to predict, by using trigonometry, where these planets will be at a given point in time.

And as the observations increased in accuracy, they realized that one circle wasn't enough to predict the motion, so they put one circle on top of another, and then they put three circles on top of each other. And this is what developed into the theory of epi-cycles, which you may have heard of. But the important thing about this is that if you know the position of one of these stars at a certain time, then you can tell where they will be in the future. So if you know the position then, you know the position now.

Those things form the basis of dynamics. To do dynamics what you need is a bunch of numbers. In this case, I'm calling them x_1, x_2, x_3, x_n. Given those numbers, you're supposed to be able to tell what these numbers will be in the future.

An example of something that's not a dynamical system is the stock market, just based on the prices of the shares, because nobody has been able to figure out what the future prices of those shares will be, just given the current prices. Something else is involved in the dynamics. With just those numbers, you don't have a deterministic dynamics.

The other thing you need for dynamics is an evolution rule. Here I'm calling it $f(t)$. $f(t)$ is some function that looks at these numbers, and spits out a set just as big of numbers, and they tell you how your system has changed. Let me try and make this a little more concrete. Let's say here I have a pendulum. Here is my pendulum, and I can tell you, and this is supposed to be a rigid bar, and if you tell me to know the position of the pendulum, all I have to tell you is the angle. So if I tell you the angle with respect to the vertical, you know where this pendulum is. You also need to know the speed, because if I just show you the pendulum in the position it is right now, you don't know how fast it's going. And as Newton explained, if you know those two numbers, the position and the velocity, you're able to predict where it's able to go. So here I'll get it going.

All right. So what we do is we represent one of the variables, so the position, and some appropriate units along this axis, and we represent the velocity along this axis. Here the angle is almost zero, so the x coordinate is almost zero, but the velocity is the largest here. The v extension here is very large, so that's why the point is down here. Whereas when I take it up here, the extension, this angle here is maximum, and the velocity is zero, because it has to stop when it comes up here.

So this collection of numbers that allows you to tell where this thing is going, we call that a phase space. Just the shape of the thing we usually call configuration space. So given that, given your phase space and your rule, you have your dynamics. It still doesn't look like the dynamics you learn in textbooks, and to make the connection, you have to understand what a vector field is.

This is your phase space here, and this is where you are when you start ticking your clock. And what I've done here is I observed something as it moved, and I recorded its position as it moves in phase space. At each tick of my clock, I mark where it is. So if I do that, and I then connect the points by little arrows, then this arrow here tells me where I'll be at the next second. If I'm at this point here and I know the arrow, I know where I'll be the second afterwards. So if you know this collection of vectors, then you know how you will move in phase space. And that's what people will call the vector field.

Now, you don't usually see it represented this way. You usually see it represented as an equation. So let's see. So people will usually write, well, okay. People will usually write, they'll say that the way something changes with time is some function of just the position of where it is. So these are your ordinary differential equations that you will see in textbooks. They are usually written this way.

Now, one thing you will always see people write, is they will always write x. or dx/dt equals something. They never study higher order systems. They never study something with two derivatives or three derivatives. Are people familiar why that is so? The reason people don't do that is that you can transform any higher order system into one of these. So let me just do that here.

I'll do that. So let's say we have an equation. Let's say we have an equation that's like this, x, two derivatives, plus $5x$, one derivative, plus x equals zero. So what you can do is you can call, you can say that x derivative is equal to some other variable, let's say y, and then you can rewrite this equation up here, and then this means that x two derivative is equal to y derivative.

That means I can write this equation up here as -- And if I join these two here, I now have an equation that I can write this way. I call x and y a vector, that's a u, and u prime is equal to x prime, y prime, which is equal to -- where is my x prime? So this is equal to y, and this is equal to minus x minus $5x$ prime.

So I can write it now as u prime is equal to this function here that only depends on u, which is this one here. So you can always transform an equation that's higher order into a first order one, just like I did here. So that's why you'll see all textbooks would just handle these vector first order equations.

All right. So to get dynamics, you need to know a bunch of numbers, and you need to have a rule that tells you how to evolve it. If you know this rule, you can observe it for a very short time and get the vector field. And you can also try and do the opposite. You can have your vector field and try and get the dynamics out of that.

Now, physicists and mathematicians have spent an enormous amount of time trying to figure out how to do this. Nowadays we don't have to worry very much about that, because we have a computer. We can go into our computer, type in your equations, and you can figure out, at least for short times, how things will move.

So what I'm going to try and give you is a view of how can you collaborate with your computer to try and understand dynamical systems because there are a lot of things that the computer can do nowadays that you don't have to worry about anymore. So for example, figuring out how the solution changes for a short amount of time is easy to do on a computer. You don't have to worry about that anymore. But if you want to know how it behaves for very long times, then that's a different story.

All right. So one thing you have to know how to solve is the linear equation. So here I wrote it in one dimension. So you have one number. You have one position, one coordinate. Take the time derivative of that, and that's proportional to x. So you have to be able to solve this --

This is the way your calculus professor doesn't want you to solve it, which is you move your $x dt Ax$, you do it the way Leibniz did. I put the dt here, and then you get -- This is very easy to integrate. It's the log. This is even easier, it's just the linear. If he grumbles, you just say that you're using infinitesimal calculus.

All right. Now the interesting thing about this solution: so you notice that you get an exponential. See, the solution is basically $x f(t)$ equals the exponential of that constant. See, this a here is the same constant that was here. It's the same constant that was in the equation. And this x_0 is where you start out.

Now, this solution here also works in many dimensions. So now, x is a vector, A is now a matrix, x dot is also a vector, but the solution is still the same shape. It's just that things mean different. They mean different things here. That's one of the things mathematicians like to do. They like to make the new stuff look like the old stuff so you don't have to learn new stuff.

So in this case you just have to understand what does it mean to take the exponential of a matrix. And the hard way of doing it is what I wrote here. This is not the way you would program your computer to do it. It's not very efficient. But this is the definition of the exponential in terms of a power series.

Well, wherever you see x you put a matrix. And wherever you see multiplication, you do matrix multiplication. And if you sum up this series, you get the exponential of a matrix. Now, the idea is right. If you do it on a computer, you shouldn't do it this way, there are much smarter ways of doing it.

But this helps you think. Because if you could waste computer time, and one way you could solve this problem is you take this matrix here, whatever it is, and you try and diagonalize it. Now you can always write the matrix as some diagonal part, plus maybe some two by two parts, plus some more diagonal stuff. And if you have a matrix, once you diagonalize it, you know you're going to get the i-gon values here. So all you really need to know about matrixes are they're i-gon values.

Now, how likely do you think it is to get an i-gon value one, in a matrix? I mean if I just gave you some matrix that ran them, how likely would it be that one of these i-gon values were one? Very slim. So the only way you get i-gon values, you'll see why later on and during the talk why i-gon value one means something. But i-gon value ones are rare. If I just give you a typical matrix, you're not going to get the i-gon value of one, unless there is some special symmetry in the problem.

And in physics there is a problem that has that a special symmetry, and they're known as Hamiltonian systems. So you do get ones when you have Hamiltonian systems. But if you just pick a random dynamical system, then it's very unlikely that you get ones. So if you're trying to learn dynamics, you should at least know how to solve the linear case. So go in any OD book, and you learn how to solve the linear case. They'll actually analyze all the possible cases for these matrices. So when the i-gon values are larger than one and smaller than one, when they're complex, they'll analyze all of these cases for you. And it's important to try these out, try them out in MATLAB, Mathematica. You should all play around with these equations to get a feel for it. Now here comes a theory that explains why you don't have to worry too much about local nonlinearities.

...Suppose I gave you a vector field, and on that picture I showed you on the computer, the thing was a circle. The motion of the pendulum drew a circle on the phase space. Now what would happen if instead of measuring angles in radiants, I measured them in degrees? The scale of that plot would change, and my circle wouldn't be a circle anymore, it would be some type of ellipse. It would do something like this, because I'd have to multiply it by 50 something to change radiants to degrees, and it becomes long.

And the shape of that circle changes as I change my coordinates. So you have to imagine this vector field as stuck to some type of rubber sheet. Imagine that the vector field is drawn on a rubber sheet. And changes of coordinates means stretching that rubber sheet in all possible ways. So those vectors are going to move around as I stretch this rubber sheet in different directions. I think I have a little stretching here in my computer.

Let me show you this first. So this was the vector field for the pendulum. I actually drew the little vectors in here. As you can see, they're pretty small in here and grow off as you go into the periphery. And now what happens if I change my coordinates on this vector field? Well, my vectors move around. They change shape.

Now, if I look at any small patch of this thing here, if I look at any small patch of that vector field, then I can always take that little piece of rubber, and stretch it, and move it around, so that all the vectors are straight. That's always possible to do, if you take a small enough region. And that's known as the rectification theorem.

Now, this vector field here, corresponds to the linear equation, so we know how to solve this. So in any little patch, we always know how to solve it. Here, I actually solved it down here for you. There is some vector here with many entries. Everybody is zero, except this direction, so we know how to solve it. It's just exponential. So it's very easy to do. And what you do, then, is you go around and for every little piece of your vector field, you find a little patch where you can solve it.

Now, it may be that the way you stretch it around here is different from the way you have to stretch it around here. And I indicated that by having different functions here, h's, that tell you how to stretch it. So moving things around is a change of coordinates, and that's given by some function that tells you exactly how to pull this rubber, this rubber phase space around, so that you get it straight. So you have many of these functions here.

Now, if we could figure out how to do this for the entire phase space, you've solved your problem. There is nothing to do. And that's exactly what you do on textbook examples of the pendulum, motion of planets. It's just that there is one of these functions that transforms it into either everybody in a straight line, or the pendulum. There is always a function like this.

Rectification doesn't always work, because if it did, I'd be unemployed. Where does it fail? I'll list you three of the most striking ways in which it fails.

It fails when I have a point in my vector field where I don't know where to go, an equilibrium point. So if I'm sitting exactly at this point, you see how the arrows just point out of there? So if I'm sitting exactly there, I don't know which way to go. I just sit there.

So around there, you can imagine if I have everybody pointing out, there is no way I can take a circle of vectors pointing out, and transform them into everybody pointing in one direction. I would have to cut it, and cutting is not a form of stretching, so you're not allowed to do that.

The other way in which it can fail is the regions where I can do it become smaller, and smaller, and smaller, and smaller until it vanishes. This usually happens when your equations blow up, you know, when your numbers go off to infinity. Usually what's happening is your neighborhoods are getting smaller, and smaller, and smaller.

But the most interesting way in which it fails was mentioned this morning, which is just a periodic orbit somewhere. So now let's just think about this for a little bit. Imagine I'm in phase space, and I have some orbit that goes around and comes back to itself. So here I've drawn it on the plane so you're just seeing some projection of it. And then I put some Poincaré's section, as Jackson pointed out, and it pierces that section.

But I'm just going to draw a line here, and this is the orbit here, this is this green orbit, it's right here, and this is the section here, this blue line here. Now, let's sprinkle points, a whole bunch of points, along this blue line, and we'll follow all of these points as we go along here and come back. So one of the things that may happen is that those points start moving away. The points start moving away from the orbit, and only a few of them come back to this initial neighborhood.

So now imagine you did that, imagine you took a point here and after it went around once, it would come back here, and then after it went around once, it would come back somewhere here. Eventually it's going to bang here, or it's going to fold, or it's going to do something. It can't just keep going on forever away from it. It can't go back on it because it either goes back exactly, or it can't go back. Because remember, from every point there is a unique trajectory. So if it came back on itself, it will again be a periodic orbit.

So if I chose a coordinate, let's say of one for a point here, when it came back it would be two, and when it came back it would be three, and I would eventually run out of a way of describing it. So in the neighborhood of a periodic orbit, this theorem of rectification will fail. You can always do it on little patches of this, but globally on this thing it will fail. So that is the way the rectification fails. It fails on a global scale.

So here is a two dimensional picture of the same thing. You go around, and if I distribute the points uniformly around it, they'll get stretched like that, and they'll get stretched even further, and even further.

So now what happens if I keep stretching it on and on? For reasons of conservation of energy, or for some reason external to your dynamics, you can't go off to infinity. So if you keep on stretching, and you're limited to live in this circle here, you have to fold back. So this thing was going to have to fold back on itself. And that's the essence of most dynamics as we understand it in the lower dimensions. Dynamics is basically stretching stuff and folding it. That's all dynamics does: stretch and fold, stretch and fold, stretch and fold.

Well, I have to show you this. There is no way you have to learn this. So people decided that they could quantify this stretching and folding. So if this is my periodic orbit, this point here is some periodic orbit going through a plane, I know that from the rectification theorem I know in that neighborhood I can make it linear. If I can make it linear, I can diagonalize that matrix that tells me the linear behavior, and you're not going to get i-gon values one, because if you do, you can just change it a little bit, and they're no longer one.

So they're either bigger than one, or smaller than one, in absolute value. So the ones that are bigger than one I painted orange. So this is the orange direction is the direction of the things that, of the i-gon vector that's bigger than one, and the green ones are the ones that are smaller than one.

Now, imagine that at times zero this is my situation, and I evolved this in time. Well, these orange guys here will get further and further away from the red point, and they'll fold back and form what is known as the unstable manifold. And if I run this thing backwards in time, then my orange guy will shrink, but my stable direction will grow. Because forward in time, it used to be smaller than one, and then when I run it backwards, it's going to grow. You can just play around with your exponentials, and you'll see that that is the case. And that's how I get the stable manifold. And these guys fold on top of each other.

So here is a review of what I showed you. There is an evolution law, $f(f(t))$ that's given for you. So this can be how you think prices are moving, it can be how the density of a population in a region is changing. It can be how a pendulum is moving.

You know that for small chunks, for small little pieces of this, I can always transform it into a linear system, and my computer can solve this. But if I go to long times, then I get these periodic orbits, and things start failing, and I have to figure out what to do with the periodic orbits. So this, what's here, is more or less what you need to know about local dynamics. So these are dynamics, just look at a small neighborhood. After some time, Smale decided to study all of these periodic orbits at once.

And what he did was this. Well, he says, "I'm going to simplify this thing." Because remember, we can change things locally with this rubber dynamics. So he says, "Let me just make it really simple. Let me pretend that my squishing and folding is done in the following way." I start out with a little square. It doesn't have to be a circle; it can be a square. They're both equivalent under the stretching stuff. So I take this square here, and I squish it, so it becomes this little square. Then I stretch it out in this direction, and then I fold it on itself.

So now you're supposed to imagine that somewhere in the middle of this square there is a periodic orbit, and this is what's happening to the points in that neighborhood. So they get squished, they get stretched, and then they fold back on themselves. And we ask ourselves, you ask yourself, "Well, what if I keep doing this over and over again?" Because if you're in a periodic orbit, those points keep coming back over and over to themselves. So what happens to the points in that neighborhood? Well, I get the square, I fold it, and it comes back on itself.

Now, some of these points here, like these guys here, this little section here, this little section here, they go away from the square. They go off to some other part of your phase space that you're not studying at this time. But some of them, indicated by green here, come back into the square.

So now where were these points originally? Where were they in this original square here? Well, let's try and unfold this. So now if you pull this thing out here, this is this part here, see, it came upside down, and this part here is here. Now this guy here, remember had been squished and stretched? So now we squish this way and stretch it this way. So if I do that, I get these two bands here. I get these two bands here.

So these points here in green, when they go forward in time, they end up here. So now think of the intersection of these guys with these, the four little squares here. So inside those four little squares, I get points that come back twice. I can repeat this over and over again. So next time you're in the shower, instead of singing, think how this thing works, and you'll start realizing that this thing actually forms a fractal in there. There is an infinity of points.

All right. I'm going to mention one thing for the skeptics, which is any dynamics you make up in two dimensions is either a horseshoe or a twist, or a combination of those things. The only thing is that you have to generalize your horseshoe a little bit. You have to allow different types of folds. So for example, folds like this, or folds like this. As long as you allow that, that's the only possible dynamics you can get in two dimensions. So that's why people study it so much.

So now you know you have, so you know how to do it for short times. You know there is a theorem that guarantees that you can do it for short times, the rectification theorem. You know that things fail around periodic orbits. Periodic orbits have to be treated in terms of neighborhoods, because around that periodic orbit I get one of these horseshoes that gets an infinity of other orbits around it, and that provides a picture of the dynamics. So you can imagine your dynamics as being formed by a whole bunch of periodic orbits all over, and each of these is behaving as a little horseshoe.

Now, this is actually, you can actually make this more precise. Because the periodic orbits not only organize how points move in phase space, they also allow you to compute averages. And that's what I would like to spend my next few minutes talking about.

See, everybody, when they study dynamical systems, they are also interested in how things move, and they're usually interested in computing something about it. So you want to know what's the average energy dissipated in your pendulum as it moves, or you want to know if you're studying rabbits and foxes in a predator-prey model, you want to know what's the average number of foxes that I have in my model. Or if you come up with some model for how income moves among people, you want to know what's the average income. So you usually want to compute average of something as things evolve. And you no longer want to compute trajectory, because after all, these things are chaotic. You can't; trajectories don't mean much.

Jack Cohen: Averages?

Ronnie Mainieri: Well, averages, yes. They do mean something, and I can give you an exact statement of what they mean. If you're going to take a dynamical systems point of view, you're giving out a dynamical system. It may be that a dynamical system description for your model is not a good one, and then yes, then the whole mystery is figuring out, "What is the good description?"

Jack Cohen: That means you lose any change though.

Ronnie Mainieri: Pardon?

Jack Cohen: If you take means, your answer is a mean.

Ronnie Mainieri: No. I'm saying that if you have a dynamical system, then means are the interesting thing.

The reason is that given how means evolve in time, you actually know everything you want to know about the dynamical system. You'll see it in the next two slides.

So let me just tell people what is a mean. A mean means that you take something, any function, so you have your phase space, you have points, you have the thing hopping around here so it goes from here to here, from here to here, to here, and you take some function, and you decide you compute the average of this function.

And then usually typically two things happen. Either this average you're computing changes, depending on your initial condition, it gives a different answer for every initial condition, or you're lucky and you find out it doesn't depend on your initial condition.

When the averages don't depend on the initial condition, your system is ergodic. So that's what the word ergodic means. It means you'll get the same one, the same answer for your averages for almost every initial condition you pick. So you see that computing averages in some type of machine, or some type of program, where you spit in this function that you're interested in— like how far away I am from a point—and it gives you back a number. And it has the nice property of being linear. So if I were computing the distance from a point in meters and the average comes out to be one meter, and I decide to measure it in feet, then the average will come out to be, I don't know, what is it, 3.3 feet, or whatever it is? You can just multiply the average answer, and you'll get the result. So it's linear.

And what that tells you is that there is another way of computing averages which is you can take the function you are trying to average, and multiply it by this other one, and you'll get the average value. So you usually see the ergodic theorem stated as there exists a measure that allows you to do the average, or there is a weight function that allows you to compute the average. So instead of doing the average by evolving in time, you can do the average by averaging over space.

And this function here most of the time is not a function. It can be very weird-looking. So here, I'll show you what it is for the Hénon map. So you can see it's a very rough function. I mean, it seems to be peaked a lot. You can actually show that it's actually smooth in this direction, but in this direction here, it's a fractal. So when you plot on the computer, it looks like it's rough in both directions, but it's not. So those rows can be really weird functions.

So now I'll give you an example of computing an average. So here is a simple system. It's a physicist's version of the pinball machine. And what I do is I throw in a ball into this pinball machine, so these are the bumpers, and I ask how many times does it bump before it goes away. So one, two, three times. One, two, three times this time, and I do this many, many times.

And I make a plot of how many times it bounced, a histogram of how many times it bumped—two times, three times, four times, five times, and so on. And if I plot this on the log of how many times it bumped, I found out that it falls as a linear scale. So the slope of this line I call the escape rate. So it's some number.

So I want to compute the escape rate for this set of bumpers.

So I go off on my computer, and I do it 10^{10} times, so I do it 10 billion times, and I get a curve like this. So after I bumped it I don't know, I forget what it was, 100,000 times, I put one point, and another 100,000 times I compute the escape rate again, and I keep doing that, and it seems to be converging to something. So some number here like 7403 or something like that. And after doing 10^{10} times, you should be satisfied with your answer. The only problem is that it's wrong. The actual answer is way down here. It's actually 723 something.

So imagine you were doing a computer simulation, you would have absolutely no idea that you had the wrong answer because it seems to have converged to something. And so now how did I actually get that answer? I can't explain in the next two minutes how I got that answer, but it has to do with analyzing how averages move.

So this is, well, I actually plotted only half the value here, but to give you an idea, I did 10^{10} iterations, and I got .36. This is on a slightly different system. And if you just think of it in terms of the periodic orbit, so if you tend to average this in terms of your periodic orbits, you study the same system substantially, for a substantially shorter time, but you get a substantially more accurate result.

What happens in the neighborhood of 10^4 periodic orbit was 10^3 periodic orbits actually, and I average on the neighborhood of every periodic orbit, and that's how I got these answers. So this is a techniques known as cycle expansions, and you can use it in quantum mechanics.

(Applause)

Mike Hagar: My name is Mike Hagar. You said in the introduction something was said about applying this to social systems.

Ronnie Mainieri: Yes.

Mike Hagar: I'd be curious to hear you say a couple of words about that.

Ronnie Mainieri: See, the point of view I take on chaotic dynamics is that's really a branch of statistical mechanics. I can't make the connection here because it would take me too far in. If Mitchell Feigenbaum were here, he would have spoken about the thing that provides the link, and then I would have been able to make it. But it's just too much to do in 45 minutes.

My point of view is that in social systems you can think of them as agent-based systems. That means that you can think of them as a computer program where you have these little agents moving around, and they have a finite state. So that means they have a finite amount of memory.

And if you assume that, then these things are really a branch of statistical mechanics. And then there are all sorts of tools, and you can use the same type of tools to study them. So that's the 30 second answer. I'll be giving a course at the Santa Fe summer school, a 5-day course on exactly this topic. So I'll have 6 or 7 hours there to elaborate. (Applause)

Dan Braha, Ali Minai, Helen Harte, David Meyer, Yaneer Bar-Yam, and Daniel Miller
Panel Discussion

Dave Meyer: What we're going to do for the next 45-minute time slot is put up a panel consisting of the people who helped put together this meeting, who have a range of areas of expertise in complex systems, or perhaps it's better to say interest rather than expertise, and let them each say a couple of words about what they work on, and then throw it up into the audience to let you guys ask sort of questions to any of these people, or to the other people who have spoken today.

And you know, hopefully we'll get some interesting discussion going. We'll try to keep it to 45 minutes, and then we'll do the last scheduled talk of the day.

Dan Braha: My name is Dan Braha. My interest in complex systems is basically in the way that complex systems is reflected in engineering. In my opinion, complex systems can contribute to understanding complex engineered systems in two respects. The first role is to develop descriptive languages for describing the activity of complex systems. For example, describing grammars that describe languages in architectural design.

The second aspect is developing ways to understand the source of complexity, thus helping us to minimize these complexity aspects, and contributing to more traditional performance measures such as cost and performance.

Ali Minai: My name is Ali Minai. I'm also an engineer. I'm in the electrical and computer engineering department at the University of Cincinnati. Most of my work has been in computational neuroscience, and I've been working on a region of the brain called the hippocampus which is involved in memory and spatial cognition. So I've been doing modeling of that. I've also worked on neural networks quite a bit, and on chaos and synchronization of chaotic systems.

Recently I've been very interested in the application of complex systems, or the application of ideas from complex systems, to the organization and control of engineering complex systems like traffic networks, and communication networks, and things that we build. Then they get so complex that we no longer know how to control them.

And I don't mean this in an apocalyptic sense, but simply that we're used to centralized control, or distributed control in a limited way. And with these systems, that just isn't possible.

Let me just, at the risk of going on for a long time, let me just give one example. As we are able to make smaller and smaller communication devices, we are able to make more of them, and for example, sensor networks and things like that, where we make these devices which have very limited communication capabilities, but we can make lots of them.

And systems of these devices, for example, can certainly do things that we could not do with our old centralized systems. The question is how do we control them and organize them, or how do we let them organize and control themselves so that we get some interesting behavior out of them? That's the kind of thing that I'm interested in. Thank you.

Helen Harte: My name is Helen Harte, and I'm a nurse by trade. My work in the last 15 years has been clinical and operational quality improvement. I got interested in complex systems essentially because of this. The work looks like this, whether it's work in clinics or hospitals. This is what people have to deal with every day who go to work. Total unpredictability and uncontrollability. And we realize that our quality improvement projects that we're spending millions of dollars on throughout the country and hospitals and clinics, do not make any difference, because our management structures look like this.

The people who have to do this work have to fight through those kinds of management structures They cannot move their information, and the failures of quality of care are almost always failures of purpose, which is what patients right now are telling us. The system is in such terrible disarray.

The purpose used to be to figure out a way to care for the sick. If you have to deal with the health care system now, you realize the purpose is to figure out how not to care for the sick. So we have issues of purpose, we have issues of connections and relationships that fail, or intermittently work, which is actually worse than failing, or that information does not flow where it's needed, and if it does it's not accurate or it's not timely. I think that the world of complex systems has a great deal to teach us as far as improving or transforming our health care system.

Dave Meyer: I'm Dave Meyer. I am a mathematician/physicist by training. As far as my interest in complex systems goes, I guess I'm mostly interested in unifying features across different complex systems. Things like the orbits that Ronnie was talking about, how those occur in different kinds of systems, be they physical, or social, or biological systems.

Specifically, I've been working the past couple of years on quantum computing with the physics side of my brain, where there are lots of small interactions, which give rise to large-scale effects which are interesting. And the social side of my brain I've been thinking about questions like voting, and aggregating preferences of large numbers of people or agents. In those kinds of settings you find cycles which lead to the kind of chaotic behavior that Ronnie was talking about.

Yaneer Bar-Yam: I'm Yaneer Bar-Yam, and many of you have had e-mail communication with me for reasons which are evident today. I wanted to make three points about ways that we approach understanding complex systems.

One of the ways that is very powerful to think about complex systems is imagining a number of parts that are interacting together, and through their interactions we think about how patterns of behavior, of the collective behavior of the parts, arise. So there is a collective behavior which is arising from interacting components, and we can understand the collective behaviors through thinking about the components and their interactions.

And as usual, we always have to put a system in the context of an environment. We don't always do that explicitly, but in general there is an environment in which the system is also interacting with, and therefore the pattern arises not only through the interaction of the parts, but interaction within the context of the environment. That's one approach to thinking about complex systems.

A second approach to thinking about complex systems is the problem of description. The idea is that we don't think in the context about what are the processes giving rise to the pattern, but we simply focus on describing what is there, if you talk about static, or what is changing if we talk about dynamics. It's just a description of the system. And thinking about mechanisms for describing systems with languages, what ideas we need to use in order to describe the system and how it's behaving, is a key approach to addressing complex systems.

The third approach to thinking about complex systems is thinking about how they form. And in general, of course, the theory of the formation of complex systems is the theory of evolution and how such systems arise. It's evolution in the context also of development.

So those are sort of three strategies that we use to approach thinking about complex systems. The reason why I wanted to emphasize that in remarks today is that as one goes through the conference, very often people talk about a specific system or a particular area, but if everyone thinks about these general ways of approaching thinking about complex systems, one can generalize from the specific to understand how it can be applied in many other contexts.

Now, there was one other thing that I wanted to mention, and this goes back to Michel's comment on the edge of chaos. Michel made a statement about the edge of chaos as being order on the one side, and chaos on the other. But one of the examples that he used actually had order on one side and order on the other side, but the boundary was not described order. That's a phase transition.

So in a typical phase transition, you go from say liquid to gas. The liquid, in a sense, is very orderly. The gas is very orderly, there are no large-scale behaviors. But at the transition between them, there can be gas bubbling, or whatever. There can be lots of different behaviors.

I think one of the main contributions that complex systems is bringing, is shifts in very many basic concepts that people sort of assume. And one of the assumptions that is so basic that it pervades how we approach all fields is that it's good to be either in one extreme or another extreme. One is always thinking about, "Is it better to be here, or is it better to be there?" And many of the debates in society are is it better to do this, or to do that? This extreme or that extreme? But very often the better case is some balance of forces. And that, in a sense, is why the idea of the edge of chaos is so powerful because it in principle says that there is a balance of things that one wants to attain. Now, even the idea of balance may be too simplified, in the sense that it may be coexistent with different things.

For example in physiology there is a balance between structural order and rigidity of the skeleton, and dynamics and application of a nervous system. They can coexist in the same system, rather than strictly being balanced against each other. Those are a few of the thoughts that I wanted to mention.

Daniel Miller: I'm Daniel Miller from New York. My field is Psychology, and in a way I feel a bit of an anomaly here because there is very little about psychology and psychotherapy and consciousness per se, but I feel that a lot of the material that's being delivered here is very, very deeply connected to consciousness as a system, because consciousness is certainly a very biological system. It's connected strongly to neurophysiology and the discoveries that we have there.

It's certainly connected to environment and stress. Now to come to the edge of chaos and how consciousness can relate to that I think is not altogether that difficult. Because you find that people who come into psychotherapy come in on the edge of chaos, literally, in their emotional states. And their old value systems, their ways of behaving are no longer functioning for them, so they've got to have a new way of relating themselves to themselves, their internal system, to their biology, to their environment. And I feel that there ought to be some kind of way of relating this to complex systems theory. I think there is, actually. I've got papers out on it.

But I'd like to hear from anyone here, Yaneer, or Michel, or anyone who cares to comment on that.

Ali Minai: As the questioner mentioned, there is some work on the application of the ideas from chaos to psychology. One of the issues that arises with that—and I'm not questioning the utility of that approach, some of that work is very valuable—but one of the issues it raises is that it functions too much by analogy still, and not enough in terms of precise quantitative ideas.

As someone who works mainly at the level of neurons and systems of neurons, and I guess as a reductionist, I think that we need to understand consciousness in terms of the physical substrate where we believe it resides. And of course, complex systems ideas will be fundamental with that, because I think most of us, if not all of us, do think that consciousness is not a property of neurons. If it is a property of anything in the brain, it is the property of the whole system. Probably the most important area in which complex systems will inform psychology is through our understanding of the brain and how the brain gives rise to consciousness.

Yaneer Bar-Yam: A quick comment is that in understanding even such complex problems as psychology, there are often aspects of them that can be thought about in terms of simple dynamical system processes. For example, the dynamics of depression, which has a temporal behavior, may be thought about, and may in fact be analyzable in the context of a simple dynamical model. That may be useful for thinking about that problem, even though it doesn't describe the entire behavior of a system.

So that the system, even though it's highly complex, may simplify to the point where it can be described using what we would consider to be relatively simple dynamical systems theory, rather than understanding all of the components and how they interact.

Dave Meyer: Since you mentioned depression, let me tell one of my favorite examples about studying human beings as complex systems. It is from two centuries ago, roughly, when people first started doing statistics, and realized that roughly the same number of people commit suicide in London in every year, 12 per 100,000 or something like that.

And this was consistent year, to year, to year. People are complex systems, and a whole city of them is certainly a complex system, yet this average, which is constant year to year, tells you something very interesting, and at the time it was very upsetting to people. It sort of made it look as if there was some absence of free will. That 12 people, who knows which ones, were just going to kill themselves next year. Where was free will if that was bound to happen? And sure enough it did.

This was sort of the beginning of statistics, and it gives an example of some complicated dynamics. Who knows what the real dynamics are? We don't know. But there are things that you can study about systems like that, and here is an example where averages actually tell you something.

M: What do they tell you?

Dave Meyer: They tell you how many people are going to die next year. They don't tell you which ones, but a number.

Ronnie Mainieri: It also tells you that if there were a dynamical system, then there is a measure that tells you. So you can start thinking about dynamical processes that will generate that measure. Heavens knows what it is. I don't know what it is.

And there is an economics, there is an area of research called social choice, which explains how people make decisions and choices, given their preferences. And I have always been curious about the psychological basis for preference systems. And you know that social choices lead to chaotic behavior, so it would be interesting to see what's the relationship between psychology and social choice.

Dave Meyer: Social choice, I think, has a very important developmental basis in that we have free will, but we also have a lot of deterministic stuff within us from our own childhoods that make us feel the way we feel, that make us see the way we see, that make us act the way we act in many, many, many situations.

And I'm not saying that we're merely all determined processes, but that we certainly are open systems as well, and can take in and change what goes on internally. If not, psychotherapy would be totally useless. So I don't know whether I've answered your question or not, but it's in that vein.

I think that's a very important aspect of complex systems in general. The interesting ones are open systems. And it's important to understand how they interact with their environment, and how the environment drives them or doesn't drive them, depending on what's on the inside.

UNIDENTIFIED: I think there is a tendency among newcomers to the field, and I count myself among them -- I teach physiology. And I think in my early imagining I made an analogy between chaos and pathology and regularity and optimization. So I'm wondering whether that is a fallacy into which, or a pitfall, into which people like me have fallen. Will the panelists please comment?

Dave Meyer: I have one small comment about that. This is not work that I've done or know well, but my understanding is that actually for heart rhythms, chaos is exactly healthy, and regularity is pathology. So that analogy is, at best, too crude.

Yaneer Bar-Yam: Ary Goldberger actually will be talking about that on Friday. It depends upon, as usual, what scale you're looking at. Regularity and heart is good on the finer scale, but on a longer scale, i.e. over minutes, or hours, or days, it's good for it to be irregular.

And the main reason for it can be actually shown through a theorem in physics, which is called the fluctuation dissipation theorem, which says that a system can respond based upon how it fluctuates. If you apply a stress over a day, you have to be able to respond over a day. If you apply a stress over an hour, you have to be able respond over an hour. If you apply a stress over a minute, you have to be able to respond over a minute. The system has to have fluctuations, also, over many different time scales in order to be able to respond on those different scales.

Dave Meyer: One of the settings where this is increasingly well understood is in ecology, where the idea is usually termed "resilience." And there is an idea that if there is this ability to respond to different kinds of stresses in a non-destructive way, then the ecosystem continues in roughly its same state, with these fluctuations.

A good example of where that hasn't happened is in the forests around Los Alamos where because they've prevented burning for so long, the underbrush has built up, and the system is no longer resilient in this same sense. It's not the way it's supposed to be. And so when someone screws up and starts a fire in the middle of the windy season, you get a disaster like they got.

Ali Minai: The issue is you don't want to be, in a loosely speaking term, in a chaotic situation where you cannot achieve anything. You cannot do whatever the objectives of the system are. You don't want to be in a rigid situation, like in a psychological situation; you don't want to be obsessive-compulsive, for example, and only have one response to any stimulus. The same with an organization. The same with any system. You need to be in a state where you can achieve your objectives, but adapt to changes in environment, and indeed, changes in those objectives. And it's that part of the edge of chaos that really was part the original intent, I believe, and why, for a lot of the things you hear people discuss, including psychology, it's got this appeal. It's adaptive, but still sufficiently structured to accomplish the objectives.

Helen Harte: I was in a clinic the day before yesterday in Seattle, a large clinic—ten doctors—and I heard a conversation between the receptionist and a patient. I could only hear the receptionist's side, very nice girl, and she was very nice to the patient. But she said, "Now, what is it you're coming in for?... Oh, that's two problems. The doctor only has 15 minutes, he can only address one problem. Which problem would you like him to address, and then we'll make an appointment for the other problem." (Laughter) And that is repeating itself hundreds of thousands of times in America's physicians' offices.

UNIDENTIFIED: I want to distract you all quickly with a statistical joke, and then make a point. The statistical joke is from a friend of mine who works for the census. And he says, "Miami is a really interesting city because it's the only place where you're born Hispanic and die Jewish." (Laughter) The more serious point I'd like to make is this metaphorical mathematical issue. I am new to complexity also. Actually, I just came here from doing organizational development work with the College of Nursing and found the metaphorical use of the edge of chaos, and the relationship between dead systems, equilibrium systems, complex systems, to get folks to re-think what they were doing and how they were doing it, was extremely powerful.

The second problem to leave you with is I'm trying to figure out how to explain heroin epidemics, and I need everything from neurophysiology up to global politico-economics. The metaphorical value in the complexity literature that I take away and then go into the privacy of my own home away from the mathematicians think about is how powerfully this is as enabling me to specify interactions among different levels. And then second of all, to look at those different levels as dynamic systems that have a history that are changing over time, and try to map those historical changes across levels. It's leading to extremely interesting and powerful understandings of heroin epidemics, that, at least as far as I know, haven't been represented in the literature. So figuring out this math/metaphor divide, and when it's crossed, and when it's permissible, and when it's problematic, would be really useful for someone like me.

Gary Ohn: One of the criticisms that I've seen of complexity is that it's too fuzzy. It is unbounded, and it has a tendency to degenerate into pure relativism. When you talk about open systems where the system involves the environment, the environment involves the observer, at what point can you draw some boundary that will actually allow you to perform analysis on it? Our analytical paradigms, our basic science paradigms, are based on Newtonian principles and linear analysis. Even the talk about dynamic systems analysis is essentially trying to reduce a complex system into manageable components.

I'm heartened to hear that the physics community is trying to expand into the complexity range, because it seems like many of the other sciences look to physics with some envy as far as analysis. But in order to avoid pure relativism and get some sort of framework, I would like to hear the comments of the panel and other speakers regarding some type of guidelines upon which you can draw some sort of perimeters where you can say, "Okay, I am going to study this component with these particular tools," and such.

Yaneer Bar-Yam: I think one of the things that Michel again spoke about earlier, was the problem of progressive approximations of a curve with segments. The basic idea that one has progressive approximations is something that one has to carry over into understanding complex systems, even if we're not doing linear modeling, or smooth line approximations.

Anything that we understand is some degree of approximation. And one of the problems that one encounters with thinking of applying the ideas of science to any system is that one has to develop some recognition of how accurately you want to apply your understanding. But more directly to your question about the things interacting and when can you have them interact, or when do you need to have them interact, Herbert Simon wrote a chapter in his book, *The Sciences of the Artificial*, on what he called the architecture of complexity. And in that, he talks about complex systems as being nearly decomposable. The idea being that complex systems are hierarchically structured, and you can basically treat the parts as being separate, except for the fact that they also interact.

But the problem there is to understand how you can balance this notion of almost separate from interacting, and how you deal with that. That's something that we are learning how to do, in some sense, by progressively approximating the nature of the parts, and their interactions in different ways. So the parts have to be strongly described, and the interactions sometimes can be more casually described, if you will, less accurately described, in order to describe the system.

UNIDENTIFIED: I'm going to be talking next and responding to this directly, but as a biologist, I think there is a big rift between the physicists and the biologists. We all know that complexity is right, because it's complex; it's hierarchical; the function comes out of multiple interacting components. But I guess what the biologist, or the person that asked that question from a hospital, was probably trying to say is that the physicists can describe probabilistically, statistically, and be really excited that they describe an average or a population, even the dynamic.

The biologist wants to be able to deal with tangibles, or have tools that will allow them to deal with more complex systems. Once you get to that, then you could start applying your tools. But the physicist thinks generally, and the biologist thinks specifically. There is this idea that you're going to have one equation that's going to describe the whole biological system, where as the biologist says, "You guys are talking so generally, this will never have any use."

So I think the key is beginning to define systems that are functionally relevant, and those systems may be different. The same constellation may share elements from different systems, but it's focusing the question down to basically one molecule a problem, where you still have all the functions of that molecule, you're not yet at the atomic level. I think it's happening, but there are language barriers there.

Gary Ohn: Well, the jokes regarding statistics are a primary example of that. It seems like classical physics works because the population of atoms is large enough so the statistical mechanics will work.

Yaneer Bar-Yam: Actually, medicine relies very, very heavily on poorly applied statistics. (Laughter) But why? What do they do? They do large clinical trials, right, and the clinical trials are based upon understanding how a drug works for a large population on a particular measure of well-being, which is a very poor way to apply statistics. And in fact, one could argue that one of the main contributions that complex systems approaches, including approaches that physicists have learned in terms of improving on how one deals with averages, or knowing when averaging applies or doesn't apply. And that's really the key: knowing when averaging applies and doesn't apply, is really essential and important to medicine.

And one of the things which is absolutely incredible is that people are now talking about fitting the medicine to the genetic code. And they don't realize that the whole medical system of doing clinical trials has to break down when you do that. And the only solution to that is, in fact, understanding how the system works, which is what complex systems does.

Jack Cohen: This point that Yaneer made is absolutely central and rooted, I think, in the way in which we approach complex systems. This way of looking, at the moment, is mostly telling us what we can't do.

In my last week at the IVF clinic in 1987, we got the hormone profiles for three women. These hormone profiles were totally outrageous. The progesterone, estrogen and FSH and LH were doing all the wrong kinds of things. One woman was clearly running her cycle on inhibin and activin which are different hormones all together. And I am interested in genetics and ecology relating. I am certain that we all differ from each other at about 10% of our loci. What this means is that each of us is a different kind of machine.

I took these three charts of the endocrines of these three women to our chief gynecologist, and I said, "Look Peter, isn't this amazing?" And he said, "Well, women can be infertile for a lot of reasons." I said, "No. These are the three women who produced the good eggs this morning. " Women can be fertile and physiologically healthy for totally different reasons.

And we did a sketch, and it turned out that probably only about a third of women have cycles which fit what it says in the textbook. Therefore, when you go and you try out something and you average all of these people as Yaneer says, it's mad, because each person has a different system.

When Ronnie Mainieri was talking this morning, he was talking about a different trajectory for each child. That's fine. But if you average the children, this is the crazy business. The mean field human being is a nonsense.

Dave Meyer: You've said "mean field" several times and you're using it wrong. You probably shouldn't say it quite like that. But Ronnie, since you made this point earlier, maybe you should respond.

Ronnie Mainieri: There is the law of large numbers, so you can't use average when the moments don't exist. So that's one point.

I wanted to make a point also related to averages. At Los Alamos, one of the things we do is a large simulation of traffic. We have this huge simulation of the traffic in Dallas, and we've been trying to reproduce when you get traffic jams, and it just doesn't show up on the simulation the exact way.

You don't get the same type of traffic jams. If you look at the size of the cluster of cars in a traffic jam, if you look at how fast they're going, it just doesn't reproduce what you observe in Dallas in what's coming out of the simulation.

Well, if you sit there and do analysis, you find out that you are averaging the wrong thing. What you should be averaging is how do cars drive slowly? When they start driving slowly, how is it that they do it, and how often do they do it? So if you look at the tail of the distribution of slow velocities, you get the right behavior.

Now, this is something you couldn't have guessed. Who would have thought of looking at that type of function to average, so you have to know what you're doing. You have to know which quantity to look at, and the only way you get that is by analyzing models.

Another example is one David and I had looked at and has to do with organizations. Usually in organizations you get cliques of people, and it's been very hard to figure out what these cliques are. But if you start looking at them in terms of the tail of the distribution, you find out that there are certain general features there. So you have to know what to look at. You just can't average any quantity that comes to mind, because some quantities just lead to bad behavior. And I didn't have a chance to say here, but you can make it precise in terms of dynamical systems. Those are the ones that have i-gon value 1, and what I have shown you, and those usually lead to bad behavior.

Jack Cohen: Yaneer was saying that the doctors average, for example, the amount of hormones you've got, but that actually, I've got a 20th of the average amount of male sex hormones. I'm entirely normal in the way of behaving, but, many --

Ronnie Mainieri: Thank you for sharing. (Laughter)

Jack Cohen: -- regard this as highly pathological.

Ronnie Mainieri: Yes, and I agree completely with you. Let me try and rephrase it in my language—we have different scales of how things work. What really matters is the dynamics of how these things work. I agree completely with you. You have to figure out what are the quantities that are independent of the absolute scale of things. And there are certain quantities like that.

Fred Jennings: I'm an economist, and all of this is of great interest to me, because economics is probably worse than physics about dealing with complex systems. We are trained to put problems in neat little boxes, and not worry about anything outside of the box that's an externality.

However, a simple recognition of what I consider to be a fact, that every act every one of us makes on the world, ripples outward forever onto everyone else, puts you right in the middle of the problem the gentleman over at the other mic brought up of, "Where do you draw the line?"

It does seem to me one place to draw a line, given the unboundedness of our impact, is within the notion that I call "planning horizons" where the line is drawn by the range of those effects that we take into account before we set them in motion. And that if we can talk about planning horizons and horizon effects, we can get quite a distance through this problem. But this, in turn, gets us right to where Daniel Miller sits, and other people, because horizontal issues are inherently psychological as well as connected with other things. These are directions of thought that I think are productive for a complex systems approach.

Don Ingber
Biomedicine

Dave Meyer: This is Don Ingber who is going to speak. Don has his BA, MA, MPhil, MD and Ph.D. all from Yale, whereupon he left and went to Harvard, where he is Professor of Pathology at Harvard Medical School, and Research Associate in Pathology and Surgery at Children's Hospital, and a member of the Center for Bioengineering at MIT.

He has worked on understanding the architecture of cells using ideas from Buckminster Fuller, and I think that's what he is going to be talking about partly today. He has also worked on a number of other things, tissue engineering, cancer research. I don't know if he was the discoverer of an angiogenesis inhibitor drug that is in clinical trials now, which is pretty exciting. And he's also a founder of a company called Molecular Geodesics, which I asked him about earlier, and you guys should ask him about because it sounds kind of interesting.

Don Ingber: Okay, I'm going to take a big shift here. I've been interested for 20 years in complexity, without maybe using the word. But I work on the ultimate example of complexity, which is the living organism, how it forms, how it's regulated, and more recently, how it evolves.

I take a more wide view of complexity because of the questions asked. I'm trained in medicine as well as biology, and we need to know more than describing the system and the likelihood that things are going to happen, the averages, trajectories. I want to know how it works. I want to know how the machine actually works. And that involves structure as well as informational probability. So let me give you that as background and take it from there.

I think you know that medicine is currently undergoing a major paradigm shift. Up to now we have been interested in what life is made of: the DNA, the RNA, the proteins, genes, etc. However, very soon we're going to have all the genes laid out before use like all the car parts on the table, and then we're really going to be challenged with the ultimate question, which is really understanding how complex behaviors emerge from collective interactions among all these different components.

This challenge to understand or explain how the biological whole is greater than the sum of its parts is what I call biocomplexity, and I think it fits in with the themes mentioned earlier. Now, I think those of you who are physicists trying to get into biology, or complexity people, think purely in terms of genes, and networks, and that's biology. And the fact is that right now we have techniques so that we are literally drowning in data.

However, this is not going to explain all of biology. A simple example: probably 99% of you don't know experiments done 30 years ago. They took the nucleus with all the chromosomes and DNA out of a tumor cell. They took the nucleus out of an egg. They put the nucleus in, and you get a normal, healthy organism. That genome had all the gene defects you hear about, but it's rebooted, it's reprogrammed by structure that's in the rest of this cell. This happens again, and again. So this is only the beginning. This is basically the inventory. In development it's critical for switching which parts are put out, but it's not the whole answer.

Now, for that reason people are moving towards structure, or what they call proteomics. However, again, understanding 3-D structure doesn't explain it all, because the same protein has very different functions when it's part of what we call a macromolecular complex, what you guys would call a three-dimensional network.

It's also a different function where it is in a living cell. There are proteins that, again, make a normal cell tumorous or cancerous, and there are enzymes that get turned on such that they never shut off. That's called constitutive. But in certain cases if that enzyme lacks a targeting sequence that brings it to a tiny region of the cell, it's active all of the time, yet you don't get transformation. Spatial issues come into effect.

Now, people are beginning to get into the realm of modeling, and this is actually from George Church at Harvard. They've developed a simple model of the complex metabolic pathways of the red blood cell, and other people have done bacteria, and they have all the different kinetic parameters for the literature. And for a lot of things, they could basically put some input change like a glucose input and see how lactate changes at the other end.

However, this is not where the future lies. The future lies in trying to understand, say, how a red blood cell in somebody with sickle cell anemia starts clotting up a capillary when they have low oxygen. And that is a mechanical change for stiffness of the shell of the red blood cell.

And the other point is that in most cells, most of the metabolic pathways, the chemical processing pathways, don't occur in solution—which is the way we study it, publish it in the literature, and the way we make models—it occurs in what I've called in the literature a solid state. The enzymes and substrates are immobilized on scaffolds. You have much higher efficiencies, and you channel reactions. And all of that information is absent.

Now, this becomes even more true when you deal with cells, such as the ones I work with, that are in complex tissues, that have the ability to contract, have a complex internal structure, and move, and grow.

And so the future really is the challenge in the field, and I think the challenge that I want to get at today, and that I don't see discussed in the complexity literature, is linking informational complexity, as people often model, to structural complexity, which is basically that there is information in structure. I hope at the end of the talk you realize that perhaps the greatest information is in structure, maybe all information in structure. After all, DNA is not what it's made of, it's that it's a double helix.

I'm going to give you examples of what we've done experimentally and theoretically, and I think in the process you're going to get a feel for hierarchical issues, for emergence issues, attractor issues, and everything that you guys talk about in general terms. I'm interested in how living cells structure themselves so that they exhibit their incredible organic properties, including the ability to change shape, to move, and to grow.

Now, most people, and probably all of you, tend to think of the cell as this elastic membrane surrounding a viscous cytosol, or protoplasm, or as I always like to say, a water balloon filled with molasses. And that is the way most people model it, and that's the way most people think of it.

In contrast, what we've been able to show over the last 20 years is that living cells are actually hard-wired so that they can take the same set of inputs, chemical inputs, and produce entirely different gene program outputs, or in your thinking, can switch between different attractor bases. This hard-wiring in the cell exists in the form of a continuous series of molecular molecule, nanometer-size, molecular ropes, struts and cables that are known as the cytoskeleton, the cell skeleton, that stretch from specific transmembrane molecules on the surface, to discrete contacts on the nucleus, and from there, the other filaments to individual chromosomes and genes. And furthermore, that cells can be switched between these different gene programs through mechanical distortion of the cell and its internal supporting structure.

Let me give you an example of what I mean. A few years ago, we wanted to make cell distortion an independent variable. The idea is that living cells normally stick by putting out anchors and pulling themselves against them, flattening, flattening, like a pancake.

The idea is we wanted to give the cells all of the chemical stimuli that we know stimulate growth. These are hormones; these are things that they bind to on the dish, yet change how far they can stretch. The way we did this is by adapting a microfabrication technique from the computer industry—there is my link to physics—that was developed by George Whitesides.

And I won't go into the technique, but basically the idea was we can make little islands that were adhesive, surrounded by non-adhesive teflon-like regions, and if we plate cells on them, the idea is that on a big island, they would pull and flatten like a pancake, on a little one they look like a cupcake, and on a tiny one they would be a golf ball on a tee.

And the question was would something as general and as generic as a distortion be able to regulate cellular function and switching between gene programs. These are cells that happen to be capillary blood vessel cells, the cells that make up the blood vessels, your small blood vessels. And normally they're very spread and triangular-looking, and irregular-looking. When you put them on these circular islands, you actually get circular cells. And if you make square-shaped islands, and you plate the same cell, you literally get a cell with 90 degree corners. And this shows you that cells, they're not just a water balloon filled with molasses, because you're not going to get this kind of 90 degree corner with that technique.

Chris Chen, a student in the lab, did this experiment, where he made shape a variable. Put the cells right next to each other. They have all the same optimal stimulation from the substrates, simulation from hormones. And he asked what happens?

And what he found was that as you increase island size and you promote self-spreading, you get an increase in DNA synthesis growth and proliferation.

Conversely, as you prevent spreading and shut off growth, you turn on a gene program that says "suicide." It's called apoptosis, and it's known as the suicide death program. More recently, we've made substrates that hold cells in the trough region. We do that by using thin lines, and these cells form cords, which over a period of hours, form tubes. These are capillary blood vessel tubes in a dish. So basically, we're switching cells between a growth program—which is a whole set of genes—differentiation, or death just by the way we've stretched them.

Just a few quickies that may be interesting from a physical standpoint. This cell is square. We give it a factor that makes it move. Cells move by putting out extensions called lamellipodia at one part of the membrane. What you notice is that these cells don't put it anywhere—they only put it out from the corners. This is a structural or geometric control of the motility response, which is critical, for where a cell moves is more important than how quickly it moves. It doesn't matter that you understand any of this. What matters is that if you're going to try to model biology, it's not all genes. There are no genes going on here in the 5 to 10 minutes that these responses are going on.

Now, one thing that we observed, we have a system here where we see symmetry-breaking with mammalian cells. Mammalian cells, when they're alone on a dish, move in Brownian motion; they're random, so, ultimate symmetry. All we do is change spatial constraints. We decrease the borders so that they have to be touching, and we get this yin/yang phenomenon, where the cells either move clockwise, or counter-clockwise. And we've actually been modeling this symmetry-breaking in very simple ways.

I don't want to go further, but just to show you that in terms of physical models, it happens in biological systems if you simplify the system so that you can address a simple question, or a simple phenomenon.

Now, the reason this is all-important is this is the essence of how you get pattern-formation in all developing tissues, whether it's mammalian, insect, or plant. The way in which you get tissues to form that have very specific patterns is to have localized differentials in cell responsiveness. In a capillary, one or two cells have to respond to all of the factors that tumors give out by branching and growing, but the neighboring ones have to not respond. Even though they're the same genetically, and they look the same, if they all responded, you get a big pile of growing cells, you get another tumor. To get pattern formation, you have local response, and then you repeat this along the side wall of the growing vessel, and again, and again, and that's how you get the fractal-like patterns that are characteristic of normal tissues in all systems.

And this local control of response to multiple stimuli by mechanical distortion is what I'm suggesting may be going on. And it's well-known that in the embryo, cells pull on each other. They do the folding/unfolding that you just heard from a mathematical standpoint, but they do it by physicality and applying forces, and the forces are generated in other cells.

Sui Hwang will talk on Thursday a little bit about this, but he's been actually doing modeling with Boolean networks with this sort of system, and showing, in fact, that this idea that something as non-specific as a distortion can actually hook into the same gene pathways in something as specific as a hormone that binds a specific receptor. There is strong support for the idea that there are really attractor bases for growth and depth and differentiation, and that perhaps if we look at genetic networks using these approaches, we'll begin to pick out things that are much more linked from an information-processing way, than just seeing things that are going on and off, which is the current approach.

In terms of how this works, I won't even go into this. Let me just tell you that shape hooks into the same machinery, as I just intimated, that growth factors and hormones use, but it does it many hours after the early signal networks that people usually think of in terms of growth control. It does it ten hours later, but it still has equal control to determine whether the cell will grow or not.

The one thing that was interesting was that we found that by using a drug that doesn't affect the cell shape at all but just affects the level of tension in the cell, we can get the same blockage that's causing the cell to round. And I have to clarify, I didn't say this, but cells generate tension through molecular filaments called actomyosin filaments. They're exactly the same as a muscle. When they're triggered, they shorten, and they create tension. And so by changing the tension in the cell, we get dissipating tension. We get the same effect as rounding.

So the question is how could changing shape or tension affect dynamic regulatory networks of a biochemical nature? And I think here is the case where we're never going to understand this if we view this cell as a water balloon filled with molasses, or as a bunch of membranes.

In contrast, what we've been able to show is that cells actually use a very particular form of architecture that comes out of the Buckminster Fuller world of geodesic architecture. It's known as tensegrity. It comes from the words tensional integrity. It basically differs from most manmade brick-upon-brick systems which depend on continuous transmission of compressive forces for their stability, because it depends on continuous transmission of tensile forces. And just to give you an example, this is a geodesic dome in the cytoskeleton of one of our human endothelial cells. Each strut here is made out of actomyosin filaments oriented in ??, and those are billionths of a meter wide.

Now, the simplest way to build one of these structures is like this. This is going to come up again and again, but to me, this is the simplest model of a complex system.

This structure, if I take apart the elements, is basically a bunch of wood, sticks and strings that individually have their own properties. However, when I connect them by tension, the sticks don't touch at all. So you have these big sticks, and they're rigid, and now I just, through architecture, pull them together, and now this has emergent properties. This structure can hold itself into form. I'll show you later that it has very characteristic mechanical properties. But I started using this as a graduate student at Yale as a model for the cell. And it wasn't so weird because like a cell, when this is not attached, it takes on a round form. And when cells are passed from one dish to the other, they're round. And if I attach it to a rigid foundation, maybe I can -- Can people see this if I flatten it here? Basically, it spontaneously flattens out, just to minimize stresses and strains.

And if I were to now clip the anchors, which is what we do with chemicals that degrade the molecular anchors when we pass cells from dish to dish, this thing bounces off like that. That's exactly what cells do in a dish.

And literally, just as an aside, I was taking a sculpture course where I saw these when I was trypsinizing cells, and that's how the last 25 years of life took a right angle.

The other thing is that if you attach this to a rigid dish, you can't see tension because it's isometric, it's stiff. But if you attach it to a flexible foundation like here, it will pull it up into wrinkles.

And in fact, if you plate cells on thin, flexible substrates, like the ones you wanted to theoretically think about—these are real thin flexible silicon rubber. You put a cell on it, and they pull it up into wrinkles just as predicted in the model.

Now, I then built hierarchical models. We talk about hierarchy all the time. That is one of the inherent properties in these systems. This structure, the struts could be made up of a mass that is a tensegrity structure on a small scale. That's one beautiful example of hierarchy. The other is that I put a nucleus that was one tensegrity sphere, where the sticks and strings don't touch. I had a big one of these models that was separate, the same properties and rules. Then the trick was I attached it by black elastic thread, tensile elements that you can't see because of the black background. And now, when it's stuck and stretched, both stretch in a coordinated manner.

This, to me, is the ultimate example of hierarchical integration over different spatial scales using one fundamental guiding principle. Furthermore, it predicted what we see in living cells, that when they spread and flatten, the nucleus spreads in a coordinated manner.

This system is also used at higher macroscales in the hierarchy of life. If you look at a human skeleton in an anatomy lab, you see every bone has to be wired together and hung from a stick to look like us. In reality, we're 206 compression-resistant bones that are pulled up against the force of gravity, and stabilized through connection with a continuous series of tensile muscles, tendons and ligaments, just like this cantilevered structure is built by having huge steel girders attached by high-tension cables using this system. And furthermore, the stiffness of my arm, the shape stability, is due to the tone in my muscles, or the tension, or the pre-stress. And that is one of the fundamental definitions of these structures. The pre-stress determines the stiffness.

Now basically, my group and others have shown that cells use this type of system to stabilize themselves. [...] actomyosin filaments that form a network that go everywhere and pull, and then there are other elements that resist compression locally, and one of those is called a microtubule, which is actually a hollow molecular filament, so it has a higher second moment of inertia, it's going to be stiffer. And what I'm going to show you is local compression in a living cell here.

This is a technique that biologists now have, ways to genetically splice a little sequence from an algae that is autonomously fluorescent. If you take that little sequence and put it into any protein, you can see your protein under fluorescent light. So this is a molecular filament called a microtubule with time-lapsed photography over minutes to hours. What you see as it grows, it hits an obstacle ?? and it buckles, slips by strings, hits the cortex of the cell, and there is a buckling wave.

If I were to fix it at any instant in time, it would look like this, which is how we always see them. And that means that at any moment in time these things are locally compressed, yet we know the whole cell is under tension—if you cut it anywhere, it pulls apart. So that is tensegrity at the molecular level: global tension, local compression.

These are not the only compression elements. This is a nerve cell from the neuro people here. When nerve cells move out, as all cells, they have a leading edge that puts out little exploratory fingers or arms that move like this. What they are are polymers of these acting microfilaments that are linked together like basically taking soda straws and putting tape around them. They get very stiff. And the way they push out the membrane to lead forward motion is exactly the way this tensegrity building by ?? pushes out its membrane using compression ??.

Now, the most important thing about these structures is that they do more than explain pattern formation; they explain mechanics. If I pull on a 2"x4" in one of our houses, I'm going to get local bending and breakage. If I pull on one of these structures, all of the interconnected elements globally reorient. Here there is a weight right here. All of them globally reorient to take the load. And as more stiff elements come into the alliance of the applied stress, the structures get progressively stiffer, and stiffer, and stiffer. This is known as linear stiffening behavior.

What's amazing is a few years ago we developed a way to measure the mechanics of the cytoskeleton in living cells, non-invasively. And just for the physicists in the crowd, the way we do it is we take small beads—and by small, I mean 1 to 10 microns—that are ferromagnetic. We coat them with molecules that only bind to specific molecular receptors on cells, and then we twist them with twisting magnetic forces. We apply a torque or a sheer stress, and we can measure how far the beads rotate so we can measure angular strains, so we can do stress-strain curves over specific molecular linkages.

And what we find is that living cells actually mimic this linear stiffening behavior that these simple models express, and the effect is mediated by these different interacting cytoskeletal elements. Whereas, if we pull on other receptors that go across the membrane but are not linked to the skeleton, that are just in the bi-layer, they are very floppy, and you don't get this kind of response.

What's real important is that this linear stiffening behavior, if you go to bioengineering textbooks, is also a fundamental property of almost all living tissue. So if you want to talk about universal principles, muscle, skin, mesentery, cartilage, all give you the same behavior. And in the textbooks it says it's universal, but no one can explain it from first principles. You could curve-fit it, but we can't explain it.

And recently, Dimitrije Stamenovic, who is in biomedical engineering at BU, has been able to use a simple model like this. He has developed a micro-mechanical model, and he can now explain this linear stiffening behavior, starting from first principles. And I think this is a really nice example of emergence. You get a linear stiffening response out of nonlinear elements, and the two key elements are architecture and pre-stress. And maybe I'll come back to this at the end, but this can be thought of conceptually.

My wife is a psychiatrist for the psychiatrist ??, and we talk about there being many interacting elements in a family and their equilibration of needs that are pulling and pushing. And there can be a lot of pre-stress in that family, or there could be little pre-stress in the family. And that's going to depend on how it's going to respond to some aberration.

What's more important is that this model has come up with some *a priori* predictions that are different than current engineering models in biology that are curve-fitting models. These are totally *a priori*. Recently, one of them is that the stiffness or shear modules increase in direct proportion to the pre-stress, not applied stress, but pre-stress. And we know of data that matches that beautifully.

With a company called Molecular Geodesics, we have actually done some computer simulations. This is 36 of these little toys interconnected, held at this end, and fixed in place. And what you can see is long-distance force transfer, and an almost organic, undulating-like movement. And just to give you an idea of how organic it is, this is the leading edge of a migrating cell, which is actually made up of filaments that form networks that are not so different from this.

Now, in the biological world, people have a hard time because they really think the cell is this viscous jelly. If this is right, and cells are hard-wired—different hierarchical systems are hard-wired—then if I were to pull quickly on a receptor that couples across the membrane to the system here, boom, I should see things move in the nucleus, and throughout the depth of the cell.

This is a cell; the white dots are actually mitochondria. If you remember from basic elementary school, junior high school, they're the organelles for respiration. They're made white with this fluorescent molecule. And we have a ?? here that we're pulling with a pipette, and you can see that throughout the depth of the cell they move. This is the nucleus it's actually distorting. And somewhere in here—here it is, right here—there is a little lever arm that goes down when everything else goes up. So there are elements in here vectorizing the force distribution, much like you'd expect from something like this, as opposed to a bunch of string.

Now, other groups have shown that this works; this is going on even at smaller hierarchical scales. Hopefully you know that cells divide; they go through what's called mitosis when the chromosomes separate. And the way they separate is they form a little spindle, it's called. And a spindle is made up of those microtubules, those struts I showed you, and they formed this structure that looks like this.

What one group in Australia did is they took a very fine laser beam, they were able to break one of the struts, and when they did that, they showed the other struts buckled more. And what this shows is basically it's like a tent with flexible tent poles, and you break one, and the others buckle more, because again, there are local compression elements pushing out against the network that normally stabilizes it. And this is tensegrity at a smaller scale.

Now, you could go further. This is a virus. Most people know that viruses are geodesic, but usually you see these sort of billiard ball models, but they actually often self-assemble by putting out extensions of their protein tails that form into a geodesic scaffold. And the stability of the scaffold is dictated by the stability of the vertices. Now, if you look at the vertex at the left, this is how the biologists/biochemists/molecular biologists will model it. This is the end of a protein chain, and this is the end of another one, and this is the end of a third one. And they form a stable vertex by overlapping in consistent ways.

The stability of that vertex, and hence the whole virus, is due to the attractive pulling between these, and the stiffness of the amino acid changes so it can't be bent or buckled. What you see at the right comes from the instruction kit of tensegratory building systems, and it's exactly how you are to build something that looks like a virus. You start with many little vertices that go stick over under/over under/over under just like this, instead of having attractive, intermolecular bonds, you have tensile strings holding it together.

And I would suggest that proteins can be thought of this way because proteins have regions that are locally stiffened. They have no material properties you can measure; they're dictated. They're called alpha helix. Then they have flexible regions, and then they'll have another helix, or they'll have something called a beta strand which is a known stiffened region. And then they pull into a stable form. For example, this is a structure, it's a beta strand, an alpha helix beta strand. They pull into a stable form that is pre-stressed. So we know it's pre-stressed because if you take an enzyme and you cleave it, it splays on; it loses its shape and its function. And of course, all molecular function comes out of molecular shape. And tensegrity can be built, like, this is the same structure with all springs. It doesn't have to be elastic cords, and wood, and so forth.

Now, what does this all mean for cell regulation? Well, what's really key, and I said this in the beginning or maybe in the answer to the other question, is that what's key is that these structural systems are not just structural systems. They orient almost all of the cell's metabolic machinery, e.g., DNA synthesis, RNA synthesis and processing, protein synthesis, glycolysis, and signal transduction. Most of the enzymes and substrates that mediate these key pathways are not floating around in a liquid cytosol or in a lipid bi-layer. They are actually immobilized on the insoluble scaffolds within the cytoskeleton and the nucleus. And because you bring enzymes into juxtaposition, you could actually have what's known as channeling the reactions. The product of enzyme A is the substrate for B. And then you have whole complexes, next to other complexes.

Often, you can put a soluble molecule in the cell, and it will not compete with the same molecule that's already in the functional pathway. So there is an incredible amount of structure in cells.

Now, I think this can help explain how you can integrate mechanics and function. If you think of planet force on a macroscale, like blood pressure in a blood vessel, it's going to distort this matrix in the wall, which has cells on its attachment scaffolds, which then stretches the cells, which then stretches this internal cytoskeleton. And if cells use tensegrity like this to link all of those hierarchical levels—and I didn't go into it, but you could show the lung as a tensegrity structure with collagen bundles and elastic molecules linked together—but if it works this way, then a local distortion will result in restructural rearrangements at many points in the cell simultaneously.

Now, if biochemistry does function in a solid state, then it's really basic biophysics. If I were to pull on a filament really quickly, a molecular filament, either it breaks or it distorts. If it distorts, then each individual molecule has to distort. And if from a physical or a biophysical sense if you change molecular shape, you change molecular mechanics, you change thermodynamics, you change kinetics. If I have a spring and it vibrates here with a certain frequency, if I change its size, I change the frequency. If I change its center of gravity, I change its frequency.

Similarly with molecules, if you change their configuration, you could change their kinetics. And in fact, what this is showing is where people have identified there is a molecule in the membrane known as a stretch sensitive ion channel in all cells. And what this shows is the structural rearrangements that go when you apply stress to a cell.

And the first thing is this is a real trickle-down—excuse me George Bush—but it's a real trickle-down system where the forces trickle down. It actually goes through the cytoskeleton, over the attachment substrate, across the receptors to the cytoskeleton, back to these receptors in the membrane. And it's not like you physically hold it open; you change its kinetics, just like I was describing a minute ago.

M: Is that a mechanical, or a quantum mechanical model?

Don Ingber: This is a model from crystal structure. This discolored thing is actually from NMR, you know, crystal structure of the protein and how it undergoes --

M: Yes ?? then after it goes down it's getting deformed, right?

Don Ingber: Yes. That was done, I think, in different states, but I'm not sure. That I don't know. It wasn't our work. It was from the *Science* paper. I think this explains how you get transmission of information across receptors, which is sort of the key of how you open systems, someone said, how the cell senses its outside world. A local binding on a molecule on one side of a membrane results in global structural rearrangements through the molecule, so that there is a structural change on the other side, and that, then, induces a change in molecular binding because the configuration has changed.

Now, does this happen in cells? And the answer is yes. And I'm not going to take you through all of the data, but let me just say that the way in which cells anchor their skeleton to the outside world is through forming little spot-weld-type anchoring complexes that are called focal adhesions, and they bind through specific receptor molecules that cluster there. They're called "integrants," for integrating. What we find is that they link to the cytoskeleton, and the cytoskeleton is the scaffold for much of the signal transduction machinery that those of you who model probably come across— and I won't go into all their names—but these are what convert an input signal to some downstream response.

And what we've done recently is basically take those beads that I told you about which bind to these, and we twist them. And we ask, "Can we affect biochemistry and gene expression?" And I won't show you the data, but the answer is absolutely yes. You get stress-dependent regulation of chemical signaling pathways, as well as genes. Whereas, if you twist another receptor right at another point in the membrane, you don't get that sort of response.

So what does it all mean for tissue development? Well, I think what I'm trying to say is that I think tensegrity is basically another form of complexity analysis. That instead of just dealing with probabilistic, and statistical relationships, takes ?? structure mechanics and architecture.

So classically, we know this is a complex system where there are many interacting parts. The function comes out of interaction of all of these parts. So, you know, in the complexity approach, we're trying to figure out the network and how we get emergence, and we're just throwing every possible connection. And what we do in tensegrity is basically say that some elements, that there are forces there, and I mean, force balances come up in lots of talks in strange ways, and Yaneer mentioned it.

But instead of force balances in a two-body problem, we're talking about force balances in a 3-D network. And what happens is they come into balance across the whole system, and then the system has inherently that if you build other systems that are connected by similar rules, you get hierarchical systems that are always coupled, part and whole, but can go through -- Every time I move my arm, my skin stretches, the connective tissue, the attachment to my cells and the cytoskeleton, but nothing breaks. It's always reversible. And furthermore, if I stress it too much, everything will change its structure and chemistry to minimize the stress and remodel, and that's how our bones remodel, and how all tissues remodel.

Now, I want to end just by -- I know people who model love evolution. I found this—this is my *Mad Magazine* book from when I was a kid that I just found for my 8-year-old. This says, "The satellite seen in the upper right contains several cows. It is the first herd shot around the world." (Laughter) But I think it's a good view of evolution.

I think in terms of evolution, everybody focuses on gene mutations, on permutations, and all of this. I think my point is if we really want to find fundamental guiding principles, that genes are actually a product of evolution, and not the driving force behind evolution. It's just one manifestation of evolution.

And in fact, what's interesting is if you go into the inorganic world, you see that these same rules of geodesic tensegrity architecture dominate small systems as well as—inorganic as well as organic. So this is a recurrent theme in nature, independent of size scale. This is a carbon Buckyball. This is water in a structured state, geodesic, and this is clay.

Crystals are often fully geodesic, highly triangulated structures. What happens is that crystals are incredibly stiff, but incredibly strong and in the most economical, lowest energy state, because they're all pulling at each other, but they can't be compressed, and you basically come into a ?? of spheres.

When you hit clay -- and I learned this when I spoke at this meeting 3 or 4 years ago after Hyman Hartman who works on evolution. He had this diagram up just saying that evolutionary biologists are now thinking that evolution did not occur in a primordial soup. We think about the idea of getting everything to bind to everything else and evolutional issues, but it may have occurred in clay.

Now, clay is interesting. It is also geodesic, but instead of every tetrahedron and octahedron being close-packed as you might find in a diamond, they are packed like this. There is like an open region, and so each vertex can move, can pivot. And because it has this flexibility, it is the first catalyst, the first enzyme, if you like. And in fact, people have found that clay can catalyze polymerization of amino acids and polymerization of nucleotides. The synthesis of the substrates for the Krebs Cycle, which is the core energy cycle from which light evolved.

And so what I'm basically saying is that this idea of complexity in dealing with structural complexity may help explain how basically that there is a fundamental rule that goes from the inorganic world through the organic world, and that basically things are always trying to pull into stable forms, and in the most economical way possible.

And that once they have forms that have a little flexibility, they can explore evolutionary space, and assemble, self-assemble with other forms, and they'll not get anywhere. But all of a sudden, they'll hit a network that has its own stable form, and then these will join with others and build over time.

And so with that, that's my last slide, and I'll just leave it and hope that those of you who model begin to try to incorporate structure into the systems, because that's really how biological systems work.

And I think that's how all systems work, but we just don't think that the structure is there. If any psychiatrist or clinician doesn't think that there is a structure in a family, you know, they're wrong. One family is stable because it's a husband and wife, another is a woman and her dog, an old woman and a dog. And you destabilize either one of those, and it goes into flux, and it may come back to stable form if it has other triangulating points that can come into balance.

I'm going off, but I think you get the point. Thank you.

(Applause) I'm happy to answer questions. Yes?

M: I'm from the University of Rwanda. One problem in particular I'd like to talk about concerns the regulation of the cell cycle of a very simple bacteria, a bacteria about the size of the mitochondria that Don showed earlier on. And ?? considered rather simple things. Those really are the ?? the ?? of jelly and ?? what makes it ??. In fact, they're probably much more complicated, and bacteria ??. They were here before us. They'll be here when we're gone. ?? human beings is to make them ??.

Don Ingber: I once wrote that from a bacteria's perspective we live in its intestine. (Laughter)

M: -- in many ways superior. The thing is, bacteria do something very similar to what you've shown with the mitotic spindle separating the chromosome. They replicate their DNA; they segregate their DNA; and they divide. And we don't know how those fundamental processes ?? in E-coli, one of the best understood of all ?? all organisms. ?? how those processes are controlled, and ??. And the way I have approached this is to try and collaborate with a variety of people -- physicists and physical chemists -- to see if they can enlighten me.

So many years ago I was fascinated by the story of tensegrity, and I went to a toy shop and I bought my daughter something very similar to the thing you've got there. Of course I didn't give it to her, because I bought it for myself. (Laughter) That was a bit of ?? because my model has bells on it. So when I ?? and in your *Scientific American* article, you mentioned that maybe vibration plays a role. And over the years I've talked to a lot of physicists about how the cell might be a giant ?? oscillating and how it might converge on a frequency. And the shape changes you've shown, you can imagine that those changes in shape resulted in changes in vibration. Maybe vibration is important to DNA. The problem for me is that nonsense ??.

Don Ingber: Well, I mean, we're actually beginning to look at, for example, resonant frequencies in cells and developing ways to do that. Very little is done, but the point is tensegrity is basically an inherent harmonic coupler. And if you get two of them with their own frequencies and you put in a tension element, they become one. And then all of biology is mechano-chemical, so if you have a certain frequency and you stress some enzyme, it may change the structure, which changes its frequency response. But there is no doubt that a bone cell in your body, you take it out and if you walk around, your bone will remodel based on forces at a certain frequency, a certain rate.

If you take a cell and you put it on a plastic dish and you apply different frequencies of force, bone cells send very high frequency, low strain. Low distortion, high, high frequency.

If I look at my blood vessel, my aorta once a second, it's kind of extending, stretching a good amount, you can see it, and then coming back. If I take my cells in the dish, the optimal stimulation response frequency is about once a second, and it has to go through big strain, 10, 20, 30% strain.

So inherent in the structure is the sensitivity of these harmonic issues. Bacteria have been so hard to study. One, everybody only studies it from a genetic level because it's so powerful, and two, it's just hard to look at. But there is work going on now where physicists are, you probably know, working where there are molecules related to microtubule molecules, tubulin, and where they're going in oscillating gradients across the cell determining where it's going to divide. And I think this is where physicists are going to start coming in; they are starting to come in.

M: ?? my question wasn't a specific one ?? bacteria. My problem is a more general one which ?? explain how the ?? which is one has something very exciting ?? the regulation of bacterial cell cycle in tensegrity. But I can go and talk to a water specialist, and he'll say, "Well, I think the structure of water is very important." And water can have different structure in different regions.

Or, I can go to someone who knows about basic physics, and they could say, "Well, the distribution of lipids in the ?? of membranes is ?? heterogeneous, and we can make some very interesting models about how ?? regulate the cell cycles. And one can go on and on about how maybe it's to do with ion currents.

Don Ingber: The key problem with the complexity field that I've heard is that it doesn't have experimental testing and verification linked, and I think that's the essence. If you come up with that, and all my critics have always said 50 different things, we'd have to go and do them. And then we'd say, "Well, you're partially right, but you can't explain this and that, and we can explain both." And then we get another paper. And then they come back, and so forth. So if people come up with specifics, that's testable. If we're just doing theory, biology is so complex that it's not going to impact it.

M: Yes, but the point I wanted to make is that we talked about emergence as a very important phenomenon in the study of complex systems. And it seems to me that when we have the organizing process, if it's tensegrity or ion ?? structural information, one way we can assess their value in structuring an organism or system is whether they interact with one another in a positive feedback sense in order to generate a greater organization, which is ??.

But in the ?? general conceptual question people ask, do different organizing processes become selected because they interact with one another so as to produce organization of positive feedback? If that were the case, one could look at the bacteria and say, "Well, tensegrity doesn't really interact positively with anything else. You don't have this ?? we've got to throw this out."

Don Ingber: I mean, hopefully I was clear that tensegrity is a system for structural destabilization, not the cytoskeleton. Because I showed pictures there that one of those was a perfect dodecahedral structure. It's an enzyme complex made out of proteins; that clathrocardiped(sp?) is a higher order one. The bacteria has some structural stability. The DNA is not replicated in solution; it's off of the wall. No one has a clue what it looks like or how it's stabilized, but it has enough stiffness so you can build off of things.

And at the level of molecules, there are no compressive forces; there are only attractive forces. So I think the systems come in, but going beyond tensegrity I think that what you have in this interaction with other systems is flexibility in the system, whether it's structural or informational. You can't have things exploring and interacting, unless you can change context. If you're just stuck, you're just going to keep going and going and be inert. I mean, that's what inert is.

So I think tensegrity is one way to get that. I think the dynamic issues, you know, the chaotic issues are over different time frames. I love that explanation that you have to have a system that can respond at slow time scales as well as quick time scales. So I'm making this up. Maybe tensegrity is very good at quick, but not good at slow. And in my perspective, a lot of what I say mechanically is the instantaneous response of a tensegrity, which then evolution has evolved to remodel biochemically.

So instead of in the cell the filament stretching like this elastic string, they polymerize. So maybe it's the over-interweaving of all of these different sorts of principles. But the principles are generally right. I don't think anybody would probably argue about behavior coming out of complex parts. I mean, the individual atom has no real properties other than mass; it's really the collective, and that's true for everything. It's just certainly that more people like biologists or certainly clinicians -- it's like what can you give me that I can use that can help me in my daily life? So that's going to require a big challenge, I think—people sort of working together, and linking theorists with experimentalists. I mean that is sort of to me, if we can't do that, we're not going to get anywhere. Yes?

M: I thought I heard you say that you didn't regard the genome itself as a driving force?

Don Ingber: I said it was a driving force for evolutionary diversity after the genome came about. But basically, it was one element of a continuous spectrum of evolutionary self-organization. In other words, it wasn't just like there is the inorganic world and DNA. And in fact, there is good evidence that the RNA world probably dominated before the DNA world. But of course, once you had DNA, all the biological diversities came out of that.

M: That's part of the diversity. Between your example, I can imagine an example I was thinking of. We higher mammals who will have a high thermoregulatory, high thermogenesis, high oxidative, phosphorylation high ATP, can't one see that as a selection force upon the affinity of cytochrome C

?? for oxygen? One part of the genome exerting a selection optimizing pressure, and another part of the genome that determines cyto-architecture of a cell a yard away.

Don Ingber: I mean I think once DNA starts working, you basically increase your efficiency of giving possible -- It's like having a better computer that can throw out more numbers all the time to explore space. And yes, they're both being selected at once. I mean, I think that's the key.

(Overlapping dialogue, inaudible)

Don Ingber: Right. And DNA regulation involves higher order DNA structure, which I looked at in your picture of that folding -- Think of DNA as, once it gets bigger and bigger in a fixed nuclear space, it is the most elegantly folded and higher order folded structure you can imagine, sort of a football field length in a micron-sized nucleus. So I think all of those are being selected at once. Actually, in the end result I can turn my DNA off this instant, and I will still walk, talk, you could have sex, you could eat, you can do anything you want for a short time, and then you're going to get sick about ten hours later. So how the machine works at any instant in time is not DNA, it is the expression of DNA through proteins and assemblages. Now, how I can develop over time, or how I can respond if you give me a stimulus over time, that's -- The body evolved that system to deal with longer term responses, I think. Yes?

M: I had a comment about your data that ?? different theories. Are there any competing theories of ??.

Don Ingber: Not that we know of at this point. Basically, all the models of cells are really -- I mean, literally the dogma that is going against this are engineers' models that say the cell can be modeled as an elastic membrane with a viscous cytosol and an elastic nucleus.

But I showed you all the structure; that's not the way it's built. So you know, there are people who have modeled cells in tissues as liquids, and they model it beautifully, and it's absolutely fine as a model. But that's not the way they're constructed. So that can model one feature, and if you're only interested in that feature at that size scale, that's fine.

So if you're only interested in the whole cell, this can behave a little bit like a water balloon filled with molasses at this size. But if you want to link structured and molecular responses, you have to start taking into account real structure.

So now, our model can act like theirs at the big level, but it can go down. There are really no other models that go down. So right now there are just not any other models for the cell out there.

M: -- keep getting more data so that --

Don Ingber: That's the last three weeks. But there is also data that relates to dynamic mechanical behavior. Like that certain responses are going to be dependent on pre-stress or not. Those also hold up very nicely.

M: Thank you.

M: I have a thing related to the vibration. The cell is not vibrating, right?

Don Ingber: The cell is vibrating, actually. I mean, you know, if you watch it over -- It depends on the time scale. If you take a heart cell in a dish, it's clearly vibrating. You can see it. But other cells, people have measured that there are certain harmonic frequencies where cells put out extensions and take them back. So there are certain frequencies. It's probably hours.

M: ?? prediction. If your Q is large enough -- I don't know what the Q is -- but if the Q of the cell is large enough and you scan it as a function of frequency, or you probe it for the frequency, then what you will observe is this tensegrity model ?? then the distribution of the resonances will be very different from the distribution of resonances of a balloon.

Don Ingber: Right.

M: It's completely different.

Don Ingber: Well, people use what's called acoustic microscopy which does a little bit of sound waves. And you can map out these filaments of stiffness and material properties within it. But we're just beginning to try that. You've got to realize from the biological world, they don't care about any of this stuff. You know, it's because it's all genes and what molecule you're working on.

I mean, I love the example of the clinical situation of the nurse. Well, that's another problem on another day. From complexity theory, you know that you change one thing, it's just going to come back to that state. You have to treat the whole system. But that's not the way we're trained, and that's not the way the funding agencies go. So you have to kind of do a little of this conventional [work], and then try to do this other stuff on the side.

M: What is the ?? specifically structures in cells ?? NAV pH level fluctuates on the order of about 50 milliseconds. So there is a wave of one of the ?? cycle ?? producing a wave going throughout the cell. The vibration ??. My question was does it ring? I mean, does it has a high frequency?

Don Ingber: What do you mean ring?

M: You can hear the vibration.

M: No, no. If you probe it and send it a pulse how long do they keep on vibrating? So the question is what is the Q?

Don Ingber: Right. We actually are trying to set up equipment to do that. I had one student who wanted to play The Grateful Dead with loud speakers next to the microscope and find the resonant frequency.

M: I would like to ask a question that may go across a couple of speakers. But you mentioned hierarchical levels, and so did other speakers. One of the things that we may be missing is some empirical way to determine what is the appropriate level of structure of ?? you're looking at. I think some of the disagreements between physicists and biologists is the difference in ?? level you're talking about. And sometime you can get a parameter that's very important when you're looking at one ?? level, and when you go across ?? levels it isn't, and visa versa.

Don Ingber: I agree totally. That was my comment I got up earlier and made when someone spoke. It's choosing how you define the system. And I think from a biologist's standpoint it's relatively easy, which is, how do you have the minimum system that has the physiological function that you're interested in studying? Whether it be female, the estrogen cycle in animals, or in our case growth control in cells. By definition, most of the work goes right to individual molecules. And you know that you can find a heck of a lot about the molecule, but you're not going to answer the bigger questions.

And I should say in terms of the other models that are out there, almost all the work on cell mechanics is isolating the individual filaments I told you about. Purifying them to umpteenth level, putting them in a gel, and measuring mechanics. And then putting combinations of them together. And none of them predict, even the first graph I showed you, let alone any of the other stuff. So that's the way people think about it. It's like it's an additive effect, when it's a synergetic effect. Thank you. (Applause)

Chapter 2

Reception Session

Yaneer Bar-Yam
Session Chair
Edward Lorenz
Climate

Introduction

Yaneer Bar-Yam: It is a great honor for me to be able to introduce our speaker this evening. Ed Lorenz and his work are known very well—everyone knows of his work. And let me just say a few words, and then turn over to him for the evening presentation.

Ed is a meteorologist at MIT, and many times I know when I speak to people, and for myself, I knew that the work that he had done is so fundamental, that one thought that it happened hundreds of years ago and not so recently. But in fact, it is quite recent, and very surprising to many people. I think some of the reasons that it is so surprising were very well described earlier today in the morning. As a meteorologist, he is now Professor Emeritus, originally with the Department of Meteorology, but now it's the Department of Earth, Atmosphere and Planetary Sciences. He started as a mathematician, and in fact he started as a mathematician not far from here at Dartmouth College.

But during World War II, he became a forecaster, and actually spent time in the South Pacific in Okinawa, forecasting the weather for the military. And when he came back, he decided to continue in meteorology. And in 1959 he did the work that is known to us as the butterfly effect, or technically a sensitivity to initial conditions. But everyone knows about the butterfly effect. He has also done a lot of work on the energy balance, and energy cycle, of the atmosphere. And today, he will tell us about climate, which he defines as what you expect. (Laughter)

Edward Lorenz
Climate

Ed Lorenz: Well thank you, Yaneer. Well, it is indeed a pleasure to be here this evening to be able to talk to you about one of my favorite topics, namely climate. And the complete title of my talk, if it needs a title, is "Climate is What You Expect." Now, if there are any meteorologists in the crowd, you may have heard this before in connection with a longer statement: "Climate is What you Expect, Weather is What You Get." This has been used quite frequently to explain to the beginning student what the difference is between climate and weather. The two are sometimes confused with one another.

So anyway, the point I wanted to make is that aside from being sort of a descriptive thing of climate, the term "what you expect" may indeed be about as good a definition of climate as you could get, or as good a simple definition, without going into defining all the very many details, and so forth.

What we expect might be considered to be a set of all statistical properties of the atmosphere, and the adjacent portion of the oceans and continents. Now, in climatology we talk about the climate system. The climate system means the atmosphere, plus these other portions of the globe that come into contact with the atmosphere, including those portions which are, in turn, affected by the atmosphere. Perhaps the most important of these being the surface of the ocean.

We usually do not include in the climate system those parts of the globe which affect the climate, or affect the weather, but presumably are not, in turn, particularly affected by the weather. This would include such things as volcanism, which certainly has its effect on things. But as far as we know, there is no very great effect of the weather on the activity of volcanoes, so there is probably a little bit there. It may have something to do with the exact timing of an eruption, or something like that.

So in studying climate theoretically, this, well, I should say this climate system, if it isn't already obvious, should fit the definition of complexity pretty well. Even the atmosphere by itself is an extremely complex system, and the other things that we have to include, include all the details of the configurations of mountains and things like that, presumably all the locations of vegetation, the moisture in the soil and, as I've already said very importantly, the temperature distribution in the upper layers of the ocean.

So we can, in fact if we want, think of this climate system as a very complicated dynamical system. We think of a dynamical system as being something whose future is determined from its present, or from its present and its past. And if we say, "Well, this isn't really a dynamical system," then certainly a great many of the models of climate from which we derive much of our present knowledge of what's going on, do qualify as dynamical systems.

Now, climate models vary from things that include almost everything that we can crowd into the memory of a present day computer, very complicated things in very great geographical detail, to extremely simple systems with just a few variables. And it's one of the simplest systems that I'm going to be talking about later on. In fact, this one will be so simple, that I think you'd be justified in saying that it isn't a climate model at all.

What it is is a little model that shares certain features that come with climate, and it's good for testing some of the hypotheses about climate that have been offered, even though it doesn't really describe the climate too well. Now, the atmosphere, or the atmosphere as far as the climate system goes, is most certainly a dissipative dynamical system. That is, there is dissipation going on there.

In mathematical terms, a dissipative system would be one where if you consider all conceivable states of the system as filling up a volume in some sort of phase space, in which each point of the space represents one conceivable state, then as time evolves, this volume will continually decrease. That is, if you take a subset of these states that fills a finite volume, this volume will continue to increase, and usually it will approach zero. And the set of states that it finally approaches, although having a zero volume, will certainly not be an empty set, any more than a plane, which has –

[...] generally known as a set of attractors. In some cases it will be *the* attractor. This is a fundamental property of most dissipative dynamical systems. What happens then is these states converge upon a smaller set of states, which form the attractor.

And I'm going to make the point that to a first approximation, maybe we can consider the climate to be nothing, more or less, than the attractor of the dynamical system; those states which will ultimately be encountered, or approximately encountered again and again, as opposed to those states which will never be encountered. A state, which as far as we know, will never be encountered is one, say, in which the temperature at Nashua reaches 150 degrees some afternoon.

If you have a numerical model, it's perfectly possible to plug in initial conditions in which the temperature here is 150 degrees, and a good model will handle that pretty well. And you'll find that pretty soon the temperature won't be 150 degrees, it will never get back to that again. In fact, it is rather seldom that we'll even get over 100 degrees Fahrenheit, although it will do that once in a while. So that's an example of a state that is not on the attractor.

Another one, which would be familiar to meteorologists is one in which the winds blow the wrong way around the low pressure centers. In the northern hemisphere, they invariably go counterclockwise around them. And you can easily plug into a numerical model, a computer model, an initial state in which the winds are blowing the wrong way around all the systems. This will be a highly unstable situation. The thing will change very abruptly, and pretty soon you'll find it's behaving somewhat the way it should be. And if you wait a month or so, you may find it's behaving pretty much the way it should be.

So anyhow, those are states that show states that are not on the attractor, and we say that climate is essentially the attractor of the system.

So here we have a mathematical basis, perhaps, for talking about climate. We say it's governed by a dynamical system. A lot of work has been done on attractors, and so we're just talking about the attractor of a system, so we get back into one special problem and the theory of dynamical systems.

Now, one special feature of this particular dynamical system and a great many other ones complicated or simple, is that they are chaotic systems. That is, that there is a sensitivity on initial states.

That is, if in our model at two separate times you plugged in almost the same but not quite the same initial states and let the model run according to the equations governing the model, according to the equations which presumably represent the physical laws governing the systems, you'll find that in due time the states might not look much alike. In that case, we say we have chaos, and the weather is most certainly a chaotic dynamical system.

Now, this brings up the interesting question as to whether the climate is a chaotic system or not. And there have been a number of climatologists who have studied it this way, and have come to usually no conclusion as to whether it is. The question I would ask is whether the climate is even a dynamical system at all, if we mean by the climate something whose variations include only those that we think of as climatic variations, and not the day-to-day weather variations. Though it may be that the chaos is confined to these shorter period fluctuations. So with that, let's go to asking how we can define climate quantitatively. Well, maybe this is a definition, the set of all statistical properties of atmosphere, and adjacent portions of the ocean and continents, but what do we really mean by the set of all statistical properties, or any particular statistical property?

And as far as climate is concerned, we can assume that we know what statistical properties that we're interested in, and concentrate on any particular statistical property, or any particular set of statistical properties. The annual rain fall in Washington, D.C., the mean July temperature in Nashua, anything like that. These are specific statistical properties. And the question is, can we even define these things?

Well, here are some alternative concepts. One would be to say that by this we simply mean the infinite term "statistics." In other words, you take the average from now until infinity, which we can't very well do in the real system, but which we can do for practical purposes in some models. In fact, we can really do it in some models if the model is so simple that we can write an analytical solution to it, because then we can integrate. Or, we could calculate the average from the present time to minus infinity, and for some dynamical systems at least, these would be the same thing, that they should average out the same way.

This is very nice, incidentally, for theoretical studies of climate. Because if we're dealing with the theory of climate and have a model, the theory of infinite term statistics, of attractors, a good deal has been worked out with that, so a lot of the theory is done for us already. But there is one big objection to this, that at least it would be a big objection to most of today's climatologists, even if they might not have batted an eyelash 50 years ago, like that.

Because if we accept this definition, then by definition, climatic change is impossible. Because after all, a set of infinite term statistics isn't going to change with time. And this is the thing that many, if not most climatologists seem to be interested in these days: the topic of climatic change.

So how do we define climatic change in this case? Well, we certainly have to have a different definition of climate than infinite term statistics if we want to do this, so let's go to the second definition that I've listed here: the infinite term statistics that would prevail if the external conditions were fixed at their present values.

Now, external conditions could mean something like the solar outputs, the amount of energy received from the sun per unit time. It could mean the degree of volcanism on the earth. It could mean the geography of the earth, which over long periods of time has changed but over short periods of time has approximately constant geographical features, and so forth. So this might seem to be a fairly acceptable definition to work with.

Well, it has its problems too. For one thing, we are not yet in a position to take any system of equations which purports to describe climate, and derive from that what the climate actually is, what any particular climatalogical statistics is. There are no set of equations that we can work with and determine from it reliably what the annual precipitation ought to be in Washington D.C. We may get a fair approximation to it, but we don't do too well at things like that. So we have to be guided by our observations of the past.

Now, we don't have the ?? back to infinity, so we – In fact it was standard practice for a long time, and it still is in some places, to define climatological statistics at particular locations as the averages over the past 30 years. So all we can really look at are past averages, but what we are presumably interested in is future averages, because if we want to plan, if we want to plan a vacation next summer, if we want to start a new industry in some particular city or something, we're interested not in what the climate of that used to be, but what kind of climate it's going to have, and what the weather conditions are going to be in the next few years over decades or so, or maybe the next few weeks if it's for a vacation.

So this also has its objections. It does have the point, the good point, that it allows climatic change to take place. Because we could then say the climatic change then has occurred whenever the external conditions have occurred. And even though these climate models, as I say, won't tell us precisely what the climate is at any place, they will give us some idea, and they can give us good, qualitative ideas sometimes, of how the climate will change, whether the temperature at a certain location will warm up or drop in response to a certain change in external conditions. So we can really work with this definition in theoretical work, even though it's a little difficult to work with when dealing with observations.

Now, it does have an objection that it will not allow for climatic changes if these result from some internal process, rather than from change of external conditions. Now, there was a time when most people would have said, "Well, the only thing that can change the climate is a change in external conditions," but now it's become fairly evident that you can get extremely long period changes due entirely to the internal dynamics of the system.

And if these are not, one might decide, "Well, let's not call these climatic changes if they occur entirely due to internal dynamics." However, for looking at what has happened in the past, since we have no way of determining where ?? observe climatic changes if ?? climates are due to internal or external things. So this leads us to the logical third definition: statistics taken over a long but finite time, which is all we can do anyway if we're dealing with the past, unless we're dealing with a model, when presumably we may be able to go to minus infinity. And for statistics, this allows then for internal and external climate changes, and it seems to be a pretty good definition, until we ask, "Well, how long is this long finite time?"

And the question is we don't really know. I mentioned the 30-year time used for past statistics, but this may not be what we're interested in. And one trouble is that climatologists don't really agree among one another when a change should be considered climatic, and when it's a fluctuation which describes part of the climate.

To give an example of this, we have heard a lot in recent years about the El Niño phenomenon, or sometimes the El Niño Southern Oscillation phenomenon, originally noted as a very warm current that occurs radically off the coast of South America and causes havoc with the fishing industries and other things there. And it's become evident in more recent years that this has an effect on the weather at rather distant points from where the primary currents are occurring. And this, we seem to get an El Niño coming on the average of from two to seven years, maybe four years on the long-term average, or something like that.

So some climatologists would say, "What if El Niño comes and the climate is changed?" Then when it disappears, the climate is changed again, maybe back to just what it was before, and maybe something else. Others, including myself, would prefer to think of climate as something that endures over longer regions, and ?? say of the present climate that one of the characteristics of the present climate is a fluctuation with periods of a few years between an El Niño and what is now sometimes called La Niña, the intermediate period of that when it's off the coast of South America.

So we can take the longer term or the shorter term approach, and according to what approach you'll take, you'll come to a different conclusion as to how long this long finite term might be. So there are difficulties even with this definition.

Another difficulty with this definition, and also say it's a combined difficulty with this definition and especially with the earlier definition defined in terms of changes in external conditions, is that when the external conditions do change, there is sometimes a considerable lag before the weather features change in response to it, before the average temperatures and so forth. It could be a lag of a number of years.

In fact, it could be such a long lag that by the time it's really realized, the external conditions have changed again to something else. So this complicates, again, the definition.

So as a final suggestion, I'll put one which I tend to favor, although it is by no means obviously certainly the better one: that we define it in terms of an ensemble of possible states, and we take large ensemble statistics at a single time.

To get this ensemble, we consider one state of the system at a particular time—let's say the present weather as it exists all over the globe—together with the present conditions of the ground, and the oceans, and so forth. We can integrate that forward if we have a suitable system of equations. And as I've already said, we don't have an exact system, but we have a number of models which give reasonably good integrations.

Then we perturb this slightly, make a large number of small perturbations, and knowing that our observations of the present, of the weather and so forth are far from perfect—even now, it's not so much the instrumental error, as that there are big gaps between observing stations, and we're not sure just how we should interpolate between them at times—knowing this, we want to choose these in such a way that if the first state that we have chosen happened to be the real state, plus the inevitable errors in observation, then any one of these other states might happen to be the exact state, although presumably none of them will be it, or it might be as you ?? upon it. In other words, these are states which look very much like the present state.

Then you carry each of these forward according to the same set of equations. Since the system that we're dealing with is chaotic, these will eventually diverge from one another, and we'll get a large collection of states, all of which don't look too much like one another. But the statistics of this large collection can be considered to be the climate.

Again, the climate when? If we mean the climate right now, then we would have to go back to a past state and carry our equations forward to the present time. How far back? One year? Fifty years? We don't really know, but we can do this according to the study, according to the particular study we're trying to make there.

So what I want to do in the remaining time is to illustrate some of the advantages or disadvantages of these various definitions by introducing a very small model, to which these can be applied. It's going to be small enough so without too much computer work we can deal with ensembles of size 10,000 or something like that.

Incidentally, the various national weather services now intend to make what they call ensemble forecasts. In addition to the regular forecast that's issued, they make a number of other forecasts like that. It's certainly not 10,000, maybe they do 50 or something like that, or maybe twenty. And this will give them some idea of the certainty, the amount of confidence that they can place in a weather forecast. If they perturb it 50 times, then they all show about the same thing somewhere, then we conclude that this is probably what's going to happen. If they show a lot of different things, then we may forecast one of them to happen, but we place rather little confidence in our forecast.

Well, this will be one where we can work with big ensembles. I want to introduce a couple of definitions which most of you are probably familiar with now, but they're relevant to this. The first is a transitive and intransitive system. By a transitive system we mean essentially one that has only one attractor. An intransitive system would be one with several different disjointed attractors.

Now we don't know what the climate system is. We have one set of climatological statistics, at least according to that first definition, which applies. But if at some very distant time something else had happened, then we would have had quite a different climate which would have nothing in common with the present one.

There have been, since the middle-seventies I believe, a couple of very simple model studies, one by Sellers in this country, one by Budyko in Russia, which simplified the whole system to just a few equations, and they found two climates. One was one that you could claim, at least, resembles the climate we have. Another, which was equally possible, was the one in which the earth is completely ice-covered and remains that way forever.

So just according to the models, if somehow the earth could become completely ice-covered, it would remain that way. Now my feeling is that these particular models were over-simplified. There are some things that were not taken into account that probably would have caused the ice to melt at least in equatorial regions. And then the melting would gradually proceed to other regions, and eventually we might have a climate as we do.

I doubt very much that the climate that we have is intransitive, but this would be an example of an intransitive system, these particular climate models. And the one that is of interest here, which I found a very good study, is one that I've called "Almost Intransitive." This would be one in which there are two sets of statistics, either one of which could persist for a long time, but not forever, and the state can vary wildly within one system, but can only pass from one to the other with difficulty. But eventually it will, and then the other type of behavior will occur.

So if you look at this over a short period of time, or two different short periods, you might think that it was transitive. Well, if you look at something from observations, you'll think it's transitive anyway because if there is another thing that can happen, it won't have happened. But from some models you might think it was intransitive, unless you studied it long enough and saw that these transitions were possible.

So what I'm going to do is to say that the climate model we'll be interested in may not look much like the climate, but it will be one, and first, in which we have chaotically varied weather, or short-term variations, which will identify with weather variations. And this will be intransitive. So that as the scale takes place, eventually we will get a change from one type of behavior to another type of behavior. This will be considered to be a climatic change then. And then if we wait long enough it will change back again.

Now, there can be more than two different types. There can be tons of different types. I know of one mathematical model by Swinney at the University of Texas where he has done an experimental model too, where he has found 100 different types of behavior, so many that if you actually were doing the experiment a few times, you would conclude that the experiment was not reproducible because you get a different thing each time. But if you do it often enough, then you find every possible type of behavior, and you conclude yes, there are lots of different types of behavior, but it has to be this one, or this one, or this one, not any old thing.

So let's take a look at this model, and I'm going to base it on an old game which we used to play when I was a kid. It's called, "Bull in a China Shop." It's also available, or has recently been available commercially under the name of "Skittles." It consists of a wooden box in which there are two or more compartments separated by a narrow door.

You set up pins at each one and spin a top in one. It will go bouncing around, knocking down the pins there, but it will be only with difficulty that it will happen to hit the door, go through, and start knocking the pins down in the second compartment. You score more for getting those pins because they're harder to get.

Here is an idealized version of Bull in the China shop with the top spinning there, and it will presumably bounce around, hits a pin, but it moves all around, bounces on the sides. And in doing that it will, in due time maybe, go right through this door and start hitting the other pins there. I'll first use this as a model of an almost intransitive system because the two states will be represented by the two boxes. It can move very easily around within one box, and only more rarely, pass from one to the other.

Now this, as it stands, won't do as our model for climate. For one thing, in order to make it a simple model, we have to simplify the behavior of the top. We're not going to worry about its ?? or anything like that. In fact, we'll treat it as a particle which moves in a straight line at a constant speed and bounces perfectly from each wall. And since these compartments are rectangular, it will not behave ??. You can predict as far ahead as you wish just what its path is going to be around the rectangle. Either it would be a path that would never go through the door, or would go through at regular, predictable intervals. So we've got to change that somehow.

One way to do that is to make the boundaries curved instead. In fact, make them bulge inwards. Then if it hits in two slightly different places on two occasions, they may bounce off of quite different things there. So let's take a look at the way I'm going to modify this. Instead, we'll have a compartment like this. I haven't bothered to put any pins on this, but we have these circular arcs there, and they meet at points there, and they don't quite meet above and below the central line, which we'll call the x-axis, so that a particle moving in a straight line and bouncing from the edges, will move around very easily within one of these compartments, and only occasionally go on such a direction as to go through there into the other compartment. Then it will bounce around in here for a while, then maybe go back again, or maybe go into the next one over here since this will be an infinite strip here.

Now, this has one disadvantage as it stands: it's not a dissipative system. If we start in the central compartment, once it's gone to this compartment, there is no more reason to suspect that it will go back to this one, than it will go on to the next one. If it does that, there is no more reason to suspect that it will ever go back, then still go on. So in due time, it will tend to get further and further away from the center, and it will be a little hard to compile any statistics for its position.

So what we've got to do is to put in some dissipation, but since this is to be a model that we're going to work with mathematically, and we want it to go on to infinity, we can't have the dissipation simply being the particle slowing down as it moves; it would, in due time, come to a stop. So what we'll do is to assume that it continues at the same speed, but every time it crosses the x-axis there, coming in here like that, it will change to move more nearly perpendicularly to the x-axis than it was before it hit.

We also are going to say that it will also change at that time to move more nearly back towards the central point than it was before. We get two super-imposed modifications every time this crosses the x-axis. Here is a particle starting off with the arrow; it bounces from the center. But when it comes back to the x-axis, it crosses more nearly perpendicular to that, and bounces. And each time it crosses—you can see there is a kink in there—and the kink is such not only to make it go more nearly perpendicular, but to make it go more nearly towards the central point here, which I've called zero. And the two changes might happen to cancel each other out in some cases.

And there are two functions associated with this that we've used so that if it's coming in at an angle, theta, to the x-axis to the perpendicular, it will go out at an angle, theta prime. And the formula we've used in this model is that the tangitive(sp?) theta prime is simply one constant K, which is a positive constant less than one, times a tangitive(sp?) theta, minus some other constant L, times some particular function of x. And I write it that way because we have two different models, or one Model A where x is simply x minus x-star; x-star is some pre-chosen point that we want it to return to eventually.

The second will be a somewhat more complicated one. And these are indicated by the graphs there, because F of x against x, of course, is nothing but a straight line. This, the second case, is asymptotic to two different straight lines as x goes to plus or minus infinity, but it has this little wiggle here so that in this case it will be attracted to the two points here and here, but actually repelled from this point in the middle.

So let's take a look at Model A first and see what happens, and we'll go back to this diagram, which I showed before, which is simply Model A with a path running through there about 60 different times. Then it hits the thing before we stopped it because there isn't room to show too many lines on top of it.

And here is the same thing again but shown differently as two time series. One is for the x-coordinate of the particle, one for the y-coordinate. Here we see the x-coordinate tends to hover around zero for some time back and forth. This dip in here was this point where it almost got through the passage, a great gap, but didn't quite go through, and went back again. The second time it did get through.

y, of course, is forced to hover around zero because there is nowhere else for it to go. So here it's hovering around zero for a while, and hovering around two for a while. Right in here is what we've decided we might call a change of climate. So right in here somewhere, here we have one climate, here we have another climate here, but let's see what the time is.

We're going to identify these bounces back and forth, the chaotic bounces, as the effect of the passage of storms there, which definitely affect weather. And knowing that these storms tend to pass a given location in about three days, and knowing that it takes about two time units at the speed to hit the side that we'll call this period of 20 time units, we'll take the time unit to be three days. So this will be considered to be the analog of two months.

So here we have the climate changing after only a month. And as I say, I don't like this as a definition of climate change. The climate doesn't change from month to month as far as I'm concerned, except it may change a tiny bit. If we define it as a number, this isn't going to change at all. When does the change take place? Eventually does it change abruptly to something else, or does it gradually creep up there, in which case it might even be different tomorrow than today?

In other words, what we might have expected tomorrow might not be exactly the same as what we could have expected today if we went back. And for what we expect, of course as I said, we have to go back a ways. Basically we have to go back to a time of which we are not in a position to make a skillful weather forecast for today and then say that's what we would expect today in that case. And what we might expect today might differ by a hundredth of a degree, or something like that, for the expected temperature here, or less than that. But this is not a significant climate change.

So anyhow, having seen it in a small fraction of a second, we can carry this forward for two months. Let's carry it forward for two years and see. So this will start out to be the same picture as this, only collapsed like this so that we get two years on one line. And you will see the change taking place very near the lefthand side of the diagram.

And then, apparently having gotten into the compartment, it centered around minus two. It stayed there for approximately a year, and then went back again and stayed for the remainder of the two-year period. This only took a small fraction of a second.

So let's extend this to 25 years now, and we'll see this part collapsed up. You'll see the change minus two going on for about a year before coming up. Here is the 25-year variation. So here we have something that varies chaotically. It changes over after a month to the minus two location, and it's quite obvious which compartment it's in at any particular time. It goes up there, and stays in one compartment for quite some time.

It prefers the zero compartment, but it will get to the plus two or the minus two if you wait long enough, and occasionally even the plus four compartment and the minus four. It didn't seem ever to get into the plus six and minus six. I think it might have if we had run this long enough. Anyhow, this is an obviously almost intransitive system. It changes very easily within one compartment there, changes very easily among these values here, and only with more difficulty does this one.

So this, although certainly not a climate model, is good for examining some features of climate. In fact, what it might be good for is seeing what happens if you use these alternative possible definitions of climate on this particular thing, and see what happens. Well, before going to that, let's look at something else. We can get a probability of density function for x there, since it was sometimes moving around minus two, sometimes zero, sometimes plus two and so forth. And running it for 100 years now, we have this probability density function. What this has done is simply taken all of that as it's occurred, dividing the range into small intervals, and simply counting the number in each interval over the 100-year period.

So here we see a fair amount of irregularity in here with these spikes and so forth, which seem quite obviously just due to inadequate sampling. We would expect that the real probability density function will be something smoother like this, and we don't really know whether this peak up here is real, and this one, and whether the distinction between this and this is real, and so forth. So for good measure I ran this for 1,000 years just to see what would happen, and without running it to infinity. By comparing the two, we can more or less see what's going to happen if we run it long enough. Let's put this one back here.

Now, here is the thousand-year one, and here, for comparison, we see that it's much smoother. But apparently, particularly, it should be perfectly symmetric as the way the model is constructed. And we see this thing on both sides, so apparently this little dip in here, this one over here is something real. The dip in here between a slightly higher one right at minus two, then I'd say minus one point five here, and the peak right at zero, are real features. So here are the types of systems that can occur. And according to the first definition, the climate would be this whole set there, in fact defined by this physical thing.

The climatic mean would be zero there, but there would be other statistics than the mean that are important, the other standard deviation, and so forth, and in general the whole probability density function. And with that, we can go to the attractor and see whether the strange attractor is what you see, and you take the attractor out of the 100 years.

To get the attractor, let's recall that we really have four variables in here, or appear to have. We have the two components of the position, and two components of the velocity, the north-south and the east-west component if we think of the directions on here as the compass directions. However, since it always moves at the same speed, we have, for practical purposes, only one variable as far as motion is concerned, the direction. So we have really three variables.

Now we want to picture a two-dimensional attractor, and to do that we look at two of the variables at the time that a third one takes on a particular value. This time we'll simply look at the values of x; x is always looked at. And while I've told you the eastward component of the motion, at the time when it is crossing the x-axis—that is, at the time when y equals zero—this gives us a set of points there. And in principle, you could take any one of these as an initial condition. As far as an initial state, you could pick any point on the x-axis and have it moving with any speed, up to -- Have it moving in any direction and speed. But after a while, they converge to these points.

And what this is is called a strange attractor. What we would have if we could put enough points in here is an infinite set of discrete lines. And if you were to blow up any portion of this, what appears to be one of the lines, you would see that anything that looks like a single line here, for instance, is really an infinite set and has all the structure that the whole thing has. And no matter how much you blow it up, you will see it applying the same complicated structure here. So this certainly has all the earmarks of a strange attractor there.

We can see that the x component is centered around zero, and the u component, the east-west thing, is also centered around zero, meaning it's equally likely to be moving either way if it's here. If, however, it has gone negative, say minus two, or minus two and a half or something, the u's tend to be positive. In other words, if it is negative, it tends to be moving, at the time it crosses the x-axis, at least, it tends to be restoring itself already.

And again, this should be perfectly symmetric. It should look the same if you turn the figure upside down. And the fact that it's denser in here than in here represents the fact that 100 years is inadequate sampling.

Well, now we get to test some of the other definitions of climate. Let's take a look at the ensemble definition, and let's take a look at an ensemble of states that happens if we say we start off the initial condition not with x equals zero, as we did, but with x equals, say, 2.2.

The initial states will be those points on a small circle centered at it. That will be this small circle in here. We have 10,000 different states spaced equally around here. If we continue it for only eight days, those 10,000 points stretch out onto what appears to be a curve, here, here, down here, and up here, and here. Of course, it has to be a closed curve. And what it actually does is once it gets down to here, it doubles back on itself so close to it that you can't resolve the difference. It's continuous through here, but with only 10,000 points you don't see the points in here, and so forth. Well, this isn't much of an attractor, but eight days isn't very long either. So let's look at the same thing from one year now.

And here we see something that looks a good deal like the old attractor, except that it's centered pretty much around two in here. There are points in here. There are even a few at minus two, or even a few at minus four here; some are plus four. But it means that the big majority of those states in here, all of which were in the plus two compartment at the initial time, the big majority were still there two months later. This is out one year. Two months later they were mostly still there, a few of them ?? over and so forth. So this would be the way you might define the climate a short time ahead if you're interested in the short-period definition of climate. What's the climate going to be two months from now, or what do we expect two months from now?

Well, knowing that the present state, which we can measure, is somewhere in here, then what we expect two months from now is this. We expect it will most likely be in here, but that it can be in here, and so forth. And this is quite different from the attractor that we looked at before, which is symmetric, about zero.

Now, this brings up the point of what would happen if this x-star in here, which we said equals zero—the point towards which this restoring process tends to bring the point back to—if we had set that at plus two instead. If we set it at plus two, well then since from the way this is constructed with a symmetry, we would have exactly the same thing that we had at zero, except that the scale would be moved over the other way. So this would be what we see at plus two, and that.

And we see that these look somewhat alike, not exactly, but a good deal alike if you superimpose them so that the plus two here superimposes on the zero there. So, we can get a climate, if you wish, since this attractor ?? the other climate looking like this –

[Break in tape]

[...] With this ensemble definition, if we take a limited amount of time to extend this ensemble, take this ensemble average, but extend them over a limited amount of time, or if we take, extend them over a long time instead, we will get what looks like the other one. We get something that looks almost like this.

Now, what this means is that if a climatologist should observe a change taking place from something like this, centered at zero, to something centered at two, he would have no way of knowing whether this was due to taking this initial state there, due to an unrepresentative initial state, or whether it was due to a change in external conditions, because this quantity x-star that we've taken here is essentially an external condition. In fact, the only external condition that we're allowing to vary in this model. So, we have no way of saying whether the change effectively is external or internal simply by observing it. We can get either kind with this model. So it means that there can be problems.

I want to say just a little bit about this Model B, which is a more complicated thing. And here we have something which tends to drive it away from zero towards some other things. So if we do that, let's take a look at this. Here is a 25-year series from Model B. Now what we did is to take c equal to four, if you go back to that formula. Without going back to the formula, what I'll say is that if x is negative, then this restoring function tends to drive it towards minus four. If it's positive, it tends to drive it towards plus four. So in either case, it's driving it away from zero. So if we start off with zero, eventually, just by chance, it happened to go to the positive one to start with, the negative, and it stayed there for about 20 years.

And within this climate centered around four, we can see the individual changes here from one cell to the next. The minus two cell, the minus four cell, rather here, the minus six cell, the minus eight cell. So these changes are just like the ones that they called "climatic" in the earlier model.

But here we have a real long-time climatic change, and a much bigger one, where it suddenly starts going through the negative ones for a while. And I have extended this to a longer period. Here we have it 100 years. And what happened at this time is it stayed in this climate for about 20 years, and in this climate for maybe 25 years, and this one for maybe thirty. And this one then, now we see it going up again in here.

So if we take the attitude that climate means changes over a longer period, not the change of climate from one El Niño to the next one and in between, then we don't call these things from here, to here, to here, and back to there climatic changes, but we do call from here to here a climatic change. Now, there is another interesting thing about this thing. We can model these longer period climatic changes by putting on this additional kind of almost intransivity. But a climatologist looking at a record like this would be very likely to conclude that here is a 25-year period ?? it's oscillating quite sinusoidally in there. There is some external condition which has a 25-year period to it.

But there is no external condition in this model which has a 25-year period. I didn't select this thing to bring this out. This is the only 100-year run I made with this model, but it did happen to show this thing. Another one probably wouldn't have shown anything looking like a 25-year period, but it might have shown something looking equally like some other period.

A climatologist would expect that, so let's for good measure look at the probability density function to see whether this is what really is happening. And we can see that there are. It has the double peak in there. Presumably, if we ran this for a long time, we would see something over here that looked like the other one, then going down and having a dip, but by no means down all the way to zero at zero. Then it should be perfectly symmetric if you run this for, say, 500 years instead of 100 years, or something like that. Well, so I think I have one or two more here. Let's suppose we run a different model, this one.

Well, the gist of it is if you decide you're going to make it go back to Model A, but let the value of x-star vary over the period, then you get something that looks very much like this. And if you let it vary in such a way that puts on both types of periodic forcing, together with this almost intransivity to make it Model B, then you get something which is a combination of the two. Again, it looks like that, and you don't know which of the two it is. So again, you can't tell whether what's causing the peculiar features you see is due to periodic forcing, or due to entirely internal processes.

I never ran a probability density function for the case where we had the periodic forcing, but I suspect you wouldn't get the dip at zero; other than that it should look about like the other one. So the point I'm trying to make here is that here we have a very complicated system, some of whose properties can be simulated by taking a very simple chaotic system and building it up in such a way, adding almost intransitivity to the simple system. First adding the almost intransitivity by putting these narrow passages in there. Next, by putting on this restoring function, which restores to plus or minus four, or according to the sign, but doesn't restore to zero.

And with the almost intransitivity, together with the chaos, you could get something that looks like climate. And as I say, we don't know whether the real system we're dealing with now is almost intransitive or not. But we can produce variations which look a little like climatic variations this way. Furthermore, they can be used to test various hypotheses, provided the hypotheses apply to systems in general, and not to the earth's climate specifically. So I guess that's what I want to say. (Applause)

I thank you.

Yaneer Bar-Yam: Take some questions?

Ed Lorenz: First of all, I'd like to say that if your questions want to wander somewhat from the particular topic here, that's perfectly all right.

Yaneer Bar-Yam: So if you want to address the history of the contribution or if you want to ask more broadly about climate or weather, please feel free.

M: Hi. I was just wondering, it seems that climate was really fundamental in the initial formation of the study of chaos and nonlinear dynamics and complex systems. And I was wondering what your opinion is of the current state of affairs of the study of climate and chaos, nonlinear dynamics and complex systems, and how they can contribute to each other.

Ed Lorenz: Yes. I don't think in general there has been too much to relate the study of climate to the chaos itself. The chaos is certainly playing a part there. Chaos has played an enormous part in the study of weather, and in particular, in weather forecasting. I mentioned one of the things in standard weather forecasting recently, as I mentioned, is this ensemble forecasting, where you make an ensemble of forecasts, knowing that they aren't all going to look alike. But actually the study of climate, I think there is still a tendency to feel that the climate is more or less determined by the external conditions, regardless of the chaos.

The way the external conditions determine the climate with the chaos there will be different from the way they would determine it without the chaos. If you could somehow eliminate the chaos, you get a somewhat different climate. But I don't think this has been looked into too closely as to just what the differences would be.

M: Since there is a lot of interest now at the policy level in climate change, and as you pointed out, these issues are not being taken into account in climate models, if they were taken into account, how would that effect the policy issues and the predictions?

Ed Lorenz: Well, some of the climate models are essentially weather forecasting models extended to long periods. And whether or not you want chaos in them, the chaos will be in them if the weather forecasting part of it is any good at all. Also, this part I think is taken into account in those.

By and large I feel that they are pretty good for showing the sign and approximate magnitude of a change that would take place in response to some change in external conditions. They have been looked at less in connection with the internal climatic changes. And of course, it is possible to have an external and an internal change taking place simultaneously, and exactly, or almost exactly, canceling one another so that you don't detect it. And this is what really makes the determination of changes due to external things so difficult, because these other changes are taking place at the same time anyway. This is why we haven't yet come to definite conclusions as to what's happening. I should say, there seems to be very little question that the climate has been warming up recently; but as to the cause of this, there is still room for some speculation.

M: Could you give us a sense of how you first came upon the equations that you showed us today? Did you have an intuitive sense of how they were going to come out before you saw the graphs, or are they calculated outcomes?

Ed Lorenz: Well, no. I had often thought of this; I had tended to think of almost intransitivity as a kind of thing. First you want to ask, "Is this any different from just having a very low-frequency variation superimposed on other ones?" And the difference, as I saw it, was that with almost intransitivity it would tend to oscillate in one state for a long time, and only with difficulty, transforming from one.

And what I was trying to do was to think of a rather simple model. And a good deal of time in the last ten years or so I've tried to think of models that are understandable from a physical point of view, and one that is understandable is the motion of a particle. So I was trying to think of the motion of a particle that might behave this way, and this almost immediately suggested this narrow door to go through, and that's what was in the game. So I decided to base it on the game, and it took a while to see how to modify it to make it work here.

For instance, I didn't want discontinuities in the behavior. Now ordinarily, if you have two arcs meeting like this here, you'll get quite a difference in the behavior according to whether it hits just on one side, or just on the other side of this point up here.

Now, it turns out that you don't get any discontinuities as long as this angle is a divisor of 180 degrees. You can more or less see this by setting up two mirrors at an angle like this, looking at your face in it, and seeing what angle do you have to put them at so that your face looks like a face, instead of two faces split apart.

You'll find that any devisor of 180 degrees will work there.

So I chose these arcs to be just 60 degrees there; so that was part of the designing of it. The choice of the ratio of the distance across here to the distance across here was more or less arbitrary. So I worked on this. This is somewhat like a talk that I gave a few years ago, and I worked the model up specifically for that talk, this particular model. So it's a rather recent model.

M: I have a question about your probability distribution for your model. There were two peaks at plus and minus one. Presumably the transition state is also highly probable. Is that a generic feature of such almost intransitive models, that the transition state would also appear with high probability?

Ed Lorenz: I think a lot of a great many intransitive models would show the sort of bi-modality. There are two separate peaks, one for each of the two or more types of behavior that can take place.

This would depend some on the particular quantity that you chose to get a probability distribution for, whether it was a mean value of something, or a variance, or something else there. There are undoubtedly some quantities where you wouldn't see this type of distribution, but I would suspect that if the thing is really truly intransitive, you ought to see this for at least some of the variables of the system.

Yaneer Bar-Yam: I have a leading question. One of the things that has bothered me about the popularization of the idea of the butterfly effect is that many people think that any butterfly that flaps his wings will cause a hurricane somewhere. And the issue of difference between climate and weather is very apparent if you go to different places in the world. Some places you only get climate, and some places, like around here, you almost only get weather. (Laughter) So could you comment about that?

Ed Lorenz: Well, as I actually pointed out—and this paper I guess isn't too readily available—it was a paper that I gave at an AAAS meeting in 1971 in Washington, which wasn't published at the time. I pointed out, I think in the first or second sentence there, that this seems like a rather weird question to ask, and then I put it into context by offering two other propositions.

The first was that if a single flap of a butterfly's wings can be instrumental in generating a tornado, then it can equally likely be instrumental in preventing a tornado that otherwise would have occurred.

The second is that if this flap of the butterfly's wings can be instrumental in generating a tornado, so can every other flap of the wings of every other butterfly in the world, not to mention the more powerful species, including our own.

So what this can do, or the most a butterfly can do, is not to create new weather, but replace one possible weather by another possible weather. And of course, the thing behind it is chaos. If you take an ensemble of states, say a few weeks before, say very close together a few weeks before a certain time, and let's say that's the real time, and that a tornado actually did occur there, then presumably a certain number of these members of the ensemble will show a tornado, and a certain number will not show a tornado.

I suspect that in most cases the majority would not show a tornado, but a few of them would. What the butterfly may do is to change things from one member of the ensemble to another member of the ensemble since they can be pretty close together at the initial time. But it doesn't create new types of weather.

Yaneer Bar-Yam: Let me ask a follow-up question, which is somewhat different. In the context of thinking about sensitivity to initial conditions in any context, including social systems, corporations, and biology, the question that one wants to ask is, "What are you sensitive to?" In other words, where is the leverage point? Where is the point at which one can apply a small force, and achieve a large effect?

Now, other work that you've done recently with Kerry Emanuel over the last few years, is on where do you measure the weather in order to improve prediction, which is a related problem. When you want to predict the weather, the key thing is to know what is happening at the places to which the later times are sensitive. Could you comment about that?

Ed Lorenz: Yes. Certainly in a system as complicated as the weather, if you make a small perturbation at one point, if you wait long enough, it's going to show up almost everywhere. But in the intermediate time, which is what you're probably interested in, it's going to show up at certain places, and not at others. And these certain places might not be the place where it originally occurred as the whole thing can propagate there.

So that probably if you make a small perturbation here, it's going to be bigger here the next day. But it may be next bigger in Boston, or out in the middle Atlantic or something like that, than it is here today.

In other words, the center, or the region of maximum influence is going to propagate in addition to the growth everywhere. So you'll appreciate it more in certain places than other places. And that's one of the big problems to weather forecasting, one of the problems of this thing that you mentioned that I worked on with Kerry Emanuel. It's been suggested that there may be alternative platforms for observing the weather available for some time. These would consist of aircraft or something going on specific missions.

If you have an aircraft at your disposal and want to send it out to get some more weather information, where should you send it to do the most good? And the most good, of course, has to be defined in terms of where you want to improve the forecast the most. Do you want to improve it more locally, or do you want to improve it most on the overall average, or something like that? Whichever way you formulate it, then this becomes a well-posed question, and it's one that you can see answers to. And we have partial answers in terms of a small model.

M: Again, weather-specific question, but I think it concerns the people in other areas. One issue is what will the weather be like at some particular place and time in the future? The people worried about putting money aside for emergencies and resources of other sorts. An issue is how many catastrophes are going to happen at what sort of time scale, catastrophes or unusual events which are going to need emergency resources, or stuff like that. Now, we know that the statistics of that are weird. Isn't that as equally an important outcome from your model as exactly what the weather will be?

Ed Lorenz: Well, this is, of course, a very big problem in climate, and particularly in decision-making in connection with climate. Because if it's an extreme catastrophe, the chances are that a modest amount of computing won't pick it up. You need an extremely long run, and chances are you may not do that. So I don't know whether I've really got at this or not –

M: I think it's clear.

Ed Lorenz: You may miss the catastrophe all together.

M: It's clear to us that predicting the exact time and place for catastrophe, probably we'll never manage it, like the same for earthquakes and volcanoes --

Ed Lorenz: Predicting where the catastrophe will take –

M: Being able to predict the exact time and place of the catastrophe –

Yaneer Bar-Yam: I think the concern is can you predict the frequency of climate change, as opposed to just when it will occur is hard.

Ed Lorenz: Presumably yes, it is predictable, but again, it's like these other things. To predict something of that sort, you need a pretty good climate model to get it right. Any climate model will predict it, but if you want to predict it with a reasonable confidence that you made a good prediction, you need a pretty sophisticated climate model.

But yes, this should predict the frequency over a certain time. But if the frequency of a catastrophe is, say, once in 50 years, you don't really have a very good idea of how many times it's going to occur in any particular 50-year period. Maybe the most frequent number will be one, but there will be a lot of 50-year periods where it wouldn't occur at all, a lot where it would occur several times.

And here is something that would best be studied with an ensemble again. You'd make an ensemble of things, and you'd say over these in each case how many occurred in this particular 50-year period. And you could get at it that way, but again, with even a moderately sophisticated climate model as they stand today, you might very well get the wrong answer by a factor of two, or even ten, I would think.

M: I would value your opinion of the following idea: that complexity and predictability are sort of inversely proportional to each other. The more complex a system is, the less predictable. And they play a role similar to complimentarity of momentum and position in quantum mechanics.

Ed Lorenz: Well, if you mean physically complex, or complex in the sense of being only describable by many variables and so forth, there is no necessity why a complex system has to be chaotic at all, or unpredictable. And there is no reason why a very simple can't be chaotic if it isn't too simple. Or, I guess the theorem says if it's governed by ordinary differential equations, all you need is three to get chaos. If it's governed by a noninvertable difference equation, all you need is one.

But I would think, my feeling is that you deal with a specific kind of system, and it has a greater likelihood of being chaotic if it's more complicated. But it's not a very simple relation between the two.

M: We're getting to the point where we can build some fairly big butterflies if we want to. Can you speculate on the amount of hardware it would take, computers, and sensors, and large airplanes, or cloud seeders or whatever, to where we could tell where to flap to create a tornado or prevent a tornado?

Ed Lorenz: Well, that type of thing—I don't think with butterflies—but that type of thing has been suggested. The difficulty here is getting your observations. I think with the rate computers are going now, they'll reach the point before the observations do. But if you're expecting to do anything by adding a flap of a butterfly's wings, then you have to know what all the rest of the butterflies in the world are doing. Or if you're going to do it by exploding a ton of TNT, then you have to know what the other explosions in the world are going to be.

M: Can you describe your ideal sensor net, just something outrageous but reasonable?

(Laughter)

Ed Lorenz: I really don't know what it would be for that type of thing. Are you thinking of a sensor unit that would allow, say, one person to just wave his arm or something like that, and bring about the change, or something that fairly good human effort could do in time?

M: Something like fairly good human effort.

Ed Lorenz: Yes.

M: Like, you know, flying a fleet of helicopters and have them hover for an hour.

Ed Lorenz: Yes, right.

M: I guess what I'm asking is, if you could put up a state-of-the-art monitoring station every kilometer, every mile, every 100 meters, how close would you need them? How much how much equipment would it take?

Ed Lorenz: Well, let's see. This effort itself extends over what area, perhaps? If it's splitting up a bunch of helicopters, maybe it completely changes the atmosphere over an area of 20 miles square or something like that. You certainly need your observations as close and as dense as that. And you need to know what's happening, at present, at least to a degree of accuracy of the change that you propose to make. Because if you propose to make a change, and the change you make is less than the uncertainty in the present state, it's not going to do any good at all.

Also, if the effect is supposed to come not almost immediately but at a later time, then you have to take into account the amount that the error will increase during that time.

In other words, you would have to have extreme accuracy. You would have to be so accurate that the presently observed state, plus its errors, these errors, will not grow to any bigger size than what you're planning to do to it. So it becomes a very difficult thing, but I think the big difficulty here is getting the observations, rather than the computer --

M: Sir, I'm interested in following a little bit on the questions that have been asked about the butterfly effect. But as the ideas of complexity and chaos theory are being taken up increasingly in the business world, people are looking for answers, leverage points. As if they could identify the right butterfly and stamp on it, or control its wings, they could make a fortune. It's just a question of searching for the right butterfly.

But suppose you could inform all the butterflies about the rules of the game. Would that make it better or worse?

Ed Lorenz: Well, that's hard to say. You inform all the butterflies about the rules of the game, then you expect them to follow the rules, or you expect them to trip you up? (Laughter) I don't know that much about butterfly behavior. I assume that what you mean is that you expect them to follow the rules and make it easier for you. I suppose there would be a certain gain that way; I don't know how much.

XX: Maybe the butterfly wants to make a fortune.

M: Well, in business we're all the butterflies. That's the problem.

Yaneer Bar-Yam: If you could inform them, then that would be part of your leveraging activity, right?

M: Ah ah.

M: You mentioned a model of climate with two different attractors, one in ICH(sp?), one in normal climate. The stochastic resonances we now propose mechanisms by which the climate can switch from one to the other. So I'd like to know your views about the connection of stochastic resonance with the kind of almost intractable systems that you're studying currently.

Ed Lorenz: Yes. Well if you're thinking of stochastic effects here, some climate models I believe—not as many as should, perhaps—but some of them do have stochastic effects specifically in them.

In this little model here, I suspect if I had put in stochastic effect in these, if I had let it be a rectangular box that should give you regularity, but let the bouncing off be at an angle determined stochastically, it could have reproduced that. In other words, in a great many cases it doesn't really matter, as far as the outcome is concerned, whether certain subfeatures behave stochastically or chaotically.

A notable case in the atmosphere is turbulence where the turbulence does effect what's going on. Whether or not we think the turbulence is stochastic, as it has often been treated, or whether we can somehow show that it is chaotic and is deterministically governed but very highly susceptible to initial conditions, I don't think this would make any difference as to what the ultimate result of the presence of turbulence would be.

Chapter 3

Emergence

Michel Baranger
Session Chair
Stuart Kauffman
Herbert A. Simon Award Lecture: Emergence
Eugene Stanley
Correlations and Dynamics
Simon Levin
The Ecology and Evolution of Communities
David Clark
Emergent Dynamics of the Internet

Stuart Kauffman
Herbert A. Simon Award Lecture: Emergence

MR. KAUFFMAN: I would like to talk about the single strangest intellectual adventure that I've had in my scientific life. I went out to my office knowing that there were lots of conceptual strands that were running through my head, and I couldn't make sense of them. I hoped, all of a sudden, that if I were to make a notebook and write in it every day, whatever it was that I was thinking about would—after a month or 10 months or two years—start to fall into place.

I sat down and I wrote, "Investigations, Stuart Kauffman, December 4th, 1994." Then I chickened out and said, "Confidential, lots of BS below."

I proceeded to produce it, spending every day writing for 10 months, and the book that resulted, <u>Investigations</u>, is due out this fall.

In December of '95 I had produced enough writing to give a series of lectures on it. By September of '96 I had something that could come out in a rough form in the Santa Fe Institute.

As you'll see, the first half of what I have to tell you is science, albeit a bit in the future. The second half gets progressively stranger.

Let me begin by telling you what the first problem is. Think of a bacterium swimming upstream in a glucose stratum. I think we're willing to say that it's going to get food; that is to say, we're willing to say that the bacterium is acting on its own behalf in an environment.

This a system that can act on its own behalf in an environment—an autonomous agent. I do not mean that it's alone, I mean that it's acting on its own behalf in an environment which may include other autonomous agents.

But the bacterium is just a physical system. It's a bunch of molecules, and so in its content form, it has the property that anything it does seems profound. The question becomes: "What must 'the physical system' mean, such that it can be an autonomous agent?"

Through an obscure line of reasoning that I cannot reconstruct, I came to an answer. An autonomous agent is a self-reproducing system that does at least one thermodynamic work cycle. Notice that all free-living cells will fill this definition. I don't think you can deduce this definition, but we can find out whether it's truthful or not. We'll get to that in due course.

Let's begin with views of the origin of life. Since Watson and Crick came up with their famous double-stranded structure for DNA, RNA, where the Watson strand specifies the nucleotidase that will occur along the Crick strand by Watson/Crick pairing, everyone has thought that life must be based on the template-replicating properties of some polynuclear-type molecule, probably RNA or cousins of RNA.

The short summary of 35 years of hard work by very able chemists is that it hasn't worked yet, for all kinds of chemical reasons. Nobody's made the single-stranded RNA molecule or DNA molecule that can line up three nucleotides on the opposite strand, knit them together with phosphodiester bonds to a three-prong, five-prong, and melt the two strands in cycle. It just doesn't work. It may work one day, but it hasn't worked yet.

Does that mean that cell reproduction is impossible? The answer is, "No, not at all." First of all, Günter von Kiedrowski created the first self-reproducing system. What Günter first did is he took a hexamer of DNA—GGC DCC—and he reasoned that it would line up two primers—CCG and CGG—by Watson/Crick base pairing. Then he hoped that the hexamer would serve as a specific ligase to glue together the two primers into a second copy of the original hexamer.

If you read it from a three-prong or five-prong direction, the two sequences are identical. It works perfectly well, so Günter made the first self-reproducing system. Notice that it's not adding free nucleotides at the end of a growing chain. It's like meeting two primers, so it's acting as a ligase.

The view that life must be based on template-replicating polynucleotides, that is to say, that self-reproduction must be based upon polynucleotides, was blown apart a couple of years ago by Reza Ghadiri at the Scripps Institute.

Reza took a protein that's 32 amino acids long. I think it comes from the zinc finger. The molecule folds into a helix and then coils back on itself to make a coiled coil. He reasoned that if you were to open out the 32-amino-acid sequence, it might be able to line up two sub-fragments of a second copy of itself, 15 mirror to 17 mirror, all alongside one another and then catalyze the formation of a proper peptide bond, making a second copy of the 32-amino-acid sequence. It works like a charm. Reza made the world's first self-reproducing protein. "It's a done deal," as they say in America.

I want to raise a question. If you can have one protein that catalyzes its own formation, what about two proteins, each of whom catalyzes the formation of the other?

This is going to be our first example of what I will call a "collectively autocatalytic set." Here's two fragments of "A"—A prime and A double-prime—which together form "A," and two fragments of B which together form "B," but now "A" catalyzes the formation of B and B catalyzes the formation of A.

Günter von Kiedrowski has already made this with DNA molecules. Making a "collective autocatalytic set" is already a done deal, and Reza is not far from it. He may have already succeeded with peptides.

I want you to notice something about this collectively autocatalytic set. There is no molecule in it that catalyzes its own formation. Autocatalysis is a collective property of a whole, and a non-local property. There is no point that you can look at and say, "Here's where a collective autocatalysis happens." It happens in the whole system.

If you can do it with two molecules, why couldn't you have 10 or a hundred or a thousand molecules which are collectively autocatalytic? In fact, every free-living cell is collectively autocatalytic. There is no molecule in a bacterium, and there is no molecule in a eukaryotic cell that copies itself. DNA gets itself replicated by polymerases, but the polymerases are proteins and, to make proteins, you need RNA synthesis and you need the charging of transferring the molecules by a terminal law, a synthetasis, so that the whole system is collectively autocatalytic. Cells are, in fact, collectively autocatalytic systems.

I want to go back to the notion of autonomous agents and say something about them. It is an astonishing fact that autonomous agents do, every day, manipulate the universe on their own behalf. We all just went in and had breakfast. We changed the arrangements of piles of muffins and danishes that were sitting there on the table. Of course, somebody put them there, too. Nowhere in physics and nowhere in chemistry does the raw fact of agency appear, yet, here we are doing it. It would be my contention that asking the right question, I hope in this case, about autonomous agents, is a fruitful way of proceeding. Often finding the right question is essential to proceeding.

With Norm Packard and Doyne Farmer at Los Alamos a number of years ago, I posed the following question: If you make a sufficiently complicated soup of chemicals, what is the chance that it will contain a reflexively collectively autocatalytic set of molecules? The result is that there's a phase transition.

As the diversity of molecules in the system goes up, where the molecules can be substrates and products of reaction, but they can also be candidates to catalyze the reactions—like proteins or RNA molecules can—then there is a critical diversity at which almost certainly you get autocatalytic sense. The field of molecular diversity now puts us into position to test this. Many of you may know about this—it's called "combinatorial chemistry" or "molecular diversity." The idea is to make trillions of random DNA and random RNA, random proteins or hundreds of thousands to millions of small molecules.

We can begin to ask of such systems, do they in fact attain collectively autocatalytic sense? It's a long, wonderful story. It is now the main means of trying to find drug leads for drugs or vaccines. I happen to own patents in that area since I started thinking about it a long time ago.

The hypothesis is: if you have sufficient diversity of chemicals sitting in a pot, there is a phase transition from non-catalysis to collective catalysis.

I want to remind you of something. The struggle in the last century was to make the distinction between spontaneous and non-spontaneous physical processes, and the whole notion of entropy has to do with trying to make the distinction between spontaneous and non-spontaneous processes. You all had the exam where you put the ink droplet in the middle of the petri plate, and it diffuses outward. It's not going to reassemble itself. There's all sorts of examples of irreversible processes.

In a chemical reaction, here's free energy and here's a reaction core, and you know if I've got X and Y, X converts to Y, and Y converts to X, and when the abundance of Y and the abundance of X are such that the rate of conversion of X to Y exactly balances the rate of conversion of Y back to X, you have chemical equilibrium. That's written as X over X plus Y at equilibrium, and here's equilibrium.

Physicists and physical chemists like to draw this in terms of a reaction core, and where if you have at the beginning only X, it will run downhill to the equilibrium ratio of X plus Y. Running downhill is called "exergonic."

You're giving off free energy. If you start with only Y, then you begin to make X and you run downhill, the same equilibrium concentration of X over X plus Y. That's also exergonic. It's a spontaneous chemical reaction.

Only spontaneous things happen, but you can drive them. Suppose you want to drive this reaction when you start with X, such that you make more Y in your equilibrium. You have to add some outside source of energy, such that as you get down to the equilibrium, you pump the reaction beyond equilibrium endergonically. You're putting energy into the system, like pushing on a piston, and you push the reaction uphill in energy, and make excess Y—that's endergonic. It's a non-spontaneous process. You have to add free energy.

Let's ask about Reza Ghadiri's absolutely wonderful autocatalytic peptide. Does it perform a work cycle? Does it link spontaneous and non-spontaneous processes? It does not. This is a purely exergonic reaction. It's going to go downhill in free energy.

You start with excess 15 mirror and 17 mirror and some enzymes—the 32-amino-acid sequence. The 32-amino-acid sequence catalyzes the ligation of the 15 mirror and 17 mirror, but there's a back reaction in which the 32 amino acids seem to breakdown at that particular bond. There's an equilibrium ratio on the 32-amino-acid sequence and it's two fragments, the 15 mirror and the 17 mirror, and this reaction will just proceed to equilibrium. It's going downhill in free energy.

As beautiful as is Reza's system, it doesn't fulfill my tentative definition of an autonomous agent. It does no work cycle. There's no cyclic link there to spontaneous and non-spontaneous process, it's purely exergonic. In fact, so is Günter von Kiedrowski's system, both his single molecule and his collectively autocatalytic set. Several of the examples that I came up with regarding Doyne Farmer and Norm Packard are also purely exergonic. So, it doesn't fulfill my definition.

Here's an example of a system that does fulfill that definition. I have faked the chemistry—I am making it up. Once I've shown it to you, you'll see how to make one in principle, and with my method we will be making autonomous agents, given my definition of what "autonomous agent" is, sometime in the next 30 or 40 years.

Here's how it works: I have Günter von Kiedrowski's system, a hexamer and the two primers that together form the hexamer. I've drawn chemical reactions the way biochemists do it, with forward and back lines, and I'm missing a line. I've got the line that says that this hexamer can break down to form the two primers, but I am lacking the forward reaction that just has these two guys spontaneously ligating, which is exactly Günter's reaction.

I want to drive excess synthesis, excess replication of a hexamer uphill energetically, compared to its equilibrium, so I need a free energy source. I am going to invoke pyrophosphate, PP, which I have present in high abundance compared to its equilibrium with its breakdown products, monophosphate. This is an exergonic reaction. So I am driving uphill excess synthesis of a hexamer by the free energy released by breaking down the pyrophosphate.

Now, I'm going to invoke an analysis. I'm going to say that this molecule catalyzes that linked reaction. I'm driving excess synthesis and excess replication of the hexamer. I've broken down pyrophosphate to monophosphate, and I want this little device to cooperate cyclically, so I'm going to have to resynthesize pyrophosphate. I'm going to invoke another outside energy source.

By the way, this is a perfectly legitimate open thermodynamic system. It has sources of food in the form of the two primers and the outside energy source here. I'm going to invoke a photon. The photon hits an electron and drives it to an excited state, E* , and then the excited electron falls back to its lower energy state, and I'm going to couple that broth in free energy—that exergonic process, with the endergonic synthesis—the energy requiring the synthesis of pyrophosphate back up to its original level by this reaction. Try to imagine that this particular primer happens to catalyze that reaction.

Phosphate is going to be a positive activator of this enzyme, just like phosphotryptokinase in the metabolic pathway. The product of the enzyme feeds back to activate the enzyme. The consequence of that is you have the substrate around, and then it slowly leaks through, and then when there's enough product, the enzyme is activated, and you flush the substrate to the product. The same thing should happen here and, in fact, it does for the right kinetic constants in the differential equations for the system. You slowly leak through, and then "bam," you turn on the enzyme and you flush substrate to product, and all of a sudden, you get a lot of synthesis of the hexamer and the monophosphate.

But I want this thing to act cyclically. I want this reaction, and then I want the back reaction to follow it. To get it to follow it, I use pyrophosphate to inhibit this reaction, and so the reaction doesn't proceed backwards until the pyrophosphate's broken down to monophosphate, then the back reaction goes.

Well, it works. Mainly, we did the differential equations for it, and the thing makes excess hexamer and, in fact, under the appropriate condition, it actually goes into a cyclic operation called "the lemon cycle."

I want to point out a few things about this. First of all, this is a perfectly legitimate open chemical reaction system. It just so happens that I have coupled two things, each of which were objectively studied, namely, something that does a work cycle, which I'm going to persuade you of in a moment, and something that does autocatalysis, and I put them together. Nothing prevents us from making these things.

Let me show you the work cycle. If you were to watch a particular phosphate, it would rotate this way around this reaction. Now, that neck rotation tells you that there's a chemical motor at work. You cannot get rotation from substrates around cycles, and in some system it's an equilibrium because of detailed balance. It's just another example of a motor.

Things like this must be able to be built. People at the Whitehead

Institute in Boston are trying to make self-reproducing molecular systems. People all over the place are trying to make self-reproducing molecular systems. We just have to put it together in the work cycle.

The next thing I want you to notice is, every free living cell does this. Every free living cell is both self-reproducing and does at least one work cycle. You can ask whether or not I have stumbled upon a definition of life. I actually think I have. I won't insist on it because you'll attack me. Anybody who says, "I've found a definition of life," is going to get his rear-end handed to him, and I don't intend to have that happen to me before 10 o'clock in the morning. But I actually think that I have found a definition of life.

The next thing to notice about this is, it's the beginnings of a new technology. Fifty years from now we will be making self-reproducing systems that actually can build things. This builds something. It builds excess hexamer, but we can make this thing build all kinds of things.

Another thing to notice about my autonomous agent is that in order for it to do a work cycle, the system must be displaced from equilibrium. You cannot do a work cycle at equilibrium, so the idea that an autonomous agent is necessarily a non-equilibrium concept.

The next thing to notice is that it stores energy in the form of excess hexamer. Phil Anderson, a Nobel Laureate and friend, pointed out to me that once you've stored energy somewhere, later on, you can use it to correct mistakes. This is just what happens in DNA replication and DNA repairing. We'll make these things within the next 20 to 30 years.

The next thing to tell you is they're capable of evolving. We made the equations for it, Peter Wills and some of his students and I, and this is just to show you that in the 13th parameter, there's 13 kinetic constants in the appropriate set of differential equations, and this just looks at the efficiency of replicating, and this shows you that there's a hill.

That's the end of the science that I'm confident of. From here on, it either goes uphill or downhill, depending upon your mood. In any case, it's strange. I'm going to show you Maxwell's Demon. There's some very strange things about Maxwell's Demon.

I want to turn now to a critique of the concept of work. I have external evidence that the concept of work is a complicated concept. It's force acting through a distance. Here's my experimental evidence. I was out there at an Indian dinner with Phil Anderson in Santa Fe, and the chapati was there, the chutney was there. Phil was about to pick up some chutney on his chapati, and I said, "Phil, work is a strange concept," and he said, "Yeah." That's it. Here's a Nobel Laureate in physics. If Phil says it's a strange concept, it is a strange concept.

Let's begin to look at it. You've got the scalar—a number for the amount of work. In any specific case, something takes a direction in which the work is being done. What picks the direction or, in other words, saying there's something that organizes the process, where does the organization come from?

Well, physicists aren't dumb. They put in the initial and the boundary conditions, and then you calculate forward, so you have the particles, the forces, the laws, the initial and the backward conditions, and Newton taught us how to do it.

And here's the cheat for which I chastise the physicists hereby—consider yourself chastised. It's a cheat in a way to put in the initial and the boundary conditions, because what we really want to understand is how did the universe get to be complex since the big bang, and that's all about initial and boundary conditions coming into existence and, certainly, that's true for a biosphere. That's one way to begin to get worried about this concept of work.

A second way is to point out that an isolated thermodynamic system can do no work, but if it be partitioned, then you could have a membrane and B could have higher pressure than A. Partition B does work on System A by bulging the membrane into it.

The puzzle is how does the system get to be partitioned, where did the partition come from? For work to happen, the universe has to be divided into two parts. How did it get to be divided into two parts? The way I find most congenial is in Atkins' book on the Second Law, where he says, "Actually, work is the constrained release of energy. Work is a thing."

Let's take the cylinder and the piston. There's the working gas, the cylinder, the piston, and the working gas does work on the piston. But what are the constraints? Evidently, they're in the cylinder and the piston. Here's the cylinder, the piston, and there's the grease in there. What are the constraints on the release of energy? It's the cylinder and the piston. That's what organizes the release of energy.

Let's ask a new question. Where did the constraints come from? Ah-ha. Ah-ha-ha-ha-ha-ha. You see, it took work to make the constraints. Somebody actually built the cylinder, somebody built the piston, somebody assembled the cylinder with a piston inside and got the grease there with the working gas on the inside. That took a whole bunch of work.

Roughly stated, I want to say: it takes work to make constraints. I'm not sure that it always does, but it very often does. And it takes constraints to make work. It takes work to make constraints, and it constraints to make work. We have here shining at us the core of a virtuous cycle that has something to do with organization.

Hang onto that, because before I delve into it, I want to show you something I'm very proud of. I want to try to describe propagating work. I do not know how to define this mathematically. In fact, I don't know how to define autonomous agents mathematically.

The cannonball has powder-like matter. You light that little thingy and it blows up, and the cannonball goes over in an arc, and it hits the ground, and it makes a hole in hard dirt. That's technical, O.K.? I want you to consider what I'm so proud of: I invented this all by myself. This is an example of propagating work. It's also a fine Rube Goldberg contraption.

I constructed a sturdy paddle wheel—you can't tell that it's sturdy, but it is—and I have dug a well, and I have put the paddle wheel over the well, and I put a bucket down in the well, and I tied a red rope—I like red rope, but it looks black—onto the wheel, and I've dug a water pipe in here with a funnel.

Now, I want to water my bean field. The cannonball comes up and hits the paddle wheel, and it spins the paddle wheel. When it spins the paddle wheel, that's an exergonic process. Once the powder is exploded, it's releasing free energy. Now it pulls on the rope, which is a non-spontaneous process, it's an endergonic process, and it lifts the bucket up, and it tips over the axle, and it dumps the water into the pipe which comes down and opens the flap valve and waters my bean field, and I make $4 an acre on it.

What's the difference between the two pictures with the cannon contraption and the cannonball and the hard dirt? The obvious difference is that there's lot of microscopic changes that propagate in the world in this case and not in the other case, and there's lots of constraints that got constructed by the paddle wheel and its rigidity and the pipe and so on that are organizing the flow of the energy. Also, notice that this thing doesn't operate cyclically very easily. If I wanted to get it to work again, I would have to sort of hook up everything and fire another cannonball.

The reason engines operate in a cycle is the system is arranged such that at the end of the cycle, it's back where it started again. That's why it's operated cyclically. This thing would be a real pain in the rear if you use it over and over again.

I'm going to show you that cells do what I said. Mainly, cells do thermodynamic work to construct constraints on the release of energy, and when that energy is released in constrained ways, it goes on, and it does more work to construct more constraints.

You have to build lipids that do thermodynamic work. The lipids, like in an aqueous environment, form bilipid membranes because they have hydrophobic tails and hydrophilic heads. So, cells just did work to construct a membrane. Now, let me persuade you that in doing so they're manipulating constraints.

I have two molecules, A and B. A and B can undergo three reactions. A and B can go through two substrate and two product reactions to make D and C or another reaction to make E or another reaction to make F and G. Each of these reactions has those reaction coordinates with substrates, potential energy barrier, and products, the kind that we saw before.

Now, when A and B diffuse into the membrane, that changes the translational, vibrational, and rotational degrees of freedom—the modes of motion of A and B—and in doing so, that changes the heights of the potential webs for each of those three reactions, and the heights of the potential barrier for each of those three reactions. That is precisely the manipulation of constraints. The cell has, in fact, done work to manipulate the constraints on how energy will be released.

Furthermore, the cell also does work to build an enzyme, and the enzyme goes over and binds to the transitional stage to take A and B to C and D, lowers that potential barrier. Just like the flap-valve-opening door in my bean field, by lowering that potential barrier the energy flows downhill to make C and D. So, energy is released in a specific way to make C and D, not E or F and G.

I'm not telling you something you don't know. You've probably just not thought about it before in this context. The cell has done work, so that it's manipulated constraints, and it's released constraints by lowering the potential barrier, and energy is flowing some specific way.

It propagates because D now goes over and gives up some of its rotational and vibrational energy to snap off to a transmembrane channel and open a channel so an ion can enter.

What a cell does is something we haven't thought about. Cells do work to construct constraints on the release of energy which, when released in a specific way, could go to work to construct more constraints which, when the energy is released, can propagate to do more work and constructing more constraints on the release of energy which propagates around the cell in some web until the cellology does something rather astonishing. It makes a copy of itself.

I want you now to think of the second closure. There's an autocatalytic closure, and there's a catalytic closure. Everything that has to get catalyzed does get catalyzed, and there's also a propagating work closure by which a cell builds a copy of itself. We do not have names for these concepts, nor is it yet mathematized, but the union of these two catalytic and propagating work closures is what a cell does.

Think of a cell. You put that one bacterium on a petri plate, then you get two, four, eight, sixteen, thirty-two, sixty-four; it keeps going. What is this propagating closure that we're talking about? Is it matter alone? No. Is it energy alone? No. Is it entropy alone? No. Is it one minus entropy? Is it sharing that information? No. It's something else. It's something that we don't have a concept for. It's what an autonomous agent does.

What we're looking at is organization. I don't know how to define it mathematically. You can try category theory. We have domains and ranges, and you can map from the domain to the range, and where you get at the range, you can control the mapping of the domain with the range, and that sounds like an autocatalytic set because the substrates go to the products, and then the products control what reactions get catalyzed. The trouble with category theories is you have to say ahead of time what all possible reactions or what all possible mappings are.

I do not think you can finitely pre-state the configurations based on the biosphere. Let me show you what I mean. Let's take the notion of a Darwinian adaptation. The function of your heart is to pump blood. Your heart also makes heart sounds, but nobody thinks your heart sounds are a function of your heart. I, a doctor, don't think that the function of your heart is to make heart sounds. That means that the function of your heart is a subset of its causal consequences. And that means that there is an inalienable holism about autonomous agents and their worlds. We have to analyze you and your world to figure out what the function of your hearts are, and the same thing's true to the autonomous agent I showed you. That's the easy part of this holism.

Darwin also talked about pre-adaptation, and the idea that a causal consequence of a part of you that is not normally of functional significance might be of functional significance in some funny environment. I'm going to tell you three stories that make this point. The first concerns Yaneer. Yaneer happens to have a dominant mutant heart. You notice if you have a piano that the strings vibrate. Well, your heart's a resonant chamber too, and Yaneer's dominant Mendelian, single-gene mutation happens to allow his heart to pick up rhythmic patterns. Yaneer's in Los Angeles, and he feels something funny in his chest, and he says, "Goodness gracious, it's an earthquake," and he does the right thing, whatever the right thing is, and Yaneer survives, while millions of people die. It's really a tragedy. But Yaneer gets on the Today Show. Now, this is the part that's necessary to the story. Women flock to his side, and Yaneer mates with many of them. Soon there are a lot of little Yaneers running around with earthquake detectors in their chests.

If earthquakes happen enough, such that this can be of selective advantage, pretty soon there will be earthquake detectors running around in the biosphere.

I'll limit myself to one more story. Do you think you could say ahead of time all of the funny context-dependent causal consequences of bits and pieces of organisms that might turn out to be adaptation to the biosphere? I think the answer to that is "no." I don't think that we could finitely pre-state all possible context-dependent causal consequences of parts of organisms that might be adaptations. That means we cannot pre-state the configurations based on the biosphere.

Why does that matter? All statistical mechanics depends on knowing the configuration space in its entirety ahead of time. You cannot do it for a biosphere. The way Newton, Einstein, Bohr, and Wilson taught us to do science, you need to know the configuration space.

I'm going to end with one more story to show you that you can't say it ahead of time. 60,000,000 years ago, there was this squirrel. She had a flock of skin that went from her hand to her ankle, and from her other hand to her other ankle. She looked so ugly that nobody would play with her and nobody would eat with her. Her name was Gertrude. She was up in a tree one day eating lunch—kind of sorrowfully—and there was an owl in a neighboring pine tree named Bertha.

Bertha looked at Gertrude, said, "Lunch," and came slamming down out of the sunlight, claws extended. Gertrude was scared out of her pants, and so she jumped out of the tree, and she said, "Ga," and she flew, and she got away from the befuddled Bertha.

Well, she became the heroine of her clan, and, she had suitors all over the place. She was married in a civil ceremony a month later to a very handsome squirrel. She had been Mendelian dominant, and that's why there's flying squirrels, roughly.

Could you have said ahead of time that the flap of skin would work as wings? Maybe. Probably not. For a bacterium that happens to have a mutant protein that happens to wiggle in a funny way when it sees a calcium current, and allows it to detect a passing paramecium and, thereby, escape? Could you say that ahead of time? I think not.

I just want to close by saying it to you again: we cannot pre-state the configuration space in the biosphere. Therefore, I think the way we've been taught to do science by the physicists needs changing and, in fact, I have to change physics, as well. Thank you.

MR. BARANGER: Are there any questions?

UNIDENTIFIED: Lets say you have a bucket of bacteria; they're autonomous agents. Dump in sugar, and they'll go for it. Now, you put in a chemical that removes their ability to self-replicate. Now, you dump in more sugar, and they still go for it. Would you say that they are or are not autonomous agents at that point?

MR. KAUFFMAN: That's a neat question. I don't know. Once you've got a central definition of something, you always have further definitions. If somebody's on a lawn mower with an umbrella over it, is that a motor vehicle? I don't know. I am content if I were given a simpler case that people will like. If they could no longer replicate, then they are dying autonomous agents.

UNIDENTIFIED: I'm interested in your definition of the autonomous agents concerning the idea that you have to have a thermodynamic cycle, because it seems like in the examples you gave that one had, like phosphoryl reaction sites moving through—it seems like you could just short-circuit that and have light coming in prior to energy or have a pump provide some external energy source, but so that the thing did not cycle. I was wondering if you'd comment on why the cycle is a necessary part, or is it just part of the analogy?

MR. KAUFFMAN: Other people have asked me that question, and I can't find an answer to it that's utterly convincing, but I will say the following piece. By having the work cycle, where you are linking spontaneous and non-spontaneous processes cyclically, somehow or other it focuses our attention on the motion of organization.

Notice that between organisms, you can do some work if it makes something that is useful. I can do some work that makes something that's useful to you, if you start having advantages of trade, which is just what happens in the roots of plants where fungi make amino acids, and plants make sugar, and they exchange them.

This cyclic linking of exergonic and endergonic reactions has to do with the building of complex linked webs, the building of complicated things, like redwood trees.

We need a notion of organization, and the work cycle forces you to think about organization. That's why I like it.

MR. COHEN: Once upon a time, I think in the 1950s, Nicol produced a book called, *The Limits of Man*. At that time I was quite interested in the agriculture of what people did. Nicol asked, "How much does it cost to make a tractor?" If you're doing your agriculture, you've got to put in the cost in energy of making the tractor, and then you work it out, and you find that in the Third World countries, a tractor doesn't reap enough food, on average, to build itself. There is an antecedent to your thinking about this.

I've been concerned, as you know, for a long time with the origin of life, life on other planets, is it going to go the same way, and precisely that question which you ask, is it possible to pre-state what kind of pre-adaptations, and what bits of function are going to be exploited?

I think the only clue we've got is from what the real world does, in which things are being exploited several times. The synthesis has been invented several times, and say, it's a default of the system. There seems to be lots of ways of jumping over to photosynthesis.

MR. KAUFFMAN: I agree with the last point that something's happened multiple times.

MR. COHEN: Yes.

MR. KAUFFMAN: I think it doesn't, and I don't think you're suggesting this either.

I don't think this argues against this odd thing in mind, but I don't think you can pre-state all of the possible --

MR. COHEN: Absolutely. I don't think you can pre-state all of them, but I think there is a likelihood they're different islands, because we have a very dense history, and we can, therefore, do some geography of that likelihood.

MR. KAUFFMAN: Yes. And that was that the eye has evolved six times.

MR. COHEN: Right.

MR. KAUFFMAN: The tractor, by the way, is a fun story, again, of Darwinian pre-adaptation. I don't know if you know this, Jack. People who are trying to make the tractor, they have this huge engine block. Every time they mounted it on the chassis, it crushed the chassis. They kept doing it over and over and over and over again.

Finally, somebody said, "You know, the engine block is so rigid, we can use the engine block as the chassis and hang things off of it." It's another Darwinian pre-adaptation.

Human technological evolution is filled with the same thing.

UNIDENTIFIED: A and B have a potential barrier against them, and you said a catalyst reduces the barrier. Is it possible that they could quantum tunnel to each other?

MR. KAUFFMAN: Sure. Given the billions of years, there is always the finite probability that the tunneling will take place. I think we should think about it. Thank you all very much.

Eugene Stanley
Correlations and Dynamics

MR. STANLEY: I'm Gene Stanley. I'm a physicist from Boston University. I assume there's at least one other physicist here. I've been given the challenging job of covering three different topics, details of which are in the abstract.

All three of these are quite complex systems where modern ideas of scaling variance, numerosity, normalization group, have proved useful in elucidating them. I'll present each of these three as a puzzle.

First, I'll start with the puzzle of liquid water. Every one of us knows that water is unusual in many respects. I'll present this, therefore, as a detective story by first posing the question: what actually is the puzzle of liquid water; secondly, why do we care about this puzzle; and, thirdly, what we actually do. The "we" is a large number of people whose names appear in the top two lines.

Consider a simple response function like the isothermal compressibility which, in thermodynamics, is the response of the volume of the system with infinitesimal pressure change. Again, a typical equation is we lower the temperature. This response function decreases, and we understand that decrease because in didactical physics, we learn that this response function in thermodynamics is related to a microscopic statistical mechanical quantity, mainly, the microscopic fluctuations in specific volume feed and, therefore, as you lower the temperature, fluctuation should get smaller intuitively and, therefore, the compressibility should decrease.

Consider, now, water. At very high temperature, water is already anomalous because this compressibility is about a factor of two larger than what you would get if you're substituting all the obvious numbers that go into these formulas.

A second anomaly occurs as we start to cool water. As we cool water, this factor of increase, roughly a factor of two, gets larger and larger, and, actually, if I had a nonlinear function to a linear function, I must have a minimum. That minimum in water occurs at the temperature of New Delhi in summer of about 46 Celsius. And below 46 Celsius the fluctuations actually increase as we lower the temperature, and this is not a smaller factpr. The lowest attainable temperature is around minus 38 Celsius.

This compressibility is more than a factor of two larger than its value of the minimum. So, these fluctuations have increased by a factor of two on knowing the temperature. The same two anomalies occur for other response functions of specific heat. Everyone knows that water is a factor of two larger than for typical liquids. This is very useful for engineers, as water is a useful coolant, and the same thing happens. Fall to about 35 Celsius, the temperature of a good hot day in Boston, this response function starts to increase.

This response function is in proportion of something microscopic. The fluctuation is in entropy, which means that as you lower the temperature, the entropy is not increasing. That would be impossible—there's a theorem against that. But the fluctuations in the entropy increase. Fluctuations are not necessarily related to the quantity, itself. And these fluctuations also increase by almost a factor of two at the lowest attainable temperature.

And the third anomaly shown at the bottom here is an anomaly that we all learned about when we were about four years old. And what is that? If we look at a third response function, the coefficient of thermal expansion, how much does the line change when the temperature is infinitesimally increased? This quantity must be positive.

Why do I say, "must?" Well, it must be because in statistical mechanics this thermodynamic function is related to the cross fluctuations of entropy and volume. And everyone knows who's been moved to a larger office, positive though that may be, that the entropy in this larger office is bigger, not smaller, and there's more ways of arranging everything, and this correlation, they must be positive, and that's true for almost all of the goods and, in fact, it's true for water.

Water is anomalous in three respects, so high temperatures, the value of this cross-correlation function is a factor of three smaller than what one would expect for it to be a hybrid. As we lower the temperature of the deviation, it gets bigger and bigger, and it's this magic number that we learned at age four and before Celsius, at this number this response function becomes negative.

And what does it mean? It means below four Celsius, when in the little region of the liquid with a larger volume, there are fewer, not more, but fewer arrangements of the particles. This seems to be very counterintuitive.

All three of these response functions are linked to three microscopic fluctuation quantities, and all three of these have very, very anomalous behavior at low temperature.

UNIDENTIFIED: About what?

MR. STANLEY: Well, 46, 35, and 4 are not terribly low temperatures.

There's a third anomaly which I haven't mentioned yet, and that is an anomaly pioneered by a Russian refusenik, Sasha Voronel and Austin Enjulin (Velyulin?)(sp?), University of Arizona. And that is, if you take any of these three response functions and at low temperature, plot the value on double-log paper—paper for people who are obsessed by scaling variance, the only kind of paper you really need—if you apply this on double-log paper where the y-axis is the function and the x-axis is temperature at this variable minus some arbitrary fitting primary temperature, it turns out, if you choose for this temperature, minus 45 Celsius or, if you prefer, 228 Kelvin. So, with the y-axis we have the longer the room of function, on the x-axis we have longer of the room of temperature, minus 228 Kelvin.

If you do that you'll find that the behavior is not curved like this, but is linear. That means there is some apparent power loss singularity.

I say "apparent" because 228 Kelvin is about seven degrees colder than you can actually get, and you never know that there will really be a singularity, but at least it looks like one.

This is the puzzle with liquid water. Why do we care? There's always two reasons to care. Scientific curiosity is a good reason, and there's no question that these three anomalies are very strange and not yet explained. And the other reason is that liquid water is a prototype of a structured tetrahedral liquid of which SiO_2, germaniums in phases of phosphorous, carbon, silicon are other prominent examples. It turns out that analogues in these phenomena occur for these other liquids, and of course, there's a practical reason. Water is ubiquitous, important for life, everything else.

Clues. Clues. Well, the principal clue was weighing out my line as a problem, mainly what I just said. Water is a tetrahedral liquid. Each water molecule can be regarded as one of these little elves which has two arms and two sensitive spots, and in low temperatures there are two arms of one elf reach out for the sensitive spots in another elf, and you can imagine what happens.

Finally what happens is that, at low temperatures, this liquid then forms a gel, a cross-linked hydrogen-bonded liquid, and such a thing is called a "gel." It's a transient gel, not like the gel that you make in the icebox. Gel in the icebox, if you pick it up and twirl it, nothing happens. But if I pick up this glass of water and pour it out, then something does happen, and that's because the lifetime of these bonds is about a picosecond and, therefore, the network at any instant of time less than a picosecond looks like this. But a picosecond later there's a slightly different network because the bonds are formed randomly.

This is an important clue, and because of this clue we can make a little topological sketch of how the water should look, and what's shown here is a two-dimensional sketch of this three-dimensional gel in which each intact hydrogen bond is shown as a line, each broken hydrogen bond is not shown at all.

80% of the bonds are intact, according to spectroscopy and, therefore, 80% of these lines are intact. You can count them if you wish. The important thing is not to check the math, but the important thing is to notice that some of these water molecules have four intact bonds, others have three, others two and, perhaps, some of the unit has one.

I was doing this on an airplane once. The stewardess came over and said, "Aren't you a little old to be playing connect the dots?" but I said, "No. This is serious stuff." And why is it serious stuff? Because a part of this network remains completely intact, and the hydrogen bond, we know, has the remarkable property that, when a hydrogen bond is intact, the bond link is very narrowly distributed around .28 angstroms. And the bond angles of this four-bonded water molecule are very narrowly distributed around approximately 104 degrees.

Therefore, the local geometry of this little patch is indistinguishable from the local geometry of a completely hydrogen-bonded cell and, therefore, the volume of local volume in this region is larger than that of the global network by the same amount approximately that the local volume of solid water, ice, is larger than the rest of the liquid and when it's floating, and we know, since the experiment was done on the Titanic, that about 10% of that ice is above the water and the other 90% is lurking somewhere underneath.

Therefore, we have a situation with the local heterogeneities, with the property that they're specific volume is 10% bigger and, of course, their entropy is lower to order. That's the cross-correlations in biomin (phonetically) specific entropy that we showed at the outset. We said that one of the puzzles of liquid water was that the cross-correlations of entropy and specific volume were negative at sufficiently low temperatures.

The contribution of these heterogeneities is to give a positive delta B and a negative delta S. Plus times minus - even a physicist knows this much - plus times minus is minus.

Further, there's a negative contribution due to these cross-correlations. And the presence of such compilations due to these hydrogen-bonded local structural heterogeneities gives rise to the increase in entropy fluctuations and entropies invial (phonetically) fluctuations in the top graph.

Furthermore, as we lower the temperature, the fraction of hydrogen bonds gently increases, and as that fraction of hydrogen bonds gently increases, the fraction of four-bonded molecules increases a little more, a little more because the fraction of four-bonded molecules is P to the 4th. So, each 1% increase in P, the fraction of the hydrogen bonds, leads to a 4% increase in P to the 4th.

What about this apparent singularity? There's a very recent result that has been published. The most convincing experimental proof has been published in *Nature* of this year, and it concerns the explanation of this apparent power loss singularly. And in light of the time, I'll tell you the explanation but not show you the evidence.

The explanation is that at low enough temperatures, these little structural heterogeneities condense out and form a separate phase. A low density liquid phase is formed out of the two—out of the uniform liquid phase, or to say it more precisely, there is a liquid-liquid phase transition in which the homogenous liquid splits into two liquids differing one from the other in their density. The presence of such liquid-liquid phase transitions have been found by our group using computational methods, and the experimental proof of such a system has only been achieved in the last year.

Before I conclude about water, I'd like to give you a little bit of a clue as to how that might happen and, in order to do that, I'll show you one transparency which might represent what you would tell your grandmother if you go home for summer vacation, and she asks you, "Why do liquids exist? Why should there be a liquid?"

It's a good question, first of all. You'd scratch your head a little bit, say, "Can we talk about it tomorrow?" Then tomorrow, if your grandmother's like my grandmother, she'd say, "Remember that question?" and you'd say, "O.K.

O.K. I'll tell you."

Every material has some interaction, however weak. And in the case of most molecules there's a region of attraction. So the energy of attraction is a function, of course, of the distance between them, and if they're far enough, that's very, very tiny, but if they're close, it's bigger.

"And what that means, Grandmother, is that if you lower the temperature enough, such that the system wants to be at its lowest energy and keep the pressure high enough, out of this homogenous gas will condense a liquid phase. A liquid phase will condense at high enough pressures and low enough temperatures."

Now, if your grandmother's really smart, she'd say, "Why did you draw such a simple little schematic like that? How did you know to draw such a smooth thing, particularly for something that's complicated as water?" Where even your grandmother probably knows, because it's been known for a long time, that whatever we know about water, we know that it's anything but a simple sphere like xenon or argon or even carbon dioxide, but rather, water is this complicated tetrahedral thing.

For example, this little tetrahedron might represent a long water molecule coordinated with its four neighbors. In other words, it might represent one of these black bullets here, and this one might represent another one, say, one of these black bullets.

You ask, "How might these come together?" And if your grandmother's real smart—and she probably is, anyway, because grandmothers are smart. These two tetrahedra can come together in more than one way because of angular degrees of freedom. In fact, they can come together in an infinite number of ways.

So, if you calculate the energy for different angular approaches, you find that there are two local minima, one in which, a deeper one, in which hydrogen bonding takes place. If you will, that corresponds to these two blue neighbors approaching each other in the same plane, say, like this, and they stop within .28 angstroms apart. And so, there's a deep minimum in the potential because the hydrogen bond, although it's a weak bond, is stronger than no bond at all.

But there's a second way they can come together, and that's a way in which one of these tetrahedra, say, is rotated 90 degrees. If it's rotated 90 degrees, not surprisingly, I can come a little closer before the hardcore part of the potential starts to set in. And that second minimum corresponds to this rotated configuration, no hydrogen bond, so it's a bit narrower, a bit shallower.

So, the potential, Grandmother, if you were right, can have two minima. And suppose it does have two minima? Then the liquid that it condensed out at low enough temperatures will discover this second minimum, so long as the pressure isn't too high. If the pressure is not too high, the pressure is below a critical threshold and the temperature is below the critical threshold, out of this liquid can condense a second liquid phase. And that second liquid phase is called a "low density liquid," and that phase was probably first discussed about eight years ago by our group and has since been found in a number of liquids, unequivocally, by experiments and confirmed by computer simulations.

That's all I want to say about this very, very complex fluid. I'm going to now turn to the second topic which is something very dear to our heart, namely our heart. So, I think that each of us in this room has a beating heart; I hope so. I hope I haven't killed anybody. And for most of us this heart will beat all our life. For three gigabeats, this heart will beat regularly. And we even think of the heart as having a little metronome because we know there are pacemaker cells that beat away, and so forth and so on.

And when we go to the doctor or the nurse, the nurse feels your wrist and counts and writes a little number down in his or her chart. And what is that number? That's the pulse, the mean number of heartbeats in a fixed time interval, and this has some diagnostic value presumed because, otherwise, she wouldn't hold your hand.

But as you saw in the previous talk, the entropy might be always decreasing when we lower the temperature, but the fluctuations increase. Fluctuations, which is what this whole conference is about, have nothing to do with mean values. So, why not study the fluctuations in this signal?

Now, unfortunately, the nurse can't measure terribly accurately the time interval between beats, but it's awful easy to measure this with any number of almost trivial devices with millisecond accuracy, and, therefore, one can obtain the time interval between each successive beat of your heart for as long as you care it to be measured. A typical measurement can last 24 hours simply by putting a little ambulatory device on your hip and walking around.

Now, why would one want to do this? Seems like a crazy thing to do, but it's true. And by the way, if any of you want to do it, I'd be happy to measure you because we're collecting data for Mitsubishi Corporation which is making a device that will enable each of us to monitor our own health or, perhaps, the health of our grandmother; in my case, the health of my grandmother, because I very much want that any problems that develop can be caught right away.

I will organize this part of the talk around three questions. First, "What is the problem?" Second, "Why care?" And, third, "What can we do?"

Under "what we'll do," I'll talk about how we enlarge those fluctuations, how we discover that they have a remarkable property, almost hard to believe at first, but it's been borne out by many other laboratories, and that property is called "long-range correlations." Long-range correlations means not short-range, but long-range.

"Short-range" means short, some fixed finite value like, let's say, a minute. You could be perfectly plausible that if I get excited, as I am now, fearing that I'll be demolished by Yaneer, having used up my time, and coming up to the front looking very menacing, responding appropriately. First, it was the police, then it's Yaneer.

These low-range correlations do not have a finite persistence, but an infinite persistence. No number characterizes the persistence of these correlations. Furthermore, this correlation is actually of the opposite sign, meaning anti-correlation. That means if the heart goes faster now, as it probably is, it will compensate by going slower in the future.

We'll talk about how we can monitor the state of the heart using two modern statistical physics tricks that, probably, most of you have heard of and perhaps even used. One is called the "weight-lift method," and the other is called "multifractality." Both of these are tricks that are necessary due to the fact that heartbeat is a non-stationary signal. In other words, its statistical properties are changed as a function of time.

Let's start at the beginning then by asking yourself, what do we mean by non-stationarity? What is this problem with non-stationarity that makes this a non-trivial problem? Because, as you see here, if you're just measuring fluctuations, people have been measuring fluctuations for years, and unlike liquid water where the measurements have to be done in a metastable region with lots of tricks, the heartbeat is just right there, the signal is right there with millisecond accuracy.

Anyone in this room can download that signal from the website of the National Resource of Biomedical Signals which was just started jointly by our university, Harvard University, and so, basically, across the river, with MIT, and we store these signals.

And why this problem is non-trivial, why every child, literally, in high school can analyze this easily, as an exercise on how to do statistical analysis, is the fact that non-stationarity rears its ugly head. And I say "rears its ugly head" because the time sequence of these intervals which, on average, we all know is about one second. We have roughly 60 beats per minute, so roughly one second. But in fact, it's changing its statistical properties of function in time.

For a while, the time interval, like right now, might be very short because I'm excited or frightened, or both, and then a little later, when I go outside and look for a whiskey, it might calm down, and the time interval will be a little longer, and the standard deviation which has one value now might have a different value later, and then I might go back on the highway and get in the same traffic jam I had now, and then all kinds of hell will break loose, and so forth and so on.

The factorial properties measured by some Hurst exponent might correspondingly change as a function of time. And, therefore, to analyze this non-stationary signal turns out to be a challenge. There are methods to do this, but some of these methods had not been applied to heartbeat intervals.

Given the time, I'll tell you primarily about of one of them, which is the multi-fractal method. This paper reported the first multi-fractal analysis of heartbeat time series, and the empirical fact, not very well-understood, that the multi-fractal properties change for the diseased heart, offering, therefore, two things, namely the scientific puzzle of "why?" and the practical option of making a device that can measure these multi-fractal properties continuously as a function of time and warn us when we get in trouble.

What is this multi-fractal business? Let's try to describe this as carefully as we can. I think most people have heard of fractals, and most of you know that the fractal dimension, as being by Benoît Mandelbrot, and has as many of them means, one of them is the Hurst exponent, and that's the terminology we'll use here for various reasons, and the heart surgeons call it the Hurst exponent.

It's a number that just simply tells us that if we open up a little window box on the signal, one size, and count how many beats we have, and then we open up a window twice as large and count how many beats we have, those two numbers will be related in a fashion that's exactly the same, independent of scale.

Suppose we color-code from, say, low frequencies to the smallest value and high frequencies to the biggest value. Then we can take a recording of the sequence. This was schematic, but this recording's for a real subject, in fact, a healthy subject, and we can demonstrate. Therefore, we can measure the Hurst exponent continuously as we go along in time. I think there are about six hours of data here. So, six hours times 3600 seconds per hour is about 20,000 heartbeats.

So, there's 20,000 numbers here, and every hundred or so the Hurst exponent is measured, and a color is laid down depending on its value. So, way we have a kind of "Jacob's Coat of Many Colors" here.

And the fact that there are many colors is an indication that what I said about non-stationarity is actually borne out experimentally. You will see with your eyes, so-to-speak, what I was claiming, namely that the Hurst exponent in one part of the signal is not the same as the Hurst exponent a little later.

This picture is a color-coded indication of how the Hurst exponent is changing as a function in time. We could, of course, just put in the one axis, the Hurst exponent, and you'll see some zigzag curve ranging from a big value to a small value, but I prefer this color-coded one because it allows me to describe what is multifractality.

What is multifractality? Multifractality means that there's more than one Hurst exponent. We can do something with this. And this is a little bit abstract, but if you'll allow me to explain it, I think you'll have the essence of this concept of multifractality that has been applied to a wide range of systems including, as you may know from reading *Scientific American*, even to Wall Street fluctuations by Benoît Mandelbrot.

What do we do with this coat of many colors? First, we can just admire it and say, "yes," that means there is a wide range of fractal dimensions, h. But we can do more than that. We can actually quantify that range, approximate range. It's one thing to say something's fractal, but it's another thing to actually measure the fractal dimension for a statistically significant sample and come up with a number and say, "Yes. That object has a fractal dimension of x."

Well, similarly here, we can quantify this distribution of color in the following way: we can put a piece of colored film over this which only lets through, say, the yellow, and then we see only the yellow patterns, and that's what you see here, and then we take a different color of film it shows only the green. What you see here is the green. And we can take another one that lets show only the red, and what you see in the bottom is the red, etc.

We can break this thing down into sub-coats, if you will, and each of these coats has only one color, and the union of all the coats is the original Jacob's Coat of Many Colors. Is that clear? I hope I'm not either over-simplifying or missing the point.

UNIDENTIFIED: It was Joseph.

MR. STANLEY: Say again?

UNIDENTIFIED: Joseph.

MR. STANLEY: Joseph, not Jacob. Oh, thank you. I had a little trouble with that book. I read it a long time ago, at age 13, but it's been awhile, so it's "Joseph's Coat." I could straighten that out. I said "Jacob's" a few times, and nobody ever corrected me, so a lot of people haven't read the book since they were 13 because I know, normally, people are not shy about correcting me.

So, we've got Joseph's Coat broken into a bunch of sub-coats and many sub-coats as there are a distinct color showing. Now, what do you think we do? We look at these the same way you might look at the bar code in the supermarket. Anyone with a background in fractals, when they see a bar code as a colony of spots, gets very excited with one natural question, the only question on the edge of your tongue when you go through the supermarket line to the person checking you out is, "Did you ever measure the fractal dimension of that bar code?" I never asked it, but I always wanted to.

Now that there's nobody to be offended by that question, we'll just go out and measure it. You measure the fractal dimension of this color; you measure this pattern. It's a fractal dimension, obviously, less than one in, it's a matter what we call a "dust," with its little pieces. It's clearly not one. Normally, the whole thing filled in, so it's something less than one. And you measure the fractal dimension this way, and you measure the fractal dimension of each of the colors. And once you have all those fractal dimensions measured, what do you think you do? There's only one thing you can do, and, that is, make a plot of the fractal dimension as a function of the color.

Remember, the color was just a measure of the Hurst exponent. So, for red was in the infrared, this is the fractal dimension of the red ones, and this is the fractal dimension of the blue ones, and these are all the colors in between in Joseph's Coat.

So, you see the wide range of fractal dimensions. Miraculously, this is always sort of a parabolic looking curve, so we have quantified Joseph's Coat of Many Colors by coming up with a parabolic curve.

Now, you might say, "Why do you want to do that?" Well, because you get into the nature of something no one's ever done before. But it's also interesting because it tells us not only that this non-stationarity is present, because people knew it was present for a long time, but moreover, the non-stationarity can be characterized as an independent set of coats each with a different global fractal dimension, capital D. And further, that this curve which was done for one patient right here can be repeated for another patient, and you'll get almost the identical curve, of course, not exactly the same—each of us is a little different—but you get almost the same. I have a graph, if you wanted to see it, where there are superimposed 22 different subjects, and they look about the same.

Well, you might say, "What is this one doing here?" Well, this one is the curve for this bottom, and this bottom is a patient who, sadly, suffers from heart disease. And with your eye you can see that Joseph's Coat of Many Colors has become a cold blue coat of predominantly one or two or three colors. There's some blues and greens, if you can see it, despite all this light.

And correspondingly, when we look at the range of colors for the sick patients, that range is very small; where there's a huge range for the healthy, a small range for the sick. Is that clear? And that means that this measurement is a way of dramatically distinguishing sick from healthy, and we've tested that on enough patients to get by three of the four referees, and then there was a fourth—hopefully, none of you in this room—who insisted on doubling the number of patients, so we had to do twice as much work. But ultimately, we beat them down. We succeeded, and the thing came up.

The summary of this topic is that the heart is a very rich and interesting instrument. It's been known to be fractal, in the sense it has some properties that change with time for a long time, but we are just now beginning to learn to quantify these properties, and the utility of the quantification at the very minimum is the opportunity to build a device.

Every one of you has heard all the hype about the Human Genome Project. I've actually given some of that hype; I believe in it; I must believe in it; after all, we spent $3,000,000 of our taxpayer money to decipher this. And we all know the hype. The hype is that in a certain number of years—they never say how many—we'll each be carrying around a little chip that will be our entire DNA, and we go to the doctor, and the doctor will pump this chip into some kind of magic reading device and say, "Oh, yes. I know what you have because you were genetically disposed to having it, and now, clearly, you're showing the symptoms and, you know, we'll go ahead with some treatment."

This is way down in the road. I think you appreciate that. We are not way down the road from having little devices that can monitor the heart. In fact, we've had them for 20 years. They're getting a lot smaller, thanks to microelectronics. So, it's completely feasible that my grandmother and yours can have the state of their heart monitored.

The remarkable thing is that it's sufficient to monitor only this one quantity to already learn a lot. Of course, they can monitor other things, but just to monitor the time series of inter-beat intervals provides enough information that, in a number of pathologies, can be recognized, not just the extreme one where the heart stops. Obviously, we all know that immediately because the time intervals suddenly become infinite.

Even in the healthy grandmother walking around, taking her daily walk or whatever, that heart and how it responds can be monitored, and the effect of a common heart disease can be monitored continuously as a function of time, and steps could be taken to prevent this heart disease from getting worse.

The last topic is the economy. It's a very complex system. I think everyone knows that the Santa Fe Institute has pioneered some years ago the study of the economy. It's pioneered this study and through a workshop convened by Phil Anderson, Kenneth Arrow, and Brian Arthur. I think Stu Kauffman's probably played a role in much of that work, also, in various ways.

And I want to tell you a little about what we're doing here at Boston University and what some of us physicists are doing. I just got back a few hours ago from Santa Fe where the second of their Institutes took place with a very remarkable mix of people, mainly 50% economists and 50% physicists. And this was directed by Doyne Farmer, as you may well know, and he insisted on this appropriate balance, and everything was balanced. And it was always easy to see who was who was so because, first of all, they count their wallets in units of mega dollars, the economists do, and I'm not joking. There were people there who had $800,000,000. That's 800 of those units. I also had 800 of the unit. The only problem is that the unit was a 10-cent coin. I did have $80 in my pocket which is more than normal.

So, what about this econophysics? It's called "econophysics" because of the 50/50 blending of economists and physicists. The question is: what on earth would a physicist do in the economy? BU had a graduation yesterday, and I found myself talking to the Vice President of the Federal Reserve Board, and she said exactly that. She said, "What on earth is a physicist doing studying the economy?"

And I tried to tell her in preparation for this talk, perhaps, that in physics in the last 10, 20 years, some interesting principles have been uncovered. One of these is scaling, which was actually first applied to the economy in 1963 by Underwood. The second is universality, and the third is universalization group. And these three colors have their counterparts in the economy.

Now, why on earth should the economy have anything at all to do with critical phenomena, you may wonder and, in fact, so did the Chairman of the Federal Reserve. I had to tell her that near a critical point, if we pick a flashlight and look at a finite mixture, we see fluctuations of all scales, and we know that experimentally the same way Andrews did in 1869, when he discovered the critical point, namely, that if you have a light of all colors, that means there must be correlated regions of all different dimensions, dimensions comparable to the wavelengths of light; otherwise, light would not scatter, but not to one wavelength of light, but to many different wavelengths of light.

And so, also, I explained to her that if there was anything, any system in nature that we know where there's all different scales, it is the stock market because there's no intrinsic scale of value. The stock markets, they're all -- when you buy a stock, you are not getting those dollars in any real thing, you're buying some expectation of eventual profits, you're buying some expectation that it won't crash or go out of business, and so forth and so on but, instead, will make you at least stay even.

So, the concept of this economy in the stock market should have no intrinsic scale, just like this critical phenomenon of color. It's not totally non-intuitive; it's not non-intuitive. In other words, it's possible.

What we've done, not surprisingly, is apply the conceptual framework that others, including our group over the last 30 years, have applied to critical-point phenomenon. We've asked, can those concepts be carried over to the economy, and, specifically, we've taken a fairly empirical approach. Why? Because this critical phenomenon field was built by, first, doing experiments and then later by formulating theories.

So, specifically, what we've looked at is to study fluctuations in economic organizations, economic organizations like business firms or countries or research budgets of universities; they all fluctuate. And it turns out that those fluctuations occur all the time, and that's what we gamble on. If there were no fluctuations, there would be no stock market. Every stock would have a value. You just buy whatever you want to buy, be equivalent to stuffing it under your mattress, and having some interest rate. But the fluctuations are everything. And what we find is that those fluctuations share two of the properties of the critical phenomenon, namely, scaling exists, and universality exists.

"Universality" is a fancy word to mean that the quantitative numbers that characterize the scaling are the same for a huge range of different situations, and this is one of the more dramatic things we found, that a huge range of economic organizations obey the identical scaling mode, the same exponent.

The second result can be stated more rapidly, namely, that the price fluctuations in finance, say, stock market price fluctuations, also obey scaling laws, and those scaling laws also appear to be remarkably universal.

I think this probably is a good place to stop. I don't really have time to summarize. I think I've gotten to the point of all three examples of complex systems. All three have in common that they represent either experiments or an interplay between experiment and, if you will, a known theoretical foundation, and this is the style of our research group. We focus very strongly on empirical facts. Thank you for your attention.

UNIDENTIFIED: I have two questions. Could you say a few words of your work on the multi-fractality of DNA, especially with regards to the usable part and the junk DNA? And the second question I have is for the economy. Andrew Lowe of MIT has recently shown empirically that the whole stock market is a non-random process, and I wonder whether the long-range --

MR. STANLEY: Yes. All right. Let me answer the second question first.

All right. Very briefly, the word "random" can be used to mean uncorrelated random, and if that's what you mean, and then a person who finds correlations would be fair to say, "I've shown that the market is not random," namely, it's correlated, and that's the sense in which I alluded to use that term.

You would certainly agree we know that Andrew Lowe has not found irregularity; otherwise, Andrew Lowe would be the richest man on the face of the earth. In other words, we know that the stock market is a correlated random process. And further, we know those correlations are so difficult to figure out that no one's figured them out, yet, enough to consistently make a huge amount of money. You just bet a little bit over the average, like throwing a coin, say, 51%. So, it's enough to get rich, but not enough to take over.

Now, the second thing on DNA in one sentence—the question from this gentleman refers to the fact that, as I think most of you know, 97% of the DNA molecule is junk DNA. It is not good for proteins. It is not what you read about when you hear the Human Genome Project described where, one by one, each of the genes or the codes for a real gene is being deciphered. And there's 35,000 genes, and each of these, one by one, we're learning exactly the sequence.

But what about the 97%? Nobody knows what that 97% is there for. There are various conjectures, one of which is from our group, that that 97% may be functioning, not as a code but as a language, and the reason we have that conjecture is very simple. It's because when one analyzes in parallel the coding part of the DNA and the non-coding using methods of statistical linguistics, one seems to see, hence, that the statistical properties are different in the following direction, but the coding part is not a language, it's a code. After all, a code is not language. Whereas, the non-coding part is more like a language, but there are many, many, many caveats, so don't write this in your notebooks, because there are so many caveats that you don't want to know, they would take another talk, but this is an example of one of the conjectures about with a non-coding in either form.

I suspect that sometime between one and ten years, we'll have a better idea of what the non-coding is there for. One last thing I'd like to mention. I've brought a few copies of this little book on econophysics, which actually sold out by Cambridge University Press in five months. While a new print is being made, if anybody is thinking about it anyway, I'll be happy to give you one at half-price because that's what I paid. Thank you.

Simon Levin
The Ecology and Evolution of Communities

MR. BARANGER: The next speaker is Professor Simon Levin from Princeton University. He's a Professor of Ecology and Evolutionary Biology, and he just finished a book called, *Fragile Dominion*, and I think he's going to talk about some of the ideas in this book.

MR. LEVIN: About a year ago I was in Japan and saw two sand bars that you see here which, according to Per Bak's theory, one of which is already in a stage of self-organized criticality. All sand falls at about the same angle, one of which is in a perfectly good state and the other of which has undergone a collapse.

This is all according to the theory as expounded in Per's extremely stimulating book, *How Nature Works*, in which he talks about a notion that's familiar to most people here, namely, the notion of self-organized criticality, the idea that if one drips sand onto a table, sand will initially build up, eventually reaching a critical angle at which the system is very subject to instabilities, a few grains of sand begin to roll down, and the correlation lengths are such that they take with them many other grains of sand creating avalanches but after which the sand pile rebuilds maintaining itself in this critical stage without real intervention from outside, as a self-organized pattern.

But Per argues from this that self-organizing systems, in general, all tend towards a self-organized critical state, the same way that, perhaps, what happened if you organized a cocktail party at your house and, as more people come, the initial small groups that are talking together break apart into a spectrum of different small groups, perhaps, exhibiting a power log distribution.

Now, this is a very attractive notion and, indeed, I think it is one of the forces that's operating in the organization of ecological systems. But there are a couple of other complexities in ecological systems, as well as in economic systems, that are exhibited in the sand pile.

The two in particular that I want to focus on are that not all grains of sand in an ecosystem are identical. There's a lot of heterogeneity; there's diversity; that's what makes it a complex adaptive system, and it's that heterogeneity that evolution acts on. And as part of that process modularity arises. And the modularity also has an important function in constraining the spread of disturbances and, indeed, selection can operate to create tightly interacting cooperative networks that are sealed off, to some extent, from the rest of the system, which doesn't happen here, and that these are important features of the system. So, those are the things I want to focus on today.

Indeed, it raises a number of questions which are, first, what is the structure of ecological systems? What is the structure of the biosphere? How does that structure affect the macroscopic properties, in particular, the capacity of the system, the capacity of the biosphere, to maintain homeostatic patterns and to maintain its structure in functioning in the face of external disturbances?

This has led to a number of controlled laboratory studies, such as David Tilman's field work in Minnesota designed to look at different ecological structures and ask: what are the resilient properties of these structures?

But coupled to this has to be a subject which has not received anywhere near as much attention but is, for us, the most interesting, which is how the systems which are self-organizing both over ecological and evolutionary time, what sorts of patterns does it develop and how these patterns affect the resiliency of the system?

In the book that was mentioned to you in the introduction, I addressed six questions and will only focus on the last few today, but those questions that we naturally have to look at when we look at any ecological system is one of the patterns that we see, and why are they there, to what extent are they the result of self-organization, and to what extent are they determined by local climatic conditions, local soil conditions. That's the second question. Are these patterns uniquely determined by the local environment, or has history had a role?

Of course, both factors are important. It depends on the scale of which we're looking at the system, and one of the things that occupies most of the attention of ecologists is trying to pull apart the degree to which the patterns that we see are determined, that the degree of which they are going to result in accidents, frozen action, self-organization.

That leads us to the third question: how do ecosystems assemble themselves? And if you take a broader, spatial scale and a longer tempo scale, how do these self-organizing patterns become reinforced over evolutionary time in shaping the evolutionary features of the species that make up the systems? What is the relationship between the structure that emerges and the macroscopic properties and, in particular, the resiliency? And when we go to the evolutionary scale, what should we expect to happen? How could evolution shape the resiliency properties of systems? Do systems, indeed, move towards less and less stability, or is stability something that can emerge in a system?

The only way to understand that, I'm going to argue, is with a bottom-up approach, and I'll come back to that. But this immediately raises questions, such as those that people who have been interested in the evolution of the biosphere have thought a great deal about, mainly, what are the homeostatic properties of the biosphere, what are the homeostatic properties of ecosystems, how they maintain, and how does evolution operate on them?

We know that the fluxes of gases through the present atmosphere are very different than they would be if there were no biota. You can see that comparison here. What we would see on a dead planet is quite different than what we see and, in fact, that fact is used as evidence if one's looking for life on other planets of whether life, as we know it, might exist there. We look at the atmospheric composition, and that provides at least *prima facie* evidence of whether there's life on other planets.

This was in mind with the statement of James Hutton over two centuries ago who said, "I consider the earth to be a super-organism, and that its proper study should be biophysiology," and it's given rise to a culture called Gaia.

Every time I talk about Gaia, people come up to me afterwards and say, "Why do you talk about that? You certainly can't believe that."

Well, I'm laying it out here in order to point out where I see flaws in it, and simply to recognize that it is a point of view that's out there and that to try to tear apart which of the aspects of it with which evolutionary biologists are comfortable, which are the aspects in which they're not, and what we have to do to try to fill in the gaps.

The original notions, this from a book of Dorian Sagan, Lynn Margulis's mother, says that Gaia says simply that the temperature and composition of the earth's atmosphere are actively regulated by the cell of life on the planet. This is not a notion that would bother many ecologists. It is, indeed, why ecologists are concerned about loss of tropical rain forests and biodiversity in general.

This is reflected in the first slide in the series I showed you, namely, that the planet and the atmospheric composition is very different as a result of life, that changing life is going to cause the climate to change, and the atmospheric composition is going to change the conditions for life.

But the theory begins to develop, and the litany I'm going to take you through is due to Jim Kirchner, in a wonderful paper in the book that Stephen Schneider and Penelope Boston edited some years ago, *Scientists On Gaia*. It was Lovelock in 1991, together with Lynn Margulis, who was most responsible for the development of these ideas. They're very stimulating ideas.

Lovelock created Gaia theories about the evolution of a tightly coupled system whose components are the biota and the material environment which comprises the atmosphere, the oceans, and the surface rocks. This is basically correct, except for the grade at which the system is tightly coupled should be argued, as I will show.

The theory gets a little more worrisome in what Kirchner calls "optimizing Gaia." Lovelock and Margulis argue temperature creation. The presence of compounds of nutrient elements have been, for immense periods of time, just those that are optimal for surface life. The Gaia hypothesis is the idea that energy is expanded by the biota to actively maintain this optimum. The notion becomes, really, of the biosphere as a super-organism operating together to try to maintain these conditions.

The difficulties with this notion are the same difficulties that one would see in thinking about the economy, namely, that the biota are made up of individual agents—different species—the species made up of different genotypes.

The notion that any genotype is going to suppress its own selfish interest for the interests of the biosphere at-large is something that's long been rejected by evolutionary biologists.

We certainly know that cooperative behavior emerges, that mutualisms develop, but they involve tight feedbacks and localized interactions. So, we have to understand how these cooperative behaviors emerge—if there's homeostatic regulation at the level of the biosphere or even at the level of ecosystems—how it emerges from the collective selfish agendas of the genotypes that make it up.

The theory becomes even more worrisome in theological Gaia theory, where the earth's atmosphere is more than merely anomalous—it is to be a contrivance specifically constituted for a set purpose.

You begin to see, I hope, why evolutionary biologists studying a microscopic approach are concerned with the notion that the biosphere is operating as a single organism. It's the same problem one finds in economic theory and worrying about competitive versus cooperative markets. The biosphere is primarily a competitive market where individuals are selected for their own game.

The questions I want to focus are how ecosystems assemble themselves and how does evolution shape those ecological assemblages? I want to emphasize two points at the beginning to set the background for what I want to say.

One is that we need a bottom-up approach. We need to think about individual agents. Fish become organized into schools, into various kinds of collective patterns. This is a wonderful picture of a reindeer herd. As you can see here, the individual reindeer at the bottom are moving to the left, those at the top are moving to the left, these here are actually moving to the right, not knowing, probably not caring, that they're eventually going to move to the left. They're simply following local clues, local cues, and the pattern that you see here is self-organized as a result of that local information and those local decisions. There are many other patterns of this sort that one sees in bird flocks and in grazing animals such as this.

Sometimes these patterns really do have evolutionary input, as there's reasons for selections to have operated on the individual rules in order to maintain the individual collections. We know, because we don't see the same patterns over and over again, that the individual patterns at the macroscopic level are not what's important, what's been selected, perhaps, is for individuals to engage in behaviors that allows them to exist within groups and, perhaps, groups of a particular size.

So we turn to local rules and individual-based laws. There have been many of this sort. This happens to be from a thesis of a student of mine, Danny Grunbaum, whose interest was in trying to understand the macroscopic dynamics of schools of fish and schools and swarms of invertebrates. He did this by beginning from the rules of individual fish that could be measured behaviorally, to try to reproduce some of the sorts of patterns that could be observed. And this can be done. Dan writes down a

Newtonian law for each individual, that is, the acceleration of an individual of unit mass is equal to the sum of the forces that are acting on that. Some of those forces are inherited from the fluid dynamics, some result from individuals paying attention to what other individuals are doing.

This can result in collective dynamics that match fairly well with what's seen in nature and, indeed, one can develop macroscopic statistical descriptions that recapture those features quite well.

The first feature is that things operate from the bottom up, and the second, not unrelated to that, deals with the heterogeneity that I mentioned and the fact that evolution is operating. We all can see similar patterns existing in natural systems, and it's tempting to use the same sorts of models to describe how those patterns emerge. Sometimes that works.

It makes a big difference, as I'm going to emphasize in the next series of slides, whether the patterns that we see have any evolutionary significance, whether they have been selected, because they convey some advantages to the organisms, or whether they have simply emerged from local physical laws. The patterns may look similar, but they will differ in details.

Here you see the well-known Bénard cells that result when fluid of a particular characteristic is heated from below. These polygonal patterns emerge, and we see the same polygonal patterns arising from a quite different mechanism, if we look at the polygons in the tundra, it's a self-organized process—local rules govern the geomorphology, and the result are these patterns that we see in the tundra.

If we look at a honeycomb, we see the same patterns. If we look at the coat of a giraffe, there's a difference. Mainly, these patterns have been subject to selection. It may be unclear exactly what the advantages of the particular coat patterns are, and there are fascinating mechanistic descriptions often using diffusion-reaction equations that can account for these patterns—the movements of morphogens, etc.

The point is that those local mechanisms have been selected because they produce reliably the patterns that we see, the same way that Stuart showed in his early work. The local rules of development have been selected in order to reproduce in a reliable way the same patterns that are developing over and over again; whereas, the Bénard cells only occur under a very special set of environmental conditions. These sorts of patterns are buffered against environmental fluctuations.

The last series I'm going to show, which will really focus the points that I want to concentrate on, has to do with the branching networks that are so dear to the hearts of many people here—the fractal-like patterns. This is an erosion pattern. It could be a river basin. The recent book by my colleague, Ignacio Rodriguez-Iturbe, found river basins as fractal networks.

These can be well-described by trunks that grow with a certain probability of branching and, as those branches grow, having a certain distribution of angles. We can reproduce these networks, but they have no evolutionary significance.

The river basin, although it may be shown to optimize certain features, has not been selected among all river basins by reproducing in order to produce given features. These patterns have emerged from local rules.

In contrast, we see similar fractal-like patterns, which could be described in the same way that Jim Brown and Brian Enquist and Geoff West have been focusing on, for example, in the bronchial networks, in lungs, or in arterial networks or in systems that have been designed to optimize the collection or the redistribution of some sort of gas or fluid. By hypothesizing certain local rules, certain constraints on the system and certain optimization principles, Enquist, West, and Brown have been able to show that these sorts of fractal structures can emerge.

These look something like the river basin, except you can see they are much more contained. They're not as spread out; they're not as diffused because they're space-filling, because they're optimizing a certain function.

There's a big difference between simply describing a branching structure that has emerged from local rules and a branching structure that has been selected because, in comparison with other organisms, other genotypes that have slightly different branching structures, it performs a task better. In the competitive framework, this type wins out. This is an optimization approach.

Well, we have to go a step farther. We have to distinguish the optimization approach from a game theoretic approach. We see the same branching structures among trees, we see a trunk, branches, branches, branches, the same sorts of self-similar patterns.

What's the difference between this and the previous example? Your bronchial structure is contained within your body. You compete with other organisms, but your bronchial structure doesn't interdigitate with the bronchial structure of another organism.

When a tree develops a branching pattern, it not only optimizes its ability to capture a light or if it's the same branching structure in the roots, its ability to capture nutrients from the soil, but it also has the effect of shading other trees, and that's every bit as important in a competitive framework as simply doing the job better yourself. Anybody who's been a medical student knows that. The problem becomes immediately a game theoretic problem which is not only how can I optimize my ability to gather nutrients, or to catch a light in an environment which I am the only organism, but how do I operate in a competitive framework in which others are trying to do the same tasks. One way to do it, of course, is by suppressing their ability to compete with me, and we see this in the economic marketplace all the time.

This is what the problem is with the Gaia framework, we've got to think about this within a game theoretic framework, we've got to think about this with a view to understanding how one's success not only depends upon what other organisms are doing, but how it also affects other organisms.

There's a great legacy in evolutionary biology and in thinking about adaptive dynamics. A lot of it was given a boost when John Maynard Smith and others introduced the notion of evolutionarily stable strategy into evolutionary biology a few decades ago and that this is simply an extension of a Nash equilibrium.

The notion that if organisms are performing a certain task that conveys a certain fitness, it may or may not be the best thing they could do. It doesn't really make sense to think about optimization in the abstract but, rather, when asked, if I perform this strategy, am I able to be invaded, in a game theoretic sense? Can any mutant strategy come along and out-compete it?

One has a problem in thinking about that problem because one doesn't know at the beginning what the feasible space is of possible genotypes. It's effectively an interdimensional space in which new sorts of innovations can come along all the time. For the most part, the theory is constrained to think about this within a given feasible set of genotypes. One can ask the question, for example, in thinking about this personally, why do sea plants disperse their seeds? One reason they disperse their seeds is that if they didn't, other plants that did disperse the seeds would come and capture the sites that these plants hold, and they would have no way to gain new sites, so there would be a net loss of sites.

Dispersal and dormancy, as well, are frequency-dependent strategies which evolve within the context of other types, and there's a large literature that looks at length-history traits like dispersal and dormancy and counters it within this notion of an evolutionary stable strategy.

There are a number of problems that we encounter with simplistic game theoretic treatments. One is that the original theory has been laid out for thinking about games that are played once between organisms. There is a payoff associated with them. The early theory is static. It doesn't have an adaptive dynamic approach to it. It doesn't tell us how to deal with the fact that individuals interact with each other repeatedly, and you know the problems in going from the conclusions of a prisoner's dilemma to what happens if you have an iterated prisoner's dilemma, you immediately change the dynamic, you begin to develop cooperative behaviors that do not arise in a single play.

Secondly, and related to this, is the fact that play is typically spatially localized. This turns out to be very important from an evolutionary point of view because spatial organization tightens feedback roots and increases the possibility of altruistic cooperative behavior emerging. It turns out it also increases the possibility of spiteful behavior because, for example, when you shade or produce toxins, the individuals that you affect, the individuals in your neighborhood, are the ones that are most likely to affect you.

It's not a good thing or a bad thing, but the spatial localization of play clearly changes the dynamic and forces one to think about things in a spatial context. Indeed, not only do the spatial dynamics affect the outcomes, but the spatial localization, itself, is something that emerges.

As I said at the beginning, it can relate to modularizations, and it can change the resilience and properties of a system. We know that's true in terms of coalitions developing and cooperative behaviors and social norms and all those sorts of things which govern the dynamics of complex systems on multiple time scales and multiple spatial scales.

The third point is that the notion of an evolutionarily stable strategy is a static one, and one can certainly have multiple evolutionarily stable strategies in a system. How do we understand why this particular strategy has emerged, what's the role of chance been, how stable is it against mutation, and what happens if there are multiples of these co-existing in the system and they come into interaction?

I've already mentioned the dynamic aspects and the point about trade-off space is we can't simply look at the features of organisms and allow them to change independently. If I want to become a superior competitor, I may do so by producing a larger seed. That affects my ability to disperse that seed, it affects the number of seeds I can produce, and so there are a variety of trade-offs between traits.

Not only don't we have a good idea of what the dimensionality of that space is in the set of all possibilities, we also don't know what the constraints are unless we really have a good understanding of the physiological and metabolic costs of particular strategies.

To give you an example, I've been doing a lot of work on the dynamics of forests, together with a colleague at Princeton, Steve Pacala. These rely on individual-based stochastic processes that grow forests. One builds these models by using measured data on how plants grow in the shade, in the light, how they affect each other, and to feed that into the model.

It turns out that there are basically four properties that are most important in understanding the dynamics of the system. They relate to how well plants do under high-light environments, how well they do under low-light environments, and that's basically it, because the dynamics of the forests are that disturbances are occurring all the time.

When a disturbance occurs, light becomes available. When light becomes available, those species that are adapted for taking advantage of that do well, but they typically are the ones with small seeds, high growth potential that eventually get out-competed by the competitive dominant. The system becomes a spatio-temporal mosaic.

The key properties are the ability of a species to grow under high-light environments, the ability of it to survive under low-light environments, the shade that it casts because that affects the way it interacts with other species, and the dispersible characteristics which are reflected in the thickness of these cylinders. Those four properties allow us to distinguish among species, but there are two things I'd like you to notice here. One is that the points are not distributed randomly throughout this aspect space, they seem to be restricted to a fairly low-dimensional—maybe one-dimensional—manifold right along here. That has to do with the constraints I was talking about.

The first question is, what accounts for this constraint surface? Why don't we find species over here? Species do tend to space themselves out to maintain diversity. Perhaps that's not too surprising. Once this niche is taken, it makes competition much less here, and it's better to try to be a high-dispersing, high-light species than to try to fight it out, and so we get spacing.

The natural question that emerges is even if we know that constraints surface, how do species under evolutionary time space themselves out? Suppose that I allow species to go into competition and put an evolutionary dynamic on the system by allowing mutations and allowing new types to come in, how do species spread themselves out?

We've begun by looking at some very simple models of this. The sorts of challenges that we need to take into account, I reiterate, have to do with spatial heterogeneity, not just spacing out, but the fact that this evolution is not going on in a spatially uniform environment. There are, for example, variations in soil characteristics.

The constraints base I mentioned—how to deal with ensembles versus individuals and how we deal with co-evolution of multiple types. The classical theory is rather simplistic on this point. It just deals with evolutionarily stable strategies, but there's no attention to what's called "adaptive dynamics."

The simplest aspect of adaptive dynamics is that it involves the same ingredients that go into thinking about evolutionarily stable strategies. Let $R(V,U)$ be the fitness of an invader which would be measured as the growth rate of an invader.

When you introduce a new type, you may be introducing it with an age structure that's favorable to it. It may grow initially, but it will quickly settle down to a stable age distribution. We call it the fitness of an invader in an environment dominated by U that it's obvious that in continuous time, the fitness $R(U,U)$ is equal to zero. The notion is that the system is in some sort of a quasi-equilibrium, such that when it reaches that equilibrium, the type can neither grow nor decline in its own environment. This simply defines equilibrium. Therefore, this is true for any type, whether it's the optimal type or not.

The notion of an evolutionarily stable strategy is one that says that $R(V,U)$ is maximized as a function of V, namely, the invader type, when V is exactly not only equal to U, but equal to that evolutionarily stable strategy. In other words, a derivative of R with respect to V is equal to zero and, $R(V,U)$ is maximized at that point.

What wasn't realized early on, which should have been, is that it may well be that one can be an evolutionarily stable strategy, and once you're established you can't be invaded. But you never get to that evolutionarily stable strategy because the putative evolutionarily stable strategy can't invade anybody else, and there are some very robust examples of that.

There's a complementary notion which is called a "neighborhood invader strategy," which is a type that can invade its neighbors, and, therefore the system can eventually move towards it. There are some intermediate notions, by the way, that I won't get into here. The neighborhood invader strategy is the complement of this, and it says that R(V,U) is minimized as a function of U at U = V = W.

These two notions together are complementary and related to each other because, if I take a second derivative along that line, I get a relationship between the two second derivatives and the mixed partial, and ultimately, this gives us a diagram which, as I put up here, is simply to show you the complexity of the problem.

Depending on what the R^2 of the U is and the $R^2(V,P)$, we get a variety of different combinations that can emerge. If, indeed, we're down here where something is both an ESS and an NIS, then we can be pretty sure that the system, at least locally, will tend towards that strategy, and once it's established, it can't be invaded.

There are many other cases in which something is an ESS, but not an NIS. Even so, it turns out that even if it's not a neighborhood invader strategy, what's really important is the trade-off between being an ESS and an NIS, and you may become established in that case, or you may not become established. ESS may never show up, or you can be an NIS and not an ESS which means that the system will converge towards you, but as you become established, the system branches away from you. You can't become established there. What happens in that case is we begin to see co-existence among types. I put this up here simply to show you the complexity of that problem.

We've looked at some simple problems involving seed size to see what the outcomes are. But it raises, not a concern, but the interesting possibility that we can go beyond evolutionarily stable strategies to situations where mobile types co-exist, and what we would really need is a collective dynamic of a whole system, because, after all, all of you have been to a forest. You know there isn't a single species there. How is it that we beget multiple evolutionarily stable types?

Together with a number of collaborators—Rick Durrett at Cornell, Ann Kinzig (who is my postdoc), Steve Pacala, and Jonathan Dushoff—I've been looking at the assembly of communities starting from things that go on at a single trophic level.

The model that's here I will simply describe to you in words. It is a model that was originally introduced in simpler form by Richard Levins and explored by my student, Alan Hastings. Most recently, it's been developed by David Tilman, the ecologist at Minnesota. Basically what it says is that species are distributed along a one-dimensional continuum. They're all competing for the same resource, namely space, and they differ in two ways: one in terms of their fecundity, namely the number of seeds its produces and, therefore, their ability to capture to new space; and, secondly, in terms of their competitive ability. And then there's a trade-off between these.

In Tilman's model, the competitive situation is dealt with by a strict hierarchy, although it's possible and important to weaken that assumption. In the hierarchy, it says that the competitively superior species never sees the other species. It gets to a site. It never sees the species that are competitively inferior to it. If it gets to a site, it takes it over no matter what's there, and once it's there, it can only be invaded by something that's competitively superior to that.

We have a continuum which is a fecundity access, and the things at the right of the fecundity access are at least competitive, and we can study this system from the left, and there are a number of sites in the system. The system's well-mixed. P supplies the proportion of sites occupied by species I. Species J likes Species J. And the dynamics here simply say that there's a natural mortality in the system and survival, initially, I'll take to be the same for all species.

There's a fecundity access X of I, and that fecundity axis is inversely related to the competitive axis. And so we have sites that become available as the competitive dominants die, and those sites are immediately taken over by the more fecund species that give way eventually to competitively superior species.

This is largely the way natural systems worked. It's closer to the way grasslands work, which is where Tilman works, than it is for forests. Primarily this is because, although there's the same competitive hierarchy there, this is what's called a "successional dynamic"—it's not an instantaneous takeover. A seed landing on a site occupied by a tree, obviously, doesn't take it over right away. It may grow up and take it over, but it takes awhile. That's what this dynamic shows.

We looked at this, and the first part of it is pretty simple. That piece of 1B, the proportion of sites occupied by Species I, this is simply a contact process, but a contact process in a well-mixed system so we don't get any local effects. F1 is the fecundity of Species I, the competitive dominant species. It doesn't see any other species. It has a mortality rate M. We can solve for the equilibrium, P1 half equals 1 minus M over F1 which, obviously, is only positive, provided the fecundity is larger than the mortality. This is not a surprising result.

When we put this on a grid, however, the fecundity has to not only be larger than the mortality, but larger by an amount that accounts for what one observes in a contact process. Namely, that a lot of local colonization attempts are on sites that you already occupy. That means the fecundity has to be substantially larger than mortality, and that's an important distinction. If you look at two species, this is the dynamic of the second species, the first we've already solved, and the equilibrium here is P2 equals 1 minus M over F2 minus something. In other words, it's less than 1M over F2 by the amount of the sites that are occupied by the dominant species.

Pictorially, this says, first of all, as Hastings and Duran showed independently, that F2 has to be not only bigger than F1, but it has to be bigger than F1 times F1 over M. Remember F1 is larger than M, so this is even bigger, and we can see this graphically. F1 has to be bigger than M, and it casts a shadow, and it's only when F2 gets out there far enough that it can come in, the fecundity of the second species, and it, in turn, will cast a shadow, etc., etc.

Here we begin to get some limitation of the number of species in the system. These are simulations we did in which we initially just threw species down on the grid. The fecundity is somewhere between .1 and .2 and only at the points that were .101, etc. What we see is that the system begins to develop a pattern and that that pattern, over a period of time, has a regular structure to it.

If you put things on a lower block scale, what we see is a straight line parallel that emerges. The longer abundance is related in a rather regular way to the fecundity. Indeed, we can explain this by the passing through a continuous limit. I won't do that here. I don't want to put any special importance on the minus 3 halves law. What I really want to do is to pull things together.

We've also looked at more complicated rules for a fecundity and mortality and how they're relating to each other, and we still get regularities that emerge. The interesting feature is not the details of the regularity but the fact that from local microscopic dynamics, macroscopic patterns begin to emerge.

This is only the beginning. It looks at the system at one trophic level. It doesn't yet address the question that I laid out at the beginning, which is, "What does this mean about the resiliency of the system?" In order to look at that, we have to subject the system to perturbations during its development. And then, once the system has reached its quasi-equilibrium situation, we have to ask the question: "If I now can challenge the system with new sorts of disturbances, how resilient is it?"

This is somewhat interesting. Indeed, we've extended the model to make it look more realistic by including explicit spatial interactions, localized dispersal, and the casting effects—indeed, by treating individuals as individuals. But the system is rather simplistic. It's a competitive system for space at a single trophic level.

My student, Lee Worden, together with some others in my group, have begun to look at the question of what happens if we introduce multiple trophic levels. What happens if you allow for the possibility of cooperative behaviors, and ultimately, the question we'd like to address with these sorts of approaches is, how do these systems self-organize over time?

If we take the system with specified kinds of interactions, allow mutations that vary particular features, what sorts of attracting states does the system tend to, and how do those attracting states affect the capability of a system to respond to perturbations?

Let me conclude by saying that the theories in the past in ecology were relatively simple. They dealt with systems of ordinary differential equations, systems that are well-mixed, and evolutionary efforts typically dealt with constant selection regime, or if they dealt with the frequency dependents, they did so in a very limited and constrained way. But there are challenges of the present as we try to ask the questions of what accounts for the regularities that we see. For example, one problem we're looking at in oceanic systems is what accounts for what are called the "red field ratios," the ratios of elements that are fairly constant in those systems. Are these properties that are adaptive in some sense, or have they simply emerged from the individual evolutionary dynamics of the components? What do they say for the resilience of these systems?

We need to be able to deal with large ensembles in which we can deal explicitly with individual and group conflicts. We need to deal with the effects of the localization of interactions and to understanding how clustering emerges, not only clustering in space, but clustering in terms of which groups of species interact with each other.

One system we've been looking at is the evolution of the influenza A virus and the cross-immunity patterns that emerge, and how quasi-species evolve in which species are exceptionally interchangeable, mutating from one to the other and operating more or less as collectives, and ultimately the co-evolution among multiple interacting types.

This, however, is what makes this system interesting, and I think that the challenges that we face are not only going to inform ecology and evolution, but are actually going to expand the mathematical theory a great deal. Thank you very much.

UNIDENTIFIED: I'm struck by the parallels in the way you stylized your investigation. I can see direct parallels of species evolving there.

Assuming that the same kind of dynamics can be attributed to the map, we have the following component. If you are now a consultant to a species, to the individuals, can you say anything sensible, as a consultant, to the individual promise in your ecological system as to how they should individually address this strategic problem, both for their individual self, but also for the betterment of the overall ecological system?

MR. LEVIN: Well, you raised a number of interesting points. If I could answer your question, I probably wouldn't be here, I would be working for BIOS which is doing exactly that sort of thing.

Obviously, I agree with your initial premise. I think, indeed, it's not an accident, ecological systems are economic systems because everybody's competing for resources, and economic systems are ecological systems with predators and parasites and all those sorts of things.

If I were advising my client, I probably would advise my client not to worry so much about the global good, unless if I were really taking my client's special interest to heart, except that I would be discounting the future.

I think one of the challenges that we're facing with the ecological systems is that we need to care about the consequences of our actions at broader spatial along the temporal scales, and indeed that part of the problem is that the individual feedback groups are not tight enough that individuals can see the effects of their action. For example, we all engage to some extent in recycling, but the degree to which we believe that this is making a difference if one bottle that we have is not going into the recycling bin, it's probably not very convincing.

Society has evolved social norms to reinforce these sorts of behaviors, and there's no guarantee that those social norms will come about, and I think we're dealing with a system that's changing on such a rapid time scale that those social norms are not developing fast enough.

There are, obviously, people who care about the environment and care about future generations, but the economists' understanding of how we discount the future means that individuals do discount the future, even in terms of their own offspring two or three generations down the line. I think that this trend is becoming worse because of globalization. I'm not taking a position here on globalization, but I think the notion of globalization means we have more rapid information flow, and we're losing local feedback groups. Whereas you can get individuals to care about pollution in their neighborhood or maybe even in their town, I think it's much harder to get them to care about global warming which is, after all, going to take 50 or 100 or probably 200 years before it has any significant effect on them. In any case, even if they believe it, they think that their own actions are a small contribution.

If I were concerned about the global situation, as I develop in the book, I would encourage governments and other organizations to find mechanisms to increase the incentives locally for tightening feedback groups by giving individuals some motivation for local preservation.

It's not so different, after all, than what Alan Greenspan does by tinkering with the discount rate in order to influence individual behaviors in ways that will achieve some global good. I would give an individual, if I wanted them to profit from the market, unfortunately the advice that the fact of the matter is it's a situation where the more exploitative you are, the better off you're likely to do.

I'd be more interested in the question of how we change incentives so at least to restrain the rate of change in the system so that social norms can deal with it.

By the way, I should say that I don't believe that all social norms are necessarily beneficial. Social norms represent the collective dynamics of large groups of people. They operate over much longer time scales, they may well resolve the maladaptive behavior because the time it feedbacks to the individual, it's no longer beneficial.

We see things, for example, such as foot-binding of women in China— something that evolved through many religious customs simply to lock individuals in place and make them agree to suppress their own initial advantages to the collective advantages of the group.

UNIDENTIFIED: I would argue strongly with your picture that the economy and the ecology evolve the same way and, in particular, you gave us an example before that you have trees shading the other trees.

MR. LEVIN: Yes.

UNIDENTIFIED: If the reason they differ is precisely the property rights are imposed in economies, and this prevents this kind of shading thing so, for example, in the sun case, the sun is shining down, and you can shade it from other people, but in the economy the analog is the consumers who are rational-behaving creatures and can favor whomever they want and will take essentially what was bid.

I'm wondering what example you have. You said, "This happens all the time in the economy." What example do you have where this actually happens?

MR. LEVIN: First of all, I didn't say the economy and the ecology operate in exactly the same way. But the point of my lecture where I made that point—the example I was thinking of is what large airlines do, for example, to deal with an upstart company in a bigger neighborhood. What they do is they lower the price for a short period of time to drive out the competitors and then raise the price back up again. The only point I was making there is that sometimes the most effective strategy is simply to drive out your competitors even if you have some short-term cost too.

Obviously, there are differences between economic systems and ecological systems. For one thing, to some extent, we have governments that are exerting top-down pressures on the system, and those are localized in decision-makers, but there are very strong parallels. In terms of the self-organization of the system, even with all of the wishes of individual governments and planners, the economies are self-organizing.

As Stuart pointed out, many of the patterns that we see could not possibly have been predicted because they involve innovations in some new direction. Although there are obvious differences between economic and ecological systems, there are very strong parallels in that both systems are, to a large extent, self-organized from the bottom up. Thank you.

David Clark
Emergent Dynamics of the Internet

MR. BARANGER: And the next speaker is David Clark. David Clark is at the Laboratory for Computer Science at MIT. He's going to talk about complexity in the Internet. He's been involved in the Internet since 1976, and I think he's one of the early developers of the Internet.

MR. CLARK: I need to explain, since this is obviously a wonderfully interdisciplinary group here, that I do not, in my own work, study the complexity of systems. What I did was build a system which had a lot of complexity in it that we had not anticipated. My physicist friends tells me that the only reason we didn't anticipate these consequences is that we were ignorant. But back in 1974, we didn't appreciate the scale of what we were going to do anyway.

I'm going to talk about the Internet in some very practical terms. I'm going to show you some pictures of interesting behavior. I won't go so far as to call it "emergent," but just interesting.

I'll show you mechanistically what's going on, and then, perhaps, we can have some conversations about analogies and different images and ways you might think about this, but as I said, I'm an engineer at heart who is interested in how this system is built.

As you're trying to understand the algorithms and the dynamics we have inside the Internet, the first thing you should understand is that the problem we're trying to solve is sharing. You can have a communication link. That's not sharing. When you have a network and multiple people offering traffic into a common resource, that is when we build something we call a "network."

Now, you're trying to solve a variety of problems here. You have some physical resource like an expensive link or a satellite or something like that, and you're trying to utilize this resource in some efficient way.

The other issue in the sharing here is that you're trying to connect everybody to everybody because who knows who you want to talk to tomorrow, or who knows who that little autonomous agent in your computer, which we call a "virus," would like to send all your e-mail to tomorrow.

The sharing strategy in the Internet is very simple. It was not obvious in the beginning how general it was, but it's turned out to be remarkably effective, which is, you take the data and you break it up into these various small sized chunks. We call them "packets." These are little bits of data, say, a thousand bytes. Think small. You put addresses on the front, you dump them in the network, and they're sent across the network in a series of hops. A "hop" is a little box called a "router" which is just processor, and the package shows up, it looks at the address on the front and says, "Oh, for this address, I'll send it that way," and these little packets go poink, poink, poink, poink, poink across the Net, right? That's it. That's the whole story. Now, you can go home, or in this particular case, eat lunch.

It helps to have some sort of visualization here, very simplistic, yet not too overly simplistic. I was thinking about the way the Internet works. Imagine that I am a sender, and I am sending data to you, and these little boxes here represent packets. What we're doing is we're sending packets across the Net, and you're putting them out sort of packed on the wire, and when they get to the receiver, these data packets trigger the return of what's called an "acknowledgement packet," a little tiny packet which basically says, "Yeah. I got it," and we number these things.

You can think of the data packets coming across and being turned into acknowledgements, and the acknowledgements flow back. And when you get an acknowledgement, you can send another data. We limit the number of packets you can send at any given time.

In this particular picture, there are 12 packets for which the sender has sent the data but has not yet received the acknowledgement. The word we use to describe this is the "window size." Don't ask why. Computer scientists explain things by taking common English words and giving them new meanings, as opposed to physicians who generate entirely new words. You know, a doctor tells you that you have ankylosing spondylitis, you don't know what it is, but you know it's really bad. But we take words like "window" and "process," and we just completely revise them to a new meaning. I apologize.

The window size of this picture is 12, and whenever you get a packet, you can send the data. If you can get a look at this picture, you can see there's something circulating here, and it's sort of a self-timed picture. You should say, "Oh, My God. It's not going to look like that. That artist's conception is beautifully stable. No, no, no. Nothing's beautiful, nothing's stable." But let's make the picture a little more complicated. Imagine two people are trying to send at the same time.

Here's one of my little boxes called a router. Again, this is an artist's conception, if it ever looked like this, we'd go home happy. But I'm going to leave the packets from Source 1 and Source 2 over here, and I've given different colors. So, each one of these now has a window size of 6, and they still fit in the same size pipe. 1, 2, 3, 4, 5, 6. These things are circulating.

Why did I say this thing gets complicated? Or to be so bold, why did I see complexity in this system? We have thousands of individual packet flows. I'm retrieving my web page, you're retrieving your web page, we're all doing this kind of cool stuff on the network, and all of our sharing is statistical.

You don't get permission before you send. You don't open a connection. There's no such thing in the Internet as a call; there's no such thing as a busy signal. You want to send? You send. You send a little bit of data, and you see whether you get away with it. If that works you send a little bit more, and you see whether you get away with it. If that works, you send a little bit more, and you gradually increase the size you're sending until, eventually, something bad happens.

You can imagine multiple inputs coming into a router, and when too many packets show up at once, which happens all the time, we queue them. Nothing surprising about this, except that a queue is a shared experience, right?

If my flow and your flow encounter the same router, then whatever experience you have in the queue is the same one I have. We're taking all of these potentially individual flows, and we're cross-coupling them by letting them have the shared congestion experience in the cues.

You have to think about the Internet with hundreds of thousands of simultaneous flows, hundreds of thousands of queues in this gigantic mush of interconnectedness, and then you begin to say, "The resulting patterns you see are remarkably complicated."

I illustrate some packets queued up in that path. If you were to count them, you'd now see that each flow had a window size of 8 instead of 6, and there's something sort of suggestive about this picture, which is, I've showed this router as being full of packets. If you send too many packets, what happens is there's no space to queue on, which means that it gets thrown away.

Now people say, "The Internet drops packets. Why does the Internet drop packets?" It's not because of cosmic rays, it's not because of shock noise, it's because the algorithms that determine how much data to send hunt for the right rate, and if you haven't dropped a packet yet, they go faster.

A typical workstation you have on your desk, properly tuned up, can go a gigabit—10^9 bits a second. It can go really honking fast.

So you say, "Why don't you keep going faster until something bad happens?" O.K. What's the badness signal in the Internet that tells you you're going too fast? Answer: we drop a packet.

People say, "That's crude. Why don't you send the guy a message? Why don't you send him back a little packet?" The guy who designed this algorithm originally was a physicist. He said that he picked packet drop as the default signal of congestion because it's the one thing you can't misimplement.

If you control a message and you put it in the network, somebody can screw up, but there's no way that you cannot have storage and put it there anyway. You see, you just can't screw up this thing. So what we do in the Internet is when the buffer fills up, we drop a packet.

We have this adaptive algorithm in the software in your computer, and everyday it tries to see how fast it can go. Once each round trip, it increases the window size by 1, so it sends 10 packets. If that works, it sends 11, and if that works, it sends 12, and if that works, it sends 13, and "whop," you drop a packet, and it cuts the window size by a factor of 2, and it hunts. It's a classic control theory algorithm. It hunts up, and then it cuts by a factor of 2. That was a great idea the day they had had it.

Then the next day, we noticed that everybody was hunting in synchrony. They were all increasing their window size, "boomp, boomp, boomp, boomp, boomp, boomp." We didn't want that because when that starts happening, all of a sudden, the entire queue drops by a factor of 2 which means it goes away, and the link is idle. Why did they synchronize? It's obvious why they synchronize because we're having a common queue experience. That is to say, if the queue is full for me, then queue is full for you. If the queue has got extra space for me, then the queue's got extra space for you.

It may be that you can think about this the way some people think about pulse oscillators which is, periodically, we have a synchronizing event which is when the queue overflows, and if it overflows for me, it overflows for you. It takes about a round trip for the consequence of a packet loss to manifest, so that one round trip later, I notice there's something missing, and I do this slowdown. But, of course, in one round trip everybody increases his window size by 1. What happens is they're all synchronized. "Oh, no, no, no, no. That wasn't what we wanted."

At that point a small riot broke out here at designland because there were two competing views as to how or how not to deal with this phenomenon. What was to isolate all the flows?

I'm going to take that one queue and break it into thousands of little queuelets, and I'm going to put my packets in my queue and I'm going to put your packets in your queue, and then I'm going to schedule them according to some rigid round-robin discipline, some sort of weighted round-robin, a generalized processor sharing or something like that.

I can tell you that if you do this—we've looked at it in simulations—you get incredibly robust isolation. You peak and just go crazy, and it doesn't really affect you, but there's tremendous implementation cost. Think order of magnitude.

Routers in the Internet, they're forwarding 10,000,000 to 100,000,000 packets a second. When you make this operation substantially more complicated, the guys whose job it is to produce low-cost boxes out of silicon, basically, are throwing it out the window. They say, "I just don't want to implement this. Go away. Unthink that thought."

The other strategy for solving this was to go to the other extreme which was to do something that, instead of being highly rigid, is highly random. And the random of this strategy is incredibly elegant. The random of this strategy which is called "random early detection" says, probabilistically, even though there's plenty of space to put the packet down, shoot it anyway.

People think about that and say, "What? You had space to queue the thing, but you dropped it anyway. What are you doing?" This is called "random early detection." This is an algorithm which today we are urging people to turn on in the Internet. We are at the point where we have simulated the hell out of this, we are convinced it works, and we're going to the router-makers and the Internet service providers and saying, "You're going to do this now, right?" We wrote a document called *The RED Manifesto* which said, "Do it. The research community tells you to do it."

We can only simulate it in a network of 100 nodes or 1,000 nodes. We wait to see what happens when we run it in the Net with 1,000,000 nodes because what we've learned is that we get our own emergent behavior.

I talked to the guy who is responsible for shipping the next software. This guy is responsible for signing off on cutting the next CD. Whenever they do TCP, the network protocols, he says, "I don't sleep for a week." He said, "It scares the hell out of me because I am terrified that I am going to be the guy who ships the TCP that's got an instability in it that causes the Internet to die and congest and collapse. Once I've shipped that CD, there are a million copies out there. I can't get them back, I can't go into some consumer store: "Excuse me, but I need to force you to do an upgrade."

This is the RED algorithm. We measure the queue length in the router. If the queue length is less than a certain amount, don't do anything. As the queue length begins to build up, we have a linear increasing probability of dropping a packet. When you get to some panic point, you basically drop everything. He said, "No, no, no, no," to try to get everybody to slow down.

What you hope is that the network will stabilize. I can do the math for you, but not in the time between now and lunch, which points out that if I tell you this hunting algorithm, I just did, increase the window size once per round trip and then cut by a factor of 2, you can actually calculate what the loss rate has to be to stabilize the Net to a new operating point. You simply hunt back and forth on until you stabilize it at the point where you're slowing the sources down enough that the whole thing is stable.

With the RED algorithm you've come in here and taken what was a synchronized hunting phenomenon—everybody was synchronized by this queue—and you've broken it apart by the imposition of the random algorithm. You just randomly pick people and say, "You, you, you, slow down by factor 2," and by breaking this apart, all of a sudden, you have broken the synchronized behavior into uncluttered behavior.

That was great. At least it was great in the simulators. We need to learn how to simulate something that's got a million flows in it. We just don't know how to simulate things that are big enough. But what we've noticed, in general, is that when you look at the traffic in the Internet, what you notice is that it crosses a whole bunch of time scales. It's clumped up into blocks, and you've got a little smooth picture of the packets flowing around.

For those of you who came to this conference last time, Walter Willinger gave a talk in which he had analyzed this and concluded that if you look at the inter-packet spacing, there's something eerie about inter-heartbeat spacing. I was thinking about here packet spacing that is multifractal and that there's a tremendous number of phenomenon inside the Net that are producing this.

Walter chose to look at this cell block as a phenomenon and analyze it. I'm an engineer, and I think that what I might actually do is tell you some of the reasons why it happens. We are noticing that whenever you look at the network, the packets are in clumps. We have an engineering question, of course, which is: is this intrinsic? Should we design for it? Could we de-clump them? Does it matter? Do we care if the packets are in clumps?

But here's a paper which you will appreciate. This is plotting instantaneous packet rates. It's called, "Pictorial Proof of Self-Similarity." This is looking at traffic on the network at successive intervals of hundredths of a second, and you get this tremendous "blblblblblblbl" sort of stuff, and then he looks at it over longer periods. This is at successive tenths of a second intervals, and it's still got a lot of noise in it.

As you plot it units over one second, unit over 10 seconds, unit over 100 seconds, you can see the pictures are not entirely the same. You're getting a sort of DC level here, you're getting a sort of average value building up which is not present here, but this is still awfully bursting. As he said over a succession of papers, he's come to the conclusion that the right way to explain this traffic is that, in fact, it is multifractal.

At long time scales you say, "What's going on?" and the answer is it has to do with the sources of offered low. We've learned some things, for example, about Web sources. The distribution of the size of Web objects that are on Web services is heavy-tailed. You say, "Why is this?" "I don't know." That's the limit of my ability to explain why.

Why do humans make webpages that have this heavy-tailed characteristic? I think, basically, it is that most of the webpages are little pieces of text, except for the ones that are images, and most of the big things are images, except for the ones that are software downloads. 20 years ago somebody analyzed the distribution of file sizes in the units file system and discovered the distribution was heavy-tailed.

The problem is that the best description of the distribution of Web sizes is: the probability that Web size is bigger than x is approximately equal to 1. It's an inverse powers. 1 over x to an alpha where alpha is 1.1.

If alpha were less than 1, the mean of the distribution goes unstable. So, why only do we have infinite variance? We're getting close to having an infinite mean, and that's the intrinsic traffic we're trying to carry. That's before the Net got its hands on it.

If you plot a picture of traffic from a particular source—this is the density is how many packets per second, and this is measured over a period of an hour. That number is 3600. And this is from a source to all destinations, and what he's done here is break it apart from the same source but to different destinations, from source to Destination 1 to 2 to 3 to 4 to 5 and on.

A very good way of modeling this traffic is a series of on-and-off periods, and when the source is on, it's a reasonably constant density of packets. You can see, you get long periods of off, sometimes long periods of on, but what you discover is that the distributions of the ons and offs are heavy-tailed, and the overlay of this, the superposition, is this jumbled traffic that we're trying to handle.

However, when you get to the short time scales, when you get to the sub-round trip time scales, and there's too many disbursements in the traffic—this has nothing to do with the offered low by the users, it has to do with something else, and what it has to do with is a little weirdness is going on inside the Net that are clumping all the traffic into clumps.

I'm going to try to do something daring here. I hope it works. I'm going to try to show you some pictures of real packet distributions, and either this will be incomprehensible or it'll work, I don't know, we're going to try.

You can try to animate the flow of packets across the Internet. You could take a snapshot of the time at which the packet reaches every router. That's too complicated. We've done it, but it's really complicated.

Normally, you get an awful lot of information just by monitoring at the source. Remember my picture of the self-timed wheel of packets. For each packet, imagine they're sequence-numbered. That's too simple. But imagine they're just numbered 1, 2, 3, 4. If you read the clock when the packet leaves and then you read the clock when the acknowledgement arrives, obviously, what you're measuring is the round-trip experience of that particular day and packet and its return and the acknowledgement. You can plot that on a graph.

You plot time on a horizontal axis, and you plot sequence number on the vertical axis. We draw an arrow for each packet showing how many bytes are in it, so that's a packet being sent, and some time later we draw this lower boundary line here, when the acknowledgements return.

This is where a packet is sent, this is where the acknowledgement comes, this is the next packet being sent, this is when the acknowledgement comes, this is the next packet being sent, "tck, tck, tck, tck, tck." Oh, we sent it again. Why do you send the same packet by the same sequence number but sent twice? That's a duplicate. We send the second time because the first one got lost. Send the packet, get an acknowledgement.

In a simple situation you get really clean pictures. Here's a picture of a file transfer across the local area net. That's exactly what you think it would be. It slopes up, right? That's because the thing runs. And I can discuss this picture in more detail than you want, but you'll notice I talked about the window, right, this upper line here is the window that the receiver offers the sender. That's the way we throttle the system back.

And you'll notice what's happening here is a straight line. This is some poor little workstation basically just sending packets as fast as it can. Nothing is following it until we get here. All of a sudden, you'll notice it just hits the window. It's sent 1, 2, 3, 4 packets, it can't send anymore.

You can plot the slope of the curve. This slope tells you how fast the sender can go. The slope here, if you were to zoom back, tells you how fast the receiver can go for a period of time. The point of a little event the receiver had, it's loaded down right away, and it took off again.

This picture is so simple that you might say, "God. There's nothing going on there." Here's a picture of a real network. I'm not going to analyze it. I just want to point out there's a lot of structure in this picture, and one of the things you might notice is you don't see any nicely spaced packets, what you see is bursts of packets being sent into the network.

Now, you say, "Why are bursts of packets being sent into the network?" This is an interesting picture. This is a real cross traffic transfer across the real Internet, a megabyte of data, and what you might look at is you'll notice the line has long periods where it goes at the same slope, and you get little inflections.

Each one of those inflections represents a packet loss followed by a rate adjustment in the center. And what you've got here is a very nice rate adaptation algorithm which is constantly hunting for the right rate. It speeds up until it gets whacked, it speeds up, speeds up. You know, each one of these is an inflection but, apparently, you have something called a "bad experience" in which the TCP crawls in a hole for a second. It says, "Oh, I'll think about it because I'm in real pain."

Now, this is a tricky picture. I need to help you read it for a little while. This is one of the few times I had put the other half of the picture up where we took a simultaneous picture of the East Coast and the West Coast, so I have the picture of the packets at 2 points in their flight. This wonderful structure, I could spend a half hour looking at this picture.

But let's start someplace. Here is a burst of packets. We zoom back to the point where the packets are just little blobs of ink on the screen, right, but we zoom back. Here's a burst of packets being sent into the Net. Sometime later you get the round trips. You can actually measure them. Now, this is the round trip in the system, right? By the way, the Internet's within a factor of 2, the speed of light. The speed of light is incredibly slow, any physicist here who thought the speed of light was fast.

Here's a packet, here's its acknowledgement, here's a packet, here's its acknowledgement, and you'll notice these acknowledgements are all nicely spaced out. That's what you'd expect. Remember by a little tiny wheel the packets are sent, and they turn into the acknowledgements, and they come back. And then there's an exceedingly long pause, and then I get all their acks. What happens when they get all their acks at once? I send a burst of data. Well, that's what happened here. Here's a burst of acks.

By the way, if you can see, you will notice they're actually out of order. People keep saying, "One of the things that happens inside the Internet is that packets get out of order." You say, "Oh, that would never happen." Oh, sure. I can tell you how it has to do with perverse engineering, but….

So, again, a burst of packets here. I got a burst of acks which produced a burst of data packets. Now, what I'm going to do very quickly is overlay on this picture what the scene on the West Coast looked like, and I have to get this right. It's like that. I'm guessing. Our clocks were not quite synchronized. You can see they're off by, 20 milliseconds or so.

And so, here's a burst of packets, and you'll notice that they come out more or less as a burst of packets but at a different slope. Why do I get a different slope? They've been squished through a bottleneck. That is the service rate of the queue.

You can tell you're looking at the receiver because the packet is instantaneously followed by the acknowledgment. I'm not modeling; we're looking at the Net right there. And these acknowledgements come out of this slope, and they come back at this slope—that's great. You can see these acknowledgements being produced here, and then there's some huge delay, and then they all come back. You say, "What is going on?"

My graduate student first found this phenomenon, a woman named Lixia Zhang. She actually produced it in a simulator. This is a phenomenon that I can produce in a simulator with two data sources, one on each side sending to the other. She called it "ack-compression." Actually, she called it "ack-smashing," but we decided that was too informal.

What was happening in the Net—and I am going to explain it as a standing phenomenon—imagine that going from west to east there's a blob of data in the queue. Now, those are datas going from west to east. I'm also trying to send datas from east to west. There's a blob of datas going this way, and my acks pile behind it.

Now, the acks were nicely paced when they hit the queue, but they slide into the back of this blob the datas, so they just sit there, and the datas ooze out as fast as data packets can and, eventually, the blob of datas vanish, and at this point all the acks come out as a burst. They have to. They're just sitting in the queue. The router just sends them as fast it can, "whup, whup, whup, whup."

All of those acks now reach the source on this side, and the host says, "Oh, I got 20 acks, I'm going to send 20 datas," and this produces a blob of datas in the other direction. What you notice is you have a rotating phenomenon in which a blob of datas here, flow trying to go this way, and one rotation later produces a blob of datas here and traffic trying to go this way.

People who do simple simulations tend to put all the data sources on one side of the screen and all the data things over here. See what happens when datas override?

My Chinese student, she put all the data sources on the other side, and I tease her for being Chinese. I said, "You're doing it from right to left," and she said, "I'll fix you up. Put half of them on each side." Instantly, this happened.

Now, my friends and physicists, please, you'll never find this phenomenon in the Internet. This is a phenomenon out of a simulator. Two days later a guy came into our office and he said, "You know, I got this weird thing. I got this cross-country link, and when I do a file transfer in one direction, it goes the full speed of the link. I do a foul transfer in the other direction, it goes the full speed of the link, but when I try to do both of them at once, each one slows down to half-speed. Explain this thing to me." I said, "Oh, you had ack compression." He went and he controlled the routers. He modified the router and put the short packets first, so the acks could jump over the datas, and everything went twice as fast, and he came back, and he says, "Oh, oh Great Magician. What do I do now?" Yeah.

So, we have this phenomenon. It's just one example, called "ack compression," and it allows us to explain, I think, a lot of the clumpiness that you see in our Internet traffic when you look at a time scale of less than one round-trip.

So, he thought about that for awhile, and we said, "Well, what do we do about it?" The obvious answer is: you put big buffers in the network so that when these bursts show up they don't overflow the buffer, and everybody said, "Well, that makes a lot of sense." You put a big buffer so that all these little bursts don't matter anymore. They just land in the buffer, and we smooth them out. Then somebody woke up and said, "Well, wait a minute. We were using the buffer full as an indicator of when to trigger congestion, and so now you're saying that you've got a buffer, but it's actually solving two different problems at two different time scales."

What you've done is coupled two different time scales in the system. In fact, the only way that I can find to rationally think about this is when we try to divide what's happening in the Internet into fundamental time scales, and one time scale which makes a lot of sense is the burst fit in a buffer. One time scale is when a blob that goes into the Net is small enough to fit in a buffer versus big enough that the buffer overflows, and another time scale is the round-trip time in the Net because you can't have feedback at less than a round trip.

To make sure that the sub-round-trip burst delivered for the buffer, you have to make the buffer big enough that it can basically hold things that are bigger than round trips. But that may not be the right size to control this slow one increase per round trip putting algorithm.

If you think about what RED does in this circumstance, and what RED does is it decouples the two things that are happening in the buffer. RED says you can have a fairly big physical buffer. I can tell by the look, right now, I lost everybody. I can just tell. Everybody got real silent. They're staring down at their feet in pain.

You can have this really big buffer, and that's so that you can absorb all of these sub-round-trip bursts that we don't seem to be able to get rid of. Remember my algorithm. I'm going to stress a word from this previous slide that I had before, "average queue length." Rather than an instantaneous queue length, we average over a number of round trips like 1 or 2 or 3.

What this does is in determining what the drop rate should be, we ignore all of the sub-round-trip fluctuations in the Net. We try not to let lows trigger dropping activity but, in fact, by correlating the drop rate to an average queue length, with an average queue length that's been averaged over enough, we can't see those bursts anymore. What we hope we have done is decoupled the phenomenon at two time scales.

The designer of this actually understood all that, but they didn't know how to tune it to make it work. And people have now spent the last four or five years saying, "This thing has a whole bunch of tuning. Now, I can turn to this, I can turn that, I can turn that, and how do I set it so that it actually is efficient in terms of control and congestion, in the sense that it controls congestion as quickly as possible, but it doesn't actually get triggered by sort of unexpected correlations of these sub-round-trip bursts?" You're saying, "Oh, God. That makes my head hurt."

What we're noticing is that there seem to be phenomena that caused the traffic in the Internet to be bursting at different time scales, and since this is an engineered phenomenon—this is an artifact we actually built—our assumption is that we could actually explain why we saw what we were seeing, that we could actually explain the behavior at every time scale. What we discovered, of course, initially, is that we couldn't. We had to go in and pry the cover off and go in and analyze and, as a result, we get this very complex picture which people like Walter and his friends explain as multifractal.

From the engineering point of view, I say, "What I'd really like to do is remove this behavior, or maybe I shouldn't." And that's one of the questions. It's actually the question that I came to ask here.

There are two ways to think about the design of a large system like this, and one of them is if we engineer it very carefully, we can eliminate this kind of behavior, and the other one is, "No, you can't." A system that's this big.

Notice, I can produce ack compression in a simple simulation where I have one congested link and two sources, one at each end. You can't have a simpler net than that.

What I've got on the Internet today is millions of end points, hundreds of thousands of routers, where we're having this sort of imposed sharing experience thousands of timed simultaneous flows, and the question is sort of intrinsically, "What level of control can we have over a system like this?"

I should tell you, my physicist friends have denied it in the sense that some of them say, "This kind of clumpedness at all time scales is just inevitable." You have to live with it. You have to put mechanisms into the system like RED that are trying very hard to separate out what happens at different time scales and control each one separately.

There's the other religion which says, "You idiots. Just do something like waiting for a queue where everybody goes in his own queue, and you can effectively separate people enough that you can control each one, and you don't have to think about any of this kind of clumping or other kind of emergent behavior because you just don't have to."

Physicist friends of mine consume quite a bit of beer, arguing about this because those of us who actually designed these mechanisms would really like to know the answer to the question, so that's why I came to ask. With that, I will stop.

UNIDENTIFIED MALE: In the routing algorithms, do you simply pass things along to the next router that's somewhere in the general direction of where you want to go, or do you try to optimize the routes?

Question two: Could you decongest if you had some kind of payment mechanism that says you have to pay more as the congestion builds up but, otherwise, you could try sharing it over to some other router?

MR. CLARK: As to the first question, we run a routing algorithm in the background which attempts to compute the best round from where you are to any destination, but that, of course, has its own time constant. And an attempt to run the routing out with too short a time constant, it leads to oscillation, and we have clearly seen what's been, people like to call, "route flapping," in which all the traffic goes over here for awhile which causes something bad to happen, so the algorithm says, "Oh, well. I'll send all the traffic over here."

This is a relatively stateless Net. We're sort of proud of the fact the Internet is stateless. What that means is you can kill a router, and it comes up again, and the flows are not disrupted. The flows continue to go. The Internet crashes all the time, but you never notice—in contrast to the telephone system, whereas, if a switch goes down, all the calls are cleared.

Because it's stateless, it's very hard to say something such as "30% of the traffic should go that way," and we're working on that. "Optimal" is a loose word. We optimize over the seconds to minutes and minutes to hours and hours to back over the time scale, but we can't optimize a sub-second.

I think the imposition of economics is absolutely the right thing to do. There's been a tremendous interest by the economists in coming in and understanding this world. Their first attempts to model this took too simplistic an approach in the Internet. They tried to associate congestion-based pricing with the length of queues. Unfortunately, since there's feedback in the systems, the queues run at a constant level, and so the algorithms didn't work.

As we've gotten more sophisticated—I've actually spent the last 10 years trying to learn how to talk to economists, so I realize that my next 10 years are going to have to be learning how to talk to lawyers, but that's not something for this meeting—that, in fact, I think economics models are very good. And there are some people who are trying to figure out -- one of these fellows talked about the aggression of species, and there's a kind of equivalent phenomenon we have which is the aggression of individual end nodes. What happens if Microsoft shifts a version of TCP whose hunting algorithms are more aggressive in that control theory space? We're terrified of having an arms race here.

There's a friend of mine by the name of Scott Shenker, who used to be at Xerox PARC. He's now at ICSI at Berkeley. He was a physicist turned computer scientist turned economist, who actually believes he has a game theory proven that you cannot use a sort of a self-interested model to keep you from becoming more aggressive, and therefore, we have to impose something external in economics like making you pay. But the right way to do that, of course, all of a sudden falls prey to what we call "marketing departments," and so we haven't got an optimal solution, I guess. Economics is the right way to regulate the congestion on the Net in the long-run.

UNIDENTIFIED MALE: It seems to me that the problem you have is an optimization problem because you have a finite number of nodes and you have a finite number of channels, and you're trying to distribute over the network a finite number of packets.

It seems to me that some years ago—this was likened to a traveling salesman—a problem where the main hypothesis was that it is incorrect to send packages directly from the sender to the receiver because it introduces lumpiness or clumpiness in the network.

There was, I believe a mathematician from AT&T, around three years ago, who formulated an algorithm so that it optimizes the problem, and the packages should not be sent directly from the sender to the receiver but distributed over the network in a very undirected way. I wonder if you know of the progress of the prescription?

MR. CLARK: It would take a long time to give you an answer that would be fine and adequate. I will be brief.

People have been trying to figure out the right way to explain optimality here and the right way to impose it almost from the beginning of the network. Again, it's an issue of time scale. Traffic sources come and go. It's not as if you had these long infinite flows, and you can sort of optimize this thing.

Look at a typical Web transfer. A typical Web transfer persists in the Net for a few round trips, and so this has been called, "The man who is being consumed by mice instead of elephants." You can see, it's much easier to optimize elephants because they're fewer and they last a long time.

You people obviously do this optimization over periods measured in days looking at aggregates of flows, and we have algorithms that basically let you take your physical capacity and divide it up into virtual pads, and then put subsets of your traffic on the virtual pads. That's what Internet operators do today, but they do that, over a long time scale.

The thing that's really hard is trying to do this over very short time scales where you're looking at the coming and going of individual flows, because if you have distributed algorithms, it always runs and orders some number of multiples of round trips, and the flows go away before you can. The optimum doesn't exist. There isn't a static optimum. But that's a very brief answer, and the answer is there's a tremendous amount of work that's been done in this area, and we found very few things that actually pass the practicality test.

MR. JENSEN: Yeah. I'm Rick Jensen from Physics and Neuroscience at Wesleyan University. Can you clearly state the question that you came here to ask so that we can work on it over lunch?

MR. CLARK: Should I assume that in a network of the size and complexity of the Internet that I have to assume that traffic will show a tendency to form into clumps, so that I have to put buffering and other strategies in the Net to absorb these clumps and sort of disentangle them? Or would it be alternatively possible to imagine, by the imposition of that engineering office net, I can sort of smooth everything out to its intrinsic smoothness, and if I could, would that actually be a good thing?

In other words, would it be simpler to build the Net that way, would it be cheaper, or is it better just to be trying to separate this? In other words, can I remove the clumpedness?

There was an observation several years ago that in the very early stages of the Internet, it appeared to be oscillating. My physicist friend said, "No. It's not oscillating, it's got a strange attractor in it, believe me," but that was over a bigger scale. Can we drive this behavior out of the network or not with a system of this size and complexity? Thank you.

Chapter 4

Description and Modeling

Jack Cohen
Session Chair
Greg Chaitin
Fundamentals of Mathematics
Kathleen Carley
Agents in Societies

Greg Chaitin
Fundamentals of Mathematics

MR. CHAITIN: I'd like to talk about something very unreal. We're covering a lot of very different areas in this meeting, and I'd like to talk about something that has as little contact with reality as possible, which is metamathematics.

Instead of talking about applications of complexity to economics, biology, physics, other important things like that, what I'd like to talk about is mathematics looking at itself in the navel trying to figure out how mathematics works, trying to figure out what the scope of mathematical methods is.

This process of self-examination is called metamathematics, and it's gone on for roughly a century. It is the brainchild of a mathematician called David Hilbert, who had very great ambitions for this subject. Unfortunately, all the great progress in this field has been with discovering that every one of his conjectures was completely wrong.

Outsiders may think that the world of pure mathematics is the world of certainty and absolute truth, and it may be that that's the case compared, say, with politics, but internally, we mathematicians are full of self-doubt. We have had tremendous controversies in this century and also in previous centuries. I'd like to tell you a little bit about this story and some of the surprising outcomes. The connection of this with the subject of this meeting is that my work has been on applying notions of information complexity to metamathematics—specifically, the question of the limits of mathematical reasoning to Gödel's Incompleteness Theorem, and the work of Turing. What I've added to this mix are ideas that are relevant to this meeting which are ideas coming from a version of information theory. It's a kind of a complexity, and so that's what the connection is.

The story begins about 100 years ago with some very controversial, wonderful work by Gregor Cantor. It's not normally mentioned that Cantor was very interested in theology, and now people omit that, and they just talk about his theory of infinite sets or infinite numbers.

We all know 1, 2, and 3, and Cantor said, "Well, let's add a number after that." He called it ω. It's the first transfinite number, and then you just keep going: $\omega + 1$, $\omega + 2$, $\omega + 3$. You get eventually to 2ω, and then $2\omega + 4$, $2\omega + 6$, and then you get the numbers like $\omega^4 + 5\omega^2 + 89$. You keep going, and you get ω as an infinite - the first number, you get ω^ω, and then later on you get $\omega^\omega{}^\omega$, you get to ω to the ω to the ω to the ω with an infinite number of ω. At that point I don't know what goes on after that, but the people in this field keep going. There is a very large number which is called epsilon-nought to its friends because you have to write an infinite amount to write this number out.

The infinite numbers that Cantor introduced up to this point involve this new number, ω, and algebra, and this number, epsilon-nought—the first number that takes more than that. It goes beyond what you can write in a finite number of symbols from ω.

Anyway, this was great fun, and it was very controversial. Some said it's a disease, and they hope mathematics recovers. David Hilbert said, "This is a paradise. No one is going to expel us from this wonderful world," and it was very controversial.One of the things that happened was it did influence 20th century mathematics. 20th century mathematics is pervasively set theoretical. That was something good that came out of this. Even though not everyone buys these infinite numbers, this is the more daring part of what Cantor did. The mathematics of the century was greatly influenced by this, even though not everyone goes off the deep end quite like Cantor did.

One of the problems that Cantor noticed is he started getting contradictions playing around with these ideas of his. These are the reasons that led to absurd results, and he didn't bother to tell the world about this. My understanding is he thought this was normal because, since he thought that infinity was an attribute of God, it's natural that there should be paradoxical aspects. He wasn't bothered by this.

Bertrand Russell, on the other hand, was greatly bothered by some of these paradoxes and started telling everyone that we have here a tremendous crisis where pieces of reasoning that looked perfectly fine lead to catastrophe, and I think Bertrand Russell really deserves the main credit for letting the secret out of the bag. He also found some additional paradoxes to the ones that Cantor discovered—or I believe discovered, but didn't tell people about.

One of these paradoxes is, Cantor had shown that for any infinite set, there's an infinite set that's larger. The way it works is this infinity is called the "power set of a set." This has a greater primality than a set. The power set of a set is the set of all subsets.

Cantor has a basic theorem that if a set is infinite, the number of elements of the sum of all subsets is a larger infinity than the original set.

All of this sounds very well and good until Bertrand also said to himself, "Well, what happens if I start with the universal set—the set which contains, by definition, everything—and I apply this argument. I'm going to end up with a set which is larger than the set of everything. How is that possible? It would be the set of all subsets of everything, it would have to be an infinity which is larger than everything, but it can't be because everything already has everything in it."

When you look at the details of Cantor's proof that this should be the case, what Russell came up with is the set of all sets that are not members of themselves. The problem with the set of all sets that aren't members of themselves is it's a member of itself only if it's not a member of itself.

Why did we get into trouble? How do you decide what sets are O.K. and what aren't? In other words, you have reasoning that looks, at first glance, perfectly innocuous, reasoning of a kind you've employed in other cases to good effect, and here you're stepping off a cliff with it, and you'd like to know before you fall off the cliff, why.

At this point there was a lot of controversy, and it's still fun to read about all of this controversy. All kinds of things were discussed, and one of the proposals by David Hilbert is known by the name, the Formalist School.

To avoid problems in these problems, he wanted to give an absolutely firm foundation to all of mathematics. He said to provide this absolutely firm foundation for all of mathematics we have to have axioms. We have to have an artificial language with a very precise grammar.

The main idea is this marvelous thing called a "proofchecker." Hilbert came up with the idea that the ultimate goal, the way to give a firm foundation for mathematics once and for all and, by the way, to avoid all of these paradoxes, would be to create a completely artificial language for doing math. So you avoid all the ambiguities of normal languages, and you write down once and for all what the axioms are.

You have a very precise grammar, and the rules of the game should be so precise that a machine can decide whether a proof is valid or not. This would

148

be an idea of absolute certainty. In other words, it shouldn't be a matter of opinion whether a proof evades the rules.

Hilbert thought that by getting to work, he and some collaborators were going to once and for all give this absolutely firm foundation for mathematics where mathematical proof should be absolutely black or white—absolutely crystal clear. Hilbert said there should be a mechanical procedure that will decide if a proof is correct or not. If we've done mathematics that carefully, then it's absolutely clear what the rules of the game are.

The reason for doing this is not because he's a fanatic. I don't think Hilbert really wanted people to work in an artificial language, and completely formal proofs tend to be very long and incomprehensible. I think what Hilbert really wanted was metamathematics. Once you've crystallized out the meaning of mathematics into this formal axiomatic system the way Hilbert wanted, it becomes a meaningless game. It's not meaningless, but you can forget that mathematics has meaning and just look at it as a game with very precise rules.

You can stand outside of the system and look down at the system and say, will this work? Will I be able to prove that zero is equal to one? Are there mathematical questions which I will be unable to settle, or can I settle all the mathematical questions? This now becomes a precise question, if you can set up mathematics in a mathematical system of the kind Hilbert wanted to.

Hilbert's goal was to once and for all formalize mathematics precisely to the point where it's mechanical to decide what is direct, and having done that you forget that it has any meaning. You stand outside, you look down at it, you study it, and you try to convince everybody that the system you've picked is good so that they'll believe in it, and they'll adopt it. This would have given once and for all absolute certainty for pure mathematics.

For about 30 years it looked good, and Hilbert and a lot of bright collaborators like the young John von Neumann worked on this until disaster struck in 1931 and 1936. Turing showed in two completely different ways that it could not be done, that it was impossible to fulfill Hilbert's ambitions.

This was very, very shocking to the mathematicians of the time. You can read John von Neumann talking about what a shock it was. Of course, there were other problems at the time—terrible problems in Europe—but Turing and Gödel showed that Hilbert's program failed in two completely different ways.

The way Gödel did it is his famous Incompleteness Theorem, a very paradoxical way of refuting Hilbert's project. What Gödel did was to construct an assertion which says, "I'm unprovable," and he made an assertion about the whole numbers. That was enormous cleverness. If you think about it, why is "I'm unprovable" a problem for Hilbert?

The reason is as follows. If you can construct an assertion—and this would have been an assertion about 1, 2, 3, 4, 5, addition, and multiplication, an assertion in elementary number theory about the whole numbers, and about the positive integers—that in some complicated way, in some very common way, says, "I'm unprovable," if you have an assertion like that in ordinary real mathematics that says "I'm unprovable," it's true if only it's unprovable.

The problem is if you can prove a statement that says "I'm unprovable," you're proving something that's false, and if you can't prove a statement that says "I'm unprovable," it's true, and you have, in mathematics, an incompleteness because you have something that's true, but you can't prove it. Either way you're in terrible trouble, and most people think that if you assume that mathematics doesn't prove false things, then it has this hole, it's incomplete. This is Gödel's famous Incompleteness Theorem—a very, very, very disturbing proof.

It skirts very close to madness, the paradox, and it doesn't give you a feeling if this is something that happens all the time or if this is something completely crazy that can in theory happen, but in practice, never occurs.

Turing had a much better way of showing that Hilbert's program couldn't work and of getting Gödel's Incompleteness Theorem as a corollary, using completely different methods. What Gödel had discovered was incompleteness, that any full set of axioms for mathematics has to be incomplete, we'll leave out some of the truth. What Turing discovered was a more fundamental and dangerous phenomenon which is uncomputability, and he deduced as a corollary completeness.

Turing discovered that there are very simple questions for which there is no way to compute an answer, and in particular the holding problem is such a question. You can't decide in advance if a program is going to hold, if you put no time limit on a self-contained computer program. This is something that no algorithm, no computer program can decide, given another computer program, whether it's going to hold or not. You can start to run the other program, and if it does hold, you realize that, but you never know when to give up. You might have let it run for a billion billion years, and you say, O.K., I guess this is never going to halt it, and it halts the next minute.

The problem is proving that a program will never halt. If a program does halt, you can always, in theory, demonstrate that just by running and in being patient until it halts. Turing showed is that there's no algorithm for deciding in advance if a computer program is going to halt.

If there is no way to calculate this, Turing, in his very first 1936 paper, did use as a corollary an incompleteness result different from Gödel's and, I think, more disturbing, which is not only that you can't calculate, given a computer program, whether it's going to halt or not, but there's no way to prove it either, because if you have a formal mathematic system which always enabled you to prove whether a program will halt or not, you can run all possible proofs in size order, check whether they're correct, and eventually find either that you've proven that it's going to halt, or you've proven it's never going to halt. That would give you an algorithm for deciding whether a program is going to halt in advance.

You can always prove whether a program is going to halt within a formal mathematic system of the kind that Hilbert wanted. There would be an algorithm for deciding whether a computer program is going to halt.

Uncomputability is a more basic problem than incompleteness, and Turing's paper is really very remarkable because it makes incompleteness look much more dangerous. With Gödel's proof it's very hard to judge whether you should take this problem seriously or not, but I think that with Turing's proof it begins to worry you more.

The halting problem almost sounds like a problem in physics. It's an itemization, there is no prime bound, but you think of a machine running, and you just ask, eventually, will it come to a stop? Will it just keep chugging along forever without coming up with a final answer? It almost sounds like theoretical physics in a way. It sounds much more complete and down to earth.

I have basically several ideas here, so let me tell you about them. There were really two thrusts. On the one hand I like information theory. The idea of information appealed to me a lot, and this was at the time when Shannon's theory was headline news, and people were very excited about it. The other idea that interested me was randomness in physics - in quantum mechanics, for example.

My approach to this was to say, well, maybe the problem is more serious. Maybe what's happening is that we have in pure mathematics and the foundations of mathematics maybe a bigger problem. Maybe this is just the tip of the iceberg. Maybe mathematics is infected with the same randomness that infects quantum mechanics; not exactly the same, but maybe it's a similar phenomenon.

There was a lot of discussion earlier this century when quantum mechanics was developed, because in quantum mechanics, even in principle, there's no way to predict the future. It's believed that some things in the atom take place at random. And it's not like in most of mechanics where you have Newtonian physics, but you use statistical arguments because you can't deal with 10^{30} particles, and so you do the Newtonian calculations but, in principle, you could.

With quantum mechanics the bedrock is a probability, it's the showing-your-way function. My thought was maybe a similar thing is happening in pure mathematics, and what would that be? It would be like this: you don't have physical unpredictability in pure mathematics, but I thought sometimes the reason mathematicians can't figure out what's going on is because nothing is going on, because there is no power nor structure nor law. The situation just has no structure, and the reason they can't figure out what's going on is because nothing is going on. That was my thought, that some questions in pure mathematics perhaps had answers which have absolutely no structure or pattern.

Part of the problem with pursuing this project is you have to be able to say more precisely, "What do you mean by that?" and to do that, I took some ideas from the Shannon information theory and modified them, and looked at the size of computer programs. That was the key idea—the size of computer programs. I defined something as having no structure or pattern in a logical sense if the most concise computer program for calculating it has the same size as it does, if it can't be compressed into a more concise description.

That was the idea, this was the project, and I've been working on this—it seems like yesterday—almost 40 years. The final conclusion of this work is that when you start looking at the size of computer programs, everywhere you turn, you end up with incompleteness and uncomputability. It sort of hits you in the face.

To give you an example, if you have this definition of randomness as algorithmic incompressibility, it turns out almost all numbers satisfy this definition of randomness, but you can never prove a number is random. You can't prove numbers are random in this sense of lack of structure. To give you the idea of the proof, it's the paradox of the first uninteresting number.

The idea of why you can't prove a number has no structure or a pattern or it doesn't stand out in any particular way is, well, that's an uninteresting number, right? A number is interesting if it stands out or has some structure or a pattern that makes it stand out from the herd. So, if you divide all of the positive integers into interesting and uninteresting ones, well, consider the first uninteresting positive integer. That's rather interesting, isn't it, the fact that it's the very first one that is not of any interest, and that's how you get into trouble. You see, if you could decide via algorithm whether a number is random or not, that would end up being a characteristic of it that might make it stand out from the herd and make it not random after you proved that it was random.

There is this paradoxical notion of randomness such that you can define logical randomness or lack of structure, but the definition has to be non-effective to avoid paradoxes. It has to be that there's no way to decide whether something's satisfied the definition or not, and I believe if that's not a flaw, it's a virtue of this definition—to avoid catastrophe.

I've gone one step further than that. That was actually the first thing I did a quarter of a century ago. I have a number, I'd like to call it Ω. It's the holding probability. I don't really have time to explain much what this is, except that it's connected with Turing's holding problem. It's the probability that a program generated by coin-tossing halts. It's kind of a statistical average over all individual instances of Turing's holding problem, but since it's a probability, it's a number, a real number, between zero and one. With this simple mathematical definition, it turns out it's maximally unknowable, this number. It's sort of a worst-case, it's the anti-Hilbertian extreme, so let me tell you why.

What Ω does is it shows rather completely and clearly that the traditional notion of mathematics is not just a little bit wrong, but in some cases is absolutely totally wrong. It's the opposite of the case. The normal notion of pure mathematics is that we can agree on a small set of axioms from which all mathematical proofs will follow, and that's why mathematics gives certainty.

What this number shows is that some mathematical questions have no structure or pattern at all, and there are some mathematical questions whose answers cannot be deduced from any principle simpler than they are. This is irreducible mathematical facts, and that's the opposite extreme from the mathematic method or from reasoning.

"Irreducible mathematical facts" means mathematical facts where, essentially, the only way to prove them is if you have them as a new axiom which means you're not using deduction. Anything can be proven by adding it as a new hypothesis. The interesting thing is that there are areas of mathematics—and Ω is such an area—where mathematical truth has actually no structure or pattern and cannot be compressed into any principle simpler or more concise than they are. That's sort of the opposite of the axiomatic method. It's a case where reasoning is completely impotent.

Why do I say this? It turns out if you take this number, this holding property, and you write it in binary, it turns out that its bits have absolutely no structure or pattern. They have a simple mathematical definition, but they cannot be distinguished from independent tosses of a fair coin. You can't compress the first n bits of this holding probability into a program that's smaller than n bits. You can't diffuse the first n bits of the binary extension of this number from axioms that are smaller than n bits. Essentially, these are irreversible mathematical facts. God is playing dice with mathematical truth here.

Let me end by stating as shortly as possible what this number Ω shows. Isn't there such a notion, "if something is true, it's true for a reason"? What is that called, "principle of sufficient reason" or something. I don't know the philosophy.

Anyway, normally, you'd think that if something is true, especially in mathematics, it's true for a reason, and mathematicians are always looking for the reason that something is true. They're looking for proofs, and that's what they publish. That's the job of a mathematician—to find the reason that things are true.

What I have discovered here with this holding probability number and looking at its bits, asking what its numerical value is, these are mathematical facts which are true for no reason, they're true by accident, and that's why they forever escape the power of mathematical reasoning.

There is no reason that an individual bit of this number is going to be zero or one. It's got to be one or the other, but it's so delicately balanced that we're never going to know. It really is an accidental mathematical fact.

Now, pure mathematicians, when I say things like this, very often, they don't understand what the hell I'm talking about, or if they do, they want to throw up. I think physicists tend to feel much more comfortable with these kind of statements than pure mathematicians.

In other words, if you think of independent tosses of a fair coin, you see each toss has got to come out heads or tails, and it's equally likely to come out one or the other. There is no reason an individual toss is going to come out heads or tails, it just does. If you know all the even tosses, it doesn't help you to get any of the odd tosses. If you know the first million tosses, it gives you no help to get the next one. And the bits of this number, even though this is an individual, precisely specified number, mirror perfectly that, you see. This is sort of a worst-case for reasoning and for the traditional philosophy of pure mathematics.

I love doing mathematics. I don't think pure mathematics now falls in a heap. I think the traditional philosophy of pure mathematics we already knew had problems from Gödel and Turing. I think this just makes it a little worse. My own personal view is I'm a supporter of a school of mathematical philosophy, a small school. There's at least one book on this school, a quasi-empiricist view of the foundation of the mathematics. The name was coined, not by me but by Imre Lakatos, and there's a very nice book that I recommend to you is called *New Directions In The Philosophy of Mathematics*. It's not a technical book. It's a Princeton University Press paperback. Thomas Tymoczko was a Philosopher at Smith College, and it's a lovely, inexpensive Princeton University Press paperback published about two years ago, and you really want that one.

That's basically what I wanted to say, and I think the moral of this story is this area is one of the first to be involved with complexity. The field of complexity is very broad, and this work on using information and complexity notions even in metamathematics is actually one of the first areas where the notion of complexity was applied. This is one of the older subjects in the field of complexity, I think it's fair to say, and I would say that there's a good reason for this. The reason is that the real world is much more difficult than the world of pure math.

This is a Turing theory, but I think it's of interest. It's like thermodynamics, not a heat engine. It's like a thermodynamical theory of axiomatic systems, in a way, what this subject is. It says the limitations of a heat engine using thermodynamic arguments, you talk about the limitations of axiomatic or formal axiomatic systems in Hilbert's style using informational theoretic arguments which do have a thermodynamic flavor, I think it's fair to say.

Physicists tend to like this stuff because it reminds them of some stuff in physics, and logicians, I think, really become nauseated by the whole thing. Thank you.

UNIDENTIFIED: There's a certain percentage of bit strings of length n that can be written with fewer than n bits. Like n zeros can obviously be compressed.

MR. CHAITIN: It's very small.

UNIDENTIFIED: Right.

MR. CHAITIN: But most n-bit strings are very close to n bits.

UNIDENTIFIED: But the fact that that number is not zero would seem to put some constraints on the first n bits of Ω.

MR. CHAITIN: A good point. You might be able to get the first few bits of Ω.

UNIDENTIFIED: Can't you get many bits to be at least a little bit constrained?

MR. CHAITIN: The precise result is that with n bits of axioms you can get $n +$ most n-place bits of Ω, and that's how it goes. And you can get up to there clearly just by putting, as your axiom, what the first n bits of Ω are.

You bring up a point which is, most numbers we deal with like the square root of 2 or pi have very concise descriptions. It turns out that most real numbers have no pattern in their bits but, in fact, all the real numbers we ever deal with normally that come up do have regularities and do stand out from the crowd; otherwise, we wouldn't see them.

It's a bit paradoxical to say the majority have no distinguishing features, and that's precisely why we never bump into them, so that the whole thing has a paradoxical quality.

UNIDENTIFIED: Can you summarize your discovery in terms of a simple word? In other words, have you discovered mathematical atoms?

MR. CHAITIN: Yeah. That's a good way to put it. David Well(sp?) put it this way: mathematical truth has no finite basis. It's simply complex. That's another good word to use.

UNIDENTIFIED: And if this is so, this connects I think to the biology because all mathematical reasoning is based on mental processes, and all mental processes occurs in the brain, and processes are based on atoms and molecules, and I'm wondering whether the logical atom and material atom is somehow connected?

MR. CHAITIN: I think it's an interesting metaphor. There's a book by Sokal and Bricmont, what is it, *Impostures Intellectuelles*, something like that? And there's one chapter which says an entire book could be devoted to what they consider to be analyses of Gödel that are completely wild and miss the mark. I don't know. I tend to, since I'm obsessed by Gödel's incompleteness theorem, whenever someone mentions it, I tend to light up.

But it's true, there are all kinds of directions one can go. Some of them are just quite a metaphor; some of them may be more than that. I don't know.

UNIDENTIFIED: I'd like to ask about intermediate levels of complexity. What about the recent developments, for example, in Fermat's Theorem and for the Collatz theorem, where it turned out the proofs were extremely complicated, involving huge computer codes. Presumably, a lot more complex than people thought they were. Where is current thinking of that?

MR. CHAITIN: My results sound too negative on the face of it. If you take them at face value, it would say mathematics is a useless enterprise, and that's not the case at all. We have spectacular recent events like the proof of Fermat's Theorem.

I think that at this point an interesting research question is, how come mathematics is doing so splendidly, in spite of all these negative results that I told you about? I have no idea how to go about it, but I agree with you, I think that's an interesting project.

UNIDENTIFIED: Greg, I think I've asked you that question before about some two years ago when you were kind enough to send me your book, and it has to do with quantum mechanics, and I'm wondering if you've changed your mind?

You mention that, maybe we should take our cue from physics, since in quantum mechanics you have some randomness, and that's sort of essential.

Only if you look at a piece of the universe will you actually find some randomness, and there's a consequence. Randomness is not essential, but rather, due to the fact of how we look at the universe.

MR. CHAITIN: Is that a generally accepted view, or is that your view?

UNIDENTIFIED: People are divided on this. Half of them say it's trivial. And the other half say, "I don't know what you're talking about."

MR. CHAITIN: Well, there's also chaos, Dave.

UNIDENTIFIED: Actually, this theory correctly predicts all of quantum mechanics as we know it.

MR. CHAITIN: Well, maybe you're doing a great parallel line. You may be right, or you may be wrong. I don't know—I'm not an expert in quantum mechanics. Your viewpoint, sounds certainly different from what I learned as a child.

I have a friend in Vienna named Karl Svozil. He's a physicist, and he's an Einsteinian classicist, even though he does quantum mechanics. He really doesn't believe in randomness in quantum mechanics. He thinks that when quantum mechanics is reformulated properly—and maybe you've already done it—there will be no randomness.

His point of view is that he doesn't believe there's randomness in the physical universe, even though a great many physicists do. He tells me that in the mindscape, in the imaginary mental world of pure mathematics that I've shown, there there is randomness for sure. Whether they've done this in the physical universe, I'll leave you physicists to do all the work on that, too.

UNIDENTIFIED: I completely agree with you. I think the randomness in mathematics is much more fundamental than the randomness in physics, which can be explained. The one that you stumbled upon is much more fundamental. That's the main point I wanted to make.

MR. BAK: Greg, I have such a great question to ask you, but then I realized it was unanswerable.

MR. CHAITIN: That's the best kind.

Kathleen Carley
Agents in Societies

MS. CARLEY: My name is Kathleen Carley. I'm going to be talking about, for lack of a better title, "Complexity of Society." When I use the term "society," I'm really talking about groups, organizations, and social systems in general—it has a very broad notion.

156

The way in which groups emerge, develop memories, their learning, and so on, is through working with a constraint-based adaptation system. In this area there's a variety of different factors and things that we're concerned with. In particular, some of the things we're concerned with is how do social and cultural constraints equally affect group behavior. Group behavior includes such things as group memory, performance, development, and cultural norms.

Then the question is: How do you design a group to think what you want it to? How do you define a group? I want to develop a new airline service. How do I set that group up so that planes don't go crashing every two minutes?

Another question is, how do groups evolve? By evolution, we mean not just the number of people in the group, but the interaction, the behavior between them and so on. How do you manage that evolution? And then, what is the impact of communication and information technology on the groups? Some of the things I'll be showing you deal with new technologies like that.

We're going to be doing this by looking at a relational analysis. How do we link agents together into groups, and how can we do that in the context of knowledge, tasks, and so on?

A variety of things have been found to affect organizational performance. This is true for groups, organizations, whatever. Basically, we know how it's affected. Put smart people together, you tend to get better performance out of the group.

"Transactive memory," is a new fancy term for a "know-who," who knows who knows what, who knows who knows who. Again, it's important for affecting the performance of the group, and design, that is, the coordination that's armed the architecture and the organization. Communication, structures, authority, and so on affect an organization. We can design optimally configured organizations.

Organizations are, as everyone says, complex. They're complex at various points. Well, ignore that. An important thing is that in organizations, there are a lot of things going on. For example, we've got agents. I purposely use the term "agents" there because, when we think about society today, we're not just talking about people in collectives, we're talking about groups acting as individual agents, we're talking about people, we're talking about avatars, about Web-bots, about robots, and so on.

Secondly, these are all connected into networks, such as a communication network, a knowledge network, and so on. Each organization has a design or a form. You can think of this as the authority structure, who reports to whom. Organizations always think they have this, but they never know what it is, O.K.? This is the organization's also tasks that they do, and those are a form of some precedence network and, of course, there are stresses: time pressures, new institutional changes, turnover, and, of course, all the technology.

All of these things come together to somehow interrelate and somehow affect and determine what goes on in the company, what affects the performance, at both the individual or workload level and at the general performance level.

When I was talking with Yaneer originally, he asked, "What exactly in your model is complex?" Well, kind of everything.

This is a long-winded answer to that. At one level, we know that organizational performance itself is a function of the variety of factors in the organization's design.

Here are utility, size, and density, and we know that there's something like a response service or performance service that says, "If you designed it this way, this is the maximum level of performance you can get with that particular design for that particular task." That's fine. The organization particularly occupies a position in this. There are some other organizations over here.

Organizations don't know what the surface looks like because they can only see where it happens to be populated, so it's that one, that one, and that one and that one. And if you're here, you're here when it was over here, and you say, "Gee, it must be higher because these guys are all high." They don't know that if they just simply increase the size, they're going to plummet down in performance.

Second thing about this is that this is only the maximum or cap on performance; that is, within this particular area, when you move to a new design, you change the number of experts you have in your company, you change who they're reporting to. You actually have a learning curve you need to go through. And so, when you move over here, depending on the expertise of the personnel, what knowledge transfers, what technologies transfers, and so on, you may be anywhere here.

The irony, of course, is if you move from here to here thinking you'll improve your performance, you may actually drop down in performance.

Then the issue is: How does that affect your learning and what it is the companies are actually learning? As if this weren't all bad enough, over time, through technological changes, environmental/institutional changes, changes in the laws governing the corporations, and so on, what performances associate with what particular configuration also changes.

The same design under two different institutional arrangements in two different countries and two different technologies is totally different. For example, the right way to organize steel companies changed dramatically when these laws of pollution control came out.

Now, within the models, not only do we model this change in the environment, technology, and with the laws and so on and the changes in corporations trying to in fact learn over time, we've got people who are agents.

One of the things that some of the earliest work done on agents has suggested is that a lot of complexity of social behavior is not a function of cognitive capabilities. It's not that it should get more capable agents—they had to exhibit more complex behavior—but it's because, as you put more limitations on cognition, more social behavior need automatically appears.

For example, the automation agent or the rational agent which economists always play with should only be able to produce these kinds of takers. But as you actually make a randomly rational or make cognitive or put emotions in it, you get more complex emergent behavior coming out automatically. And, of course, the level of knowledge also affects that.

We can talk about having a model social agent, just like we would have a model agent previously in psychology as a basis for sociality. We've got agents populating these models that are cognitively extremely restricted. They have the ability, for example, to know everything they learned in kindergarten and what they learned yesterday and forget everything in between. They're bound to be rational, they're socially rational, and so on.

All this work has led to is a new kind of a view in a variety of people. We think of it as sociocognitive physics. For the physicists among here, if I'm misusing these terms, please, bear with me and forgive me for the moment, and tell me later.

The kind of principles that are underlying this work are, first, we always see the emergent reality. That is, all social behavior, all individual behavior seems to emerge out of ongoing interactions when you have intelligent, adaptive agents.

Secondly, we have synthetic adaptation. Synthetic adaptation simply means that any unit, group, organization, etc., composed of intelligent, adaptive units, and computational agents is also intelligent, also adaptive, and also computational.

Groups, in other words, have the ability to be intelligent, adaptive agents in and of themselves and not relying on the abilities of the individuals within them or the agents within them.

We have constraint-based action, that is, that all action in here is a function of the constraints on temporal, discipline, social, and institutional constraints and the condition of the participants and adapted by a set of possible actions and, of course, none like your physical molecules. What we have here is that our atoms, which are agents, have the ability to learn, and because they can learn, learning occurs at multiple levels. It occurs at the individual levels where the individual's learning on the basis of experience, individual baselining with expectation-based learning, but we also have something called, "structural learning," which is the learning of the pattern of relationships and connections among units, and there are, of course, multiple bases for learning, expectation, feedback, and so on.

One of the ironic things about agents—and this is true for people, in general—is that they are very good at forgetting how they learn something. We learn something by saying, "I forecast what I think the future is going to be like, and I remember my result," or, "I experienced a couple of things, and I remember the result." I can't tell the difference two days from now whether I forecasted it by learning, or I actually experienced it by learning. That's the fact of the different bases of learning.

The consequence, of course, is whenever you've got all these four things happening, you always get the social world as having an underlying order, you always get history matters or history and facts, you always get shake-outs occurring. They automatically occur.

Shake-outs are when, for example, you put a lot of companies in a new industry, and then after about ten years half of them fail and dry out. Or shake-outs occur, for example, when you put a group together, and they all start interacting a lot, and after a year, they'll never speak to each other again. Those are what I mean by "shake-outs." They automatically are guaranteed to occur, and you need some kind of dramatic social change or some kind of group survival, and to get either dramatic social change or group survival you have to have either intended action where you're trying to do interventions that make the group survive or some kind of catastrophic event. Those are features that always come out of these models.

Some of the insights that come out, first, are that there's this notion of a learning ecology—the idea that learning at one level interferes with learning at another level. A very good example is in corporate America. We've got individuals learning things and gather expertise over time. One other type of learning is where the CEOs learn that they can actually save money by downsizing. Well, the tragedy of downsizing in the early '90s for American corporations is they lost all their brain power. That's the two types of learning interfering with each other.

The second of the insights is the notion of social fallout. We see a population at the organizational level, but we also see the social tie—the social level. An example is when you bring a new group together, initial ties form instantly among people on the basis of very superficial similarity.

When groups first get together, they go on the basis of: Are they the same race? Are they the same gender? Are they the same religion? After six months, those connections are falling away, and people start having to learn more about each other. More people form more long-lasting ties on the basis of deeper social knowledge.

The third insight we see over and over again is a trade-off between innovation and vulnerability. That is, if you create a corporation or a group, and you want it to be invulnerable to such things as, say, information and security issues, stealing ideas from you, and so on, and you would have high levels of innovation, well, you can forget vulnerability by having a high level of depth in your connectivity among the individuals, so you don't have high slips, because high slips are your most devastating from a group perspective. However, the higher the level of connectivity, the lower the level of creativity on average, because people need some time alone, you know, to be creative.

There's also a trade-off between common view and active corporate coordination. That is, if we want to, we can design for a particular task in an organizational form that is completely optimized to do that task with a minimum of communication and with a minimum of active coordination of the group. But if you do that and you've optimized performance, you then change what tasks they're doing, the group falls apart, because they each have no idea what the other is doing; they have no way of adapting and changing to the situation. So, again, there's a trade-off between this kind of common-view notion and active coordination.

I'm going to switch here a little bit. The specific models I'm going to talk to you about are both based on this notion of interaction as occurring within an interaction knowledge space.

These little blue dots are agents. We have some synthetic agents moving around. These might be groups, organizations, whatever, and what happens is as they're going through this process here learning something, changing their knowledge, they move in a knowledge space which changes whom they interact with, and they're constantly moving through this.

The only thing is that there's this concurrent activity because everybody's doing this all at once: not only the individuals, but the group. That's the basic fundamental conceptual idea of what's going on. That's carried out using adaptation models where you have a set of individuals who all know something; so they all have complex mental models.

Each of them then can choose somebody to interact with. Here we have these two people interacting; they communicate something; they learn something; and then they socially reposition, which changes whom they interact with in the future.

Now, your first reaction is, "I talk with you; that means, then, I'll talk with you more." But that's actually not what goes on in these models at all because this is happening for everyone all at once.

When I talk to you, I may tell you something that actually makes you more like someone over there than with me, and you spend the rest of the time talking with them instead of with me. That's the kind of basic model that's underlying that for individual learning, and then these individuals are placed in this network here. They have a knowledge network, so they know who knows what. There's information networks linking their ideas together, and then, of course, there are these organizations linking which[sic] organizations together. The simulation is used to put this all together. These are all big Monte Carlo simulations with these acclimations.

This network I just showed you is operationalized at four levels. We have a set of agents, knowledge, resources, which are physical objects and tasks. We have the authority and the precedent network linking who knows whom, the expertise, who knows what; we have what ideas are linked to what, we have their capabilities, who has access to what and who is assigned to do what task, and then who actually does what task.

This is put into this model called OrgAhead. I presumed a little bit here. I called them decision-makers, but these are your agents by skills, by what resources and what knowledge they have, to man structure tasks. We put them into the model, it goes around, turns around, and comes out predicting accuracy, workload, anything you want to know about the company.

At a more functional level, each of the agents in the model is actually modeled using experience-based learning, doing a task. There are a lot of tasks. There's a radar task. There's a minor choice task. There's a variety of different tasks that you see down here.

Each agent has access to some information, some resources. They do the task, and they get better over time. If they don't go grab, they report to somebody; and they report not all the information, but a collapse of the information. Then the group at the top, one or more people, makes the decision; then it becomes the group's decision. The group gets feedback on how well the group did, not on how well they did individually. In most companies, you have no clue as to how to reward individuals individually. You've got recommendations going up, feedback coming back.

There's a strategic adaptation component, which is basically a CEO or a change unit that has knowledge about the future and can predict/forecast a little bit; and like in chess, you can only predict a few moves ahead in the future, not very far. He knows your current performance, possible change, expected performance, who knows whom, who knows what, and so on; and he suggests actual findings.

There's a few things we know about an organization that actually operates at this level. First off, we never look at many alternatives, just one or two.

Secondly, they become increasingly risk-averse over time. That means less willing to make changes in here if they think it's going to lead to a downturn in performance, more willing to accept possibly risk-increasing moves initially.

All the features you have at the strategic level for organizations map perfectly onto an A list. Those are small and simulated. So that's the basic model. It provides tons of results on organizational behavior. I'm only going to show you a couple.

First off, one of the things it predicts is that you're going to get a general transport improvement and performance over time, even with environmental change. The colors represent environmental change.

The other thing you'll notice here is that sometimes you'll get dramatic downturns of performance, like this one. One of the things is, when you look at it, every single time you have this dramatic downturn in performance, the reason that occurs is because individual learning is clashing with expectation-based learning at the strategic level. Every time you get one of these, it's because of that kind of a clash.

The other thing you'll see is that these guys change in who is the top honcho at any particular point in time. We've looked at this data, and it looks exactly like, say, the ranking patterns of universities over time, where who's in the top ten stays about there, but who's number one changes.

Now, when you use this model, I'm going to actually use it to look at real shifts in some real organizations. What we did in order to do this is we collected data from archives of 69 different and real organizations faced with a technological catastrophe. Exxon Valdez is an example; Bhopal's another example, and so on.

For these 69 general cases we collected general organizational measures, whether they were hierarchy- or team-oriented, whether they were experience-based, or whether they used the standard operating procedures, and so on. We have exactly the same elements in the model.

We did a matched analysis where we took the performance, the characteristics of design for each of these 69 corporations, and simulated each one multiple times to their predicted performance under both normal operating conditions and under crisis conditions.

For crisis conditions we've got the kind of crisis it was and the kind of things the technology-induced era caused. Was it unavailable personnel? Was it communication errors? Now, we predicted performance. First off to see, does the model help, does it match at all, and then go on and ask the question that we can't ask with the real data, and, that is, what if the organization had done the opposite of what it really did? What if, instead of shifting its structure, it did not shift, would that have changed performance any?

This is the basic results. The blue here is the simulation; the red is the actual data; and the green is our projected "what if" analysis.

I divided this up here into experience-based organizations, organizations where the individuals are told to operate on the basis of their personal experience versus organizations where the personnel were told when to actually strike and in general followed standard operating procedures. These are more militaristic-type organizations, for example.

The first thing you should notice is you get a very good match for social data. This is unbelievable. We get a very good match between the model's predictions and the organizations predictions both for general performance and for performance under crisis conditions.

This is a three-part performance indicator. This is performance on a three-point scale. This is the predicted performance, this is the actual performance. This is for organizations under normal operating conditions, under crisis conditions, and this is under a hypothetical condition.

MR. COHEN: Several of us are puzzled. What criteria do you use? What do you measure the performance against?

MS. CARLEY: For organizations, you never get a simple measure of performance. What we're using here for normal performance is they're on some average of performance on average workdays, as whether it was better or worse or typical for that industry.

For crisis conditions, it's taken into account how bad the catastrophe was relative to how bad it could have been, and those two give us a measure of their performance. And for the simulation, it's the total average based on the act itself.

Now, one thing you'll notice here in an experience-based organization is that they tend to do better under crisis conditions than under normal conditions. Crisis conditions are such that they restrict the choice of options, and the people in the organization actually work better; they know how to deal better with that for a variety of reasons.

Secondly, what you're seeing is that in following standard operating procedures, these organizations, in general, tend to do worse under crisis conditions than they do on average. Their off-centered operating procedures just really aren't designed to deal with crises which are beyond the norm, and they're different kind of entities where the experience allows you to rely on where to get it.

The second thing you'll notice is that for these organizations here that did shift their design, if you go and talk to those CEOs and with the managers of those teams, they'll tell you, "Yes. We shifted our design. That was exactly the right decision, and we did better than we would have otherwise."

What you find out when you simulate this is, for those organizations that shifted, what if they had not shifted? What you see is their performance would have been even better. In other words, they learned just the wrong thing.

For the organizations who are following standard operating procedures, if they had not shifted, their performance would have been even worse. In both cases, they are learning exactly the wrong thing from experience. There's a fallacy to learning, in other words, when what we're trying to learn is in the changing environment for human agents.

Case Example 2 says, in the modern world we're dealing with not just human agents, we're dealing with a lot of artificial agents. In the new organizations we've got people running around, doing things, knowing things, talking to each other, and so on, but we've also got invisible digital organization behind them.

We have a set of avatars. An avatar is like an artificial agent that captures a part of your knowledge at a particular point in time and can answer questions for you. Think of it as something that knows all your papers and knows your calendar and schedules meetings automatically, and knows the most commonly asked questions by people coming in and bugging you. So, there are your avatars running around. They're based on knowledge off your web page and so on, and also your smart databases and so on.

We might ask, as a manager, am I really going to improve performance by bringing in all this new kind of technology?

Well, if I'm a technologist or I'm a computer scientist, I say, of course you are. No problem. But is that really the case?

We take this model construct—again, it's based on this kind of agent-by-knowledge basis—it tries to predict diffusion; it believes change of performance or organization; and we're going to look at exactly that question. If I put in place these smart databases or these avatars, is my organizational performance really going to improve?

How many of you think organizational performance will improve if you put in place avatars? First thing, we've got the databases here, and this is the world with avatars.

If we look at what's happening here, you see that what's happening over time is that organizations where we put in place the database or an avatar take longer for people to gain expertise. Eventually, we all get some amount of expertise; it just takes longer. So, it retards the rate at which people get trained, basically.

In terms of "know who," when you have avatars in the system, there's a slight improvement in learning who knows what and who knows who knows whom. People tend to get a little bit better with their transactive memory than they would if there weren't avatars. I will talk to your avatar if I can't talk to you. Databases actually retard things a little bit, and those are small facts; it doesn't really matter.

But now, let's look at some data facts. Here we're looking at max peak interaction and shared knowledge over time for the same organization. What you're seeing here is that the natural level of interaction that your company is ever going to achieve or your group's ever going to achieve is going to be lower and take longer to achieve when you have either databases or avatars, much longer for databases.

UNIDENTIFIED: Why are the curves going down?

MS. CARLEY: Why are they going down? Because in these models individuals interact if they think there's somebody out there who knows something that they don't know, so they go to interact with them. They're similar in interaction, and both of those drive interaction down, eventually, because you all know that you're all alike, and so you're equally likely to interact with everyone.

What's happening here in terms of shared knowledge, is when you just have people interacting, they get higher levels of shared knowledge, they reach it sooner, so greater consensus to greater share in beliefs. Avatars make that a slower process, but they still stay high. Knowledge bases -- look at this, O.K.? You're never getting consensus, you're never getting shared business.

Now, what is that doing to performance? This is what I started asking. The answer is absolutely nothing. In other words, there isn't any way to organize. You get equal performance outcomes with many different organizational schemes.

What they've done here is that the use of databases and avatars is really substituting know-who and transactive memory and things like that for shared knowledge.

In summary then, what we've talked about are a variety of different things. We talked of the proposal—basically, underlying this system there's a coalition of social and cultural events. We talked about the creation of synthetic agents, the ecology of learning, the creation of learning clashes as affective behavior, learning fallacies, equal-performance outcome, and technology as a cultural change agent.

All of those things mean that in the world of social systems, what's happening is that if you say, organizations are complex sites that do this, that management complex is where a lot of the work is being done, and we talked about power laws. Yeah. It's all over place, got it by group size, by organization size, or the number of network ties, the number of websites, number of links per website, interaction by distance, all of the power model behavior.

What we really want to know are what are the common starts for irregularities that are resulting in those same models all over the place. Why is it that the Web has the same kind of power model that we see for human interaction networks? We know there's path dependency, but what we're more interested in now is morphing, that is, can we move? Knowing you're going to go from Point A to Point B, can I create a path for you that's the easiest one to move along that helps you get there in the least costly and best way?

We know that complex behavior emerges from simple agents, and what's more interesting now is that you can get social complexity from constraints. Thank you.

UNIDENTIFIED: I've always thought that the recent advances of the economy have been related to—well, to the use of computer technology and databases—avatar-type things. There is a correlation of increased productivity with increased use of these types of technology, so that's why I voted in the favor of higher performance.

Could you explain why, in fact, there is no improvement in performance?

MS. CARLEY: One answer is that I'm looking at performance here in terms of performance accuracy, the correctness of the decisions being made. I'm not looking at terms of timeliness.

In terms of timeliness, you may see great productivity. You measure it that way. I'm looking at it as accuracy. That's one answer is that where you may be thinking of it, you're using a different method.

The second answer is that one of the reasons at least that you're getting ample performance outcomes here is because of the difference that you can compensate for, the additional access to knowledge, additional access to people, by different forms of predefined coordination skills. So, there's compensation skills.

UNIDENTIFIED: Wouldn't it be a reasonable suggestion to change the parameter from the performance to productivity, which is a more common index?

MS. CARLEY: Not necessarily. In many cases, you don't really care about performance. The right thing is to use many different performance outcomes and then look at the different effects on all of them.

UNIDENTIFIED: I'm going to ask a question that Jack might ask if he weren't chairing this session, and that is, on all of your plots and output you were apparently showing an ensemble average at each instant in time, and you weren't showing error bars or the variance.

Now, I suspect that the variances are immense, especially towards the beginning of the simulations, and they should stay immense unless you have a highly dissipated set of rules which crushes down at the end.

Can you tell us how significant we should take any differences we see in the graphs that you drew?

MS. CARLEY: For the last set of graphs, in particular, the ones that are dealing with the avatars and so on, we've run those so often that it means the variances on down.

UNIDENTIFIED: The variances are square and thin. It doesn't get any smaller in numbers.

MS. CARLEY: Yeah. Maybe you're thinking variance in a different method. We're talking about the variance, not across the mean, but when they cross two points between the two samples and the two points of time. So, I know the mean --

UNIDENTIFIED: Yes.

MS. CARLEY: -- I know the variance. The higher the number of points, the lower the variance.

UNIDENTIFIED: So, you're saying that even the small differences on these graphs are statistically significant?

MS. CARLEY: Let me give you an even more precise answer. We actually calculated the statistical significance on these for these ones, and regardless of the number of times you run it for the performance outcome these are not significantly different. These ones are significantly different starting at this first little pink dot right there. The blue is always significantly different from the green up until about here, and the pink is significantly different from the green starting about here, and then, of course, all the way up.

MR. COHEN: Are they replicated?

MS. CARLEY: Are these replicated?

MR. COHEN: Each line there comes from how many graphs and how many variances?

MS. CARLEY: Three thousand.

MR. COHEN: Three thousand.

UNIDENTIFIED MALE: And how many agents?

MS. CARLEY: These have been done across 10, 20, and 30 agents.

MR. COHEN: So, your numbers are big enough that we can rely on your variances?

MS. CARLEY: Yes. Does that give you a better answer?

UNIDENTIFIED: Yes.

MR. COHEN: Yes?

UNIDENTIFIED: When you're looking at crises did you look at the size of the crises relative to the size of the organization?

For example, Valdez is not very big in terms of Exxon, but a major earthquake that knocked out maybe two or three of their refineries would be different in terms of their response.

MS. CARLEY: No. We didn't use that. We didn't use the size of the crisis relative to the crises that were typical for the industry, and we did not look at natural disasters because those are always multiple for organizations, so that we've got another private number.

UNIDENTIFIED: How much cultural diversity was represented in the examples where you were looking at actual organization performance? Were the majority of them U.S. companies, or did you look at companies in other parts of the world?

MS. CARLEY: Those 69 cases are drawn from throughout the world. The majority are from the U.S. and Europe.

UNIDENTIFIED: So, you didn't see any significant difference in behaviors between cultures?

MS. CARLEY: At the levels we're looking at I didn't see anything that was relevant. Thank you.

Chapter 5

Self-Organization

David Campbell
Session Chair
George Whitesides
Complex Chemical Systems
Irv Epstein
From Nonlinear Chemistry to Biology
Chris Adami
Artificial Life
Duncan Watts
Small World Networks

David Campbell
Session Chair

David Campbell: I'm David Campbell. I'll be Chair of the first morning session on Self Organization. We have four very interesting talks, and Yaneer offered me ten minutes to introduce with my observations. I decided to take three of those, not more, because I want to give you a chance to hear from the experts.

We're talking about self organization this morning. In some sense, this seems like a pretty self-evident subject. You know what self-organization means: we are not allowed, as we watch a system develop, to go in and mess with the parameters ourselves, and play – no deus ex machina kind of intervention – that's sort of obvious. It's also pretty clear what kind of things do it well, namely, living things, but beyond that it gets sort of murky.

What we're going to hear about today are four ways of poking around in the dark to try to understand more about self-organization. The first talk -- I should mention all the talks will be 35 plus or minus 5 minutes, and 10 plus or minus 5 minutes, of questions. So please interact with our speakers.

The one exception is George Whiteside's talks, which will be 35 plus or minus 5 minutes, or until his voice runs out. As you'll hear, he has laryngitis, so we're hoping for the best.

George is going to start off by talking about complex chemical systems. We're familiar with information from things like the BZ reaction, and with building blocks sort of assembling together. George has played a leading role in going well beyond that, and he'll tell us about that.

The next question might be, "Well anyway, how do we go from chemistry to biology?" Irv Epstein is going to tell us about that.

Third, many of you know the early work of John von Neumann in self-reproducing cellular automata, and the descendants of that have grown into a field called "artificial life." And Chris Adami in the third talk will be telling us about that.

And finally Duncan Watts will be telling us about small world networks. A really exciting result, which explains why "Six Degrees of Separation" was such a popular play on Broadway. So with that introduction, I'll turn it over to our first speaker, George Whitesides.

George Whitesides
Complex Chemical Systems

George Whitesides: First, let me apologize for my Marlena Dietrich simulation this morning. It's the best I can do. I won't sing.

The subject of the discussion for this morning is molecule-mimetic systems. As a chemist, what I'm going to do is stand in front of you and tell you that we're too restricted in working with the periodic table and atoms. And in order to understand what's going on in this area, we need better ways of tinkering with the systems.

The folks who have done this are the people who are listed here. I know that Rustem Ismagilov and Abe Stroock are both here. I believe that Bartosz Grzybowski is going to be here, so that you can ask them details about the work. I basically am reporting secretary.

Now, I should say that in this area, we are real amateurs, and part of the reason why I'm so interested in coming to, we were also interested in coming to this meeting, this very interesting meeting, is that, in fact, we can't at the moment answer that first question. And that is, in this area of complexity and emergence as the phrase goes, is there a there there?

That is, there is no question that there are lots of interesting phenomena which have the characteristic that they are complicated, and then there is a field called complexity, and the two are not superimposable on one another.

But even given that, does complexity offer either an intellectual methodology for understanding complicated problems -- which is fine -- or, more from our point of view, does it offer a prescription for discovering new phenomena?

And one of the characteristics of chemistry is that chemists are accustomed to working in a world in which it is relatively straightforward to test hypotheses by synthesis. So we tend to go at things by synthesizing, and then observing, as opposed to simply observing.

And one of the interesting issues about complexity and complex systems is that in principle, it offers a way of looking into ways in which one might design systems which show new phenomena. And I'll show you some examples as we go along, of directions that we're taking in this, but I want to leave it on the table as a question really to discuss. Is complexity an interesting area? I mean I believe it is, but I have a hard time right at the moment unambiguously demonstrating that.

Now, chemistry has not, I think, been per se a really major player in the area of complexity, but there are a number of reasons for thinking in chemical systems, or molecular systems, that there are interesting things there.

The Belousov-Zhabotinsky (BZ) reaction we saw yesterday -- there are a series of reactions very much like the BZ reactions, which are pattern formers. These are systems in which patterns emerge by competitions between reactions, particularly catalytic reaction and diffusion.

And one understands these in general terms. In detail, they turn out to be extremely complicated, most of them, but they undoubtedly, unquestionably, are interesting.

And then there is unarguably this highly chemical subject called "life." And life, you can, without any problems at all, pick out sections of life that are chemical, and that are complex and complicated. Metabolism has that characteristic. The area on which I think people are focusing right at the moment is signaling pathways. And those of you who have met Sui Huang who is there, a student with Don Ingber, should chat with him about the issue of signaling pathways. He has done some spectacular work in that area.

Now, the characteristics of these systems are that they are systems that are typically fed back, so feedback is an essential part of understanding them -- nonlinear -- and then one of the characteristic structures in this area is what chemists call "a futile cycle" which is schematically outlined here A, B, C, D, E -- these are all chemical species of some sort.

And if one has a process in which B is being converted catalytically to C, and C is being converted catalytically back to B, that seems to get nowhere.

But, it has the nice characteristic that if you have something that's an A that modulates this, then it can shift the balance between B and C. C then modulates a corresponding process down here, D to E, and so on. That kind of cycle, which is called a futile cycle -- the conversion of A to B and B to A -- is a structure that is ubiquitous in biological systems, and used very prominently in high amplification, nonlinear feedback loops and systems. So it's a structure that's a very interesting structure.

And at the core of all of this are structures of this sort. This is an enzyme schematically. And the point of interest in it is two things: one is historical, and one begins to come to the subject of the talk.

That is that if you look at this, it is a beautiful, complicated structure. And as you all know, it's formed by taking a string of amino acids in a polypeptide, and having those fold up into the structure that is schematically represented there.

What we tend sometimes in chemistry to leave out, is the fact that it does something. You know, it recognizes a substrate, and then it catalyzes a transformation. It's the dynamic aspect that is really important, the crucial aspect.

And chemistry is making pretty good progress in understanding how to go from the linear sequence, to the tertiary structure, to the three-dimensional structure. What we are not making very good progress in is understanding the details of the catalysis. That is, we have a very hard time right now designing catalytic networks, or designing catalysis in any form.

And catalysis, which is the dynamic part of systems, is the key to many of the issues in connecting molecules to the complex systems that are obvious in life.

Now, let me raise another couple of points. One of them is really, to me, the central issue, and that is, is emergence a useful concept or is it not? There are two ways of looking at emergence. One of them is to say that emergence means the properties or behaviors of things that emerge because of collective behaviors. And that says that emergence tells you something about something that is genuinely new, and that obviously is very interesting, wherever it occurs.

The other way of thinking about emergence is to say it's just a code phrase, which means that we don't understand the physics well enough to make predictions.

Now, I happen to be very much on the side of thinking that emergence is an interesting property, but I argue, particularly with physicists, about this, and they say chemistry has always been a field with sort of so-so intellectual standards.

(Laughter)

That's what you'd expect of a chemist.

And my response in turn is if you can go by some physics-based process from DNA to the B Minor Mass then you've got me. But until you can do that, well - - (Laughter) – so somewhere along there, something interesting happens.

But I think the question is a legitimate question. And one of the subjects that I want to return to a little bit later, is this. And I'm going to show you a system – this is a system developed by Bartosz Grzybowski -- which was designed to have the characteristic that it is dynamic, so on, and so forth.

The question that we were interested in asking in this system is in a system in which we understand as much as we can, and we have several particles interacting with one another, at what point do new properties begin to emerge? That is, at what point do we begin to see behaviors that we were not able to predict, whether we understood the physics or not?

And the answer is to me, disconcertingly early. That is, some very simple systems, show some very complicated behaviors. And it may well be that this --

M: George? In defense of physicists, and to save your voice for a couple of seconds, some of them are bit more modest. There is a very wonderful article by Phil Anderson called "More is Different" in 1972 in *Science* where he takes a far more modest position that's much more in line with yours.

George Whitesides: Right. I know, and David Pines does too, and you know, but I think of these as the good physicists. So I appreciate your defense of the field.

Now, another sort of background subject, and that is that I think many of us are here in the room because in the back of our mind is the idea that life is interesting. And what is it that one needs to begin to replicate in life?

Everyone has their own check list, and the one that I use is here. Stuart Kauffman proposed another one yesterday. They all turn out to be the same thing, and you can use what happens to be easiest for you. My view is that what one is interested in is something with these four characteristics. And I don't know whether some of these can be disposed of, these are the characteristics of biological life. It's compartmentalized, and it's compartmentalized, I think, for second-law reasons. You can't have what's inside the cell diffusing over the ocean; it has to be kept in a cell, so it's compartmentalized.

Energy-dissipating. This is one of the aspects of complexity which chemistry has not been very good at dealing with, and it's an area that we're trying to learn how to design. So this is a focus of ours. Self-replicating. In fact, there are no self-replicating systems that aren't knock-offs of the biological systems. So that's an area where one has made almost no progress up to this point. And then adaptive. And "adaptive" is one of those words which you sort of don't know what it means. What it really means, if you think about living systems, is that the system tends to adapt its internal homeostasis in such a fashion that the outside world can change, but it continues to function. That is, it keeps the inside constant, while the outside changes.

The sort of gaia approach that we heard about yesterday is the other way. That's the inside changing to make the outside stay constant. That's a different issue. I think we're concerned here with systems that are adaptive, in the sense that the internal networks change to maintain a homeostasis in the system.

Now, one last bit of background. We, again, tend to think about complexity in terms of the set of words that I show up there. Thus, complexity is that area which falls in between simple linear behavior, and chaotic behavior. And one of the systems that Abe Stroock has dealt with, and it's over on the poster, is microfluidic systems, where one can --

Fluidic systems have the virtue, in general, that one can go all the way from laminar to turbulent. You can go from linear behavior to chaotic behavior, and it's wonderful from that point of view. Fluid systems are wonderful from that point of view. But this is the area where I think we'd like to focus. And then down here there are two words: one of them physics-related, one of them more chemical.

Chemistry tends to work in self-assembly. Self-assembly is the construction of components which can be molecules or particles, that form structures without external intervention. I don't get in and tinker; they form their own structures. And we're making very good progress in this kind of thing.

It's very closely related to self-organization. One can argue that the difference between these two is that the chemical self-assembling systems tend to be close to equilibrium, the self-organizing systems are a little bit further from equilibrium. But that distinction between how far are you from equilibrium is an interesting metric that one can use in thinking about this area. It hasn't been, so far as I know, very well-quantified. But obviously, the further you are from equilibrium, in general, the greater the tendency will be for something to happen that brings you to equilibrium.

Now, let me begin to talk about what we're interested in. I'm going to start with the simpler problem, which is the self-assembly problem, and there are a variety of scales of interest that are relevant here. If we start over here, this is a molecular scale, and this is simply a hydrogen bonding network, of the sort that is typically found in DNA. It's a slightly different one. So these are structures that are one to a few nanometers in size.

There are self-assembling structures at all scales. This is simply a crystal of polystyrene latex beads. Those are micron scale beads. One can see self-assembly up to quite large structures, and I will show you some quite large structures. But these are all, probably, close to equilibrium structures. And the characteristic of all of them is that they are held together, they are held, or the particles are held, at a certain distance from one another.

Now, when you say something is held close to something else, what one is thinking about is generally a bond. Now, we're all familiar with the concept of a bond in the sense of a chemical bond, and it is whatever holds two hydrogen atoms together to make a hydrogen molecule. I should point out that although everyone is very comfortable with the idea of a bond in a molecular sense, if you stop and ask what is really going on there, you very quickly find that you're in deep trouble. Because as a chemist, the older I get, the less I understand why this works.

It's not straightforward. The energetics are determined primarily by Coulomb's Law, we know that. Whether quantum mechanics is really important is another issue. And then the relationship between kinetic energy and volume of space that's available for the electronics to move in, and all of the rest of that -- it is a thoroughly non-intuitive process that just holds together.

But in a general sense, one can say that there is a bond, and we'd like to generalize that concept into a structure that will hold together other kinds of objects. And that's one of the things that I want to talk about.

Now, I'm going to show you two systems. When one thinks about a bond in general, one can think in terms of a potential function. Here the two components held together at some distance are -- what is that reflecting? It reflects a potential curve which looks like this, which has the characteristic that at this distance, the system is at its lowest energy, and hence, in a dissipated structure, it is stable at that approximate distance.

Now, this curve may represent a single force interacting in an interesting way, or it may represent competition between two forces: a repulsive force, and an attractive force. And I'll show you examples of both.

The example of this is the first part of the talk, which has to do with the assembly of two pieces into structures, or multiple pieces into structures, at the liquid-liquid or liquid-air interface, by capillarity. And let me briefly explain what capillarity is.

It is a characteristic of condensed matter that interfaces are always less stable than bulk. So every system will tend to minimize its exposed surface, and I can go into why that might be, but the general notion is that if you think just about a liquid-air interface, a molecule at the interface has other molecules around it only in half of its universe. The ones that are in the bulk have molecules around them everywhere.

And since van der Waals interactions are attractive at some distance between everything, if you have molecules around you everywhere, it's a more stable situation than if half of your universe is exposed. So these molecules, the interfacial molecules, are less stable than the bulk molecules, and the system will always tend to try to minimize its surface energy by minimizing its surface area.

And what we are going to be working with, then, is particles about a millimeter in size, that float at that liquid-liquid interface. And the sides are wetted by one of the liquids, and form menisci. And if you think about these two particles coming together in this fashion, what happens is that the liquid interfacial area is minimized by that process, and that minimization lowers the overall free energy, and that is the driving force. We call that a capillary bond.

So here is a bond that we are able to design, that holds two things together. The virtue of it relative to a chemical bond is that it is purely classical, so it's easy to understand, in principle, and it's easy to see what's going on, because the particles are big enough to look at. I'll show you some examples in a moment.

Now the second system that we work with -- and this is sort of stepping off in the direction of energy-dissipating -- is the one that's shown here at the bottom. And what this system is going to be composed of is small disks, disks that are about a millimeter in diameter, and they have a magnetic core. And we will cause these disks to spin around their axis at about 1,000 cycles per second, while floating at a relatively low friction interface, which is a liquid-liquid interface, in the presence of an average magnetic field which is dished this way.

And because the field is dished, these will tend to come together. And because they're spinning, they produce a vortex, and vortex-vortex interactions are repulsive, so they tend to stay apart. And that provides an attractive interaction here, and a repulsive interaction, which is balanced to give a characteristic distance. And the question is, what kinds of structures emerge

176

from those behaviors? So I'll talk about both of those.

Let's start with the simpler case first. I'm just going to show you a few pictures to give you an idea of what goes on in this area, what is now possible. These are structures held together by capillary interactions of objects floating at the, in this case, water-perfluorodecalin interfaces. It starts with this kind of object, and this is about a millimeter from here to here, just to give you an indication. It's a hexagonal plate about a millimeter thick, and a couple of millimeters across, arranged in such a fashion that this edge is hydrophobic, the dark edges are hydrophobic, the light edges are hydrophilic. So there are menisci here and here, and here and here, and not here.

You put a bunch of those particles together, and they form structures. So this comes together and it forms a crystalline lattice of that sort. Pieces of this sort form lines of the kind that's indicated there. So these are structures that are self-assembling. They are probably close to thermodynamic minima. They are the equivalent of millimeter-sized particles of crystallization in molecular crystals.

Now, this has developed into a fairly flexible method of doing things. It's also an interesting subject from the point of view of this conference in terms of its complexity, or its complicatedness. And the issue here is that in these capillary forces, one can fairly easily write down an equation which describes the interaction between them -- between two pieces, these objects interacting through capillary forces.

The trouble is that you can't solve this equation for anything other than a very simple case, that is parallel slabs.

So it has the characteristic that it is relatively simple -- we know what the physical interactions are -- but it's also complicated enough that we can't do an analytical solution for it. The pieces, on the other hand, can do an analytical solution. I mean, they come together. They know what they're doing. So this is, if you like, a kind of analog computer, these systems are a kind of analog computer, that solve this in a multi-particle system, and they do a pretty good job of it.

Now, there is some quite sophisticated bond engineering that one can do in this kind of system, and let me just give you a couple of examples. Here are some examples of hierarchical self-assembly. If we take pieces that look like that where the dark areas are where the menisci are, these come together quite reliably in this fashion. We take a piece that looks like that, it crystallizes into a lattice of that sort. But we now take pieces of this kind, and mix them with pieces of that sort, and there is a tendency for these preferentially to form the first trefoils of this kind, so we see a structure that looks like that.

And if we take that kind of structure, actually with this structure, these first come together to form trefoils, and then the trefoils come together, simulating that behavior, to form extended lattices of this kind. A quite high number of defects, but you can see the structures that are forming.

This is the equivalent of, if you like, the beginning of a periodic table. Things are forming selective bonds between different components in the system in a way that's under quite good control, because we can control the shapes and the shape of the menisci. And you can extend this to phenomena that begin to look like phenomena that are actually fairly complicated, in a chemical sense. And here is an example.

I showed you a moment ago that if you took these sorts of structures and put them together, they tended to form a lattice, but the lattice was defective. If you include in this an inert sphere, just something that is space-filling in this two-dimensional world, what happens is that this is included in the lattice that's formed, and you get quite large, relatively defect-free crystalline lattices this way. This is the precise analogy of something that's called an inclusion compound in chemistry. Gas hydrates are an example, methane plus water forms a crystalline solid that is quite stable. And in that particular three-dimensional structure, these things would be water molecules, and those would be methanes. So one is seeing behaviors that are very much like behaviors that you see in molecular systems. And one can begin to do engineering of this sort. I'll just give you one example.

Suppose you want to make a two-color grid spontaneously. How does one do that? What we do is to take pieces that look like this, and we make the faces chiral so-called; that is, they have a handedness. And you think about this as being, of this object, as floating at the interface, with the dark patch here being hydrophobic. Looking at the face from the side, what you would see would be a meniscus which would be tilted at a characteristic angle relative to the aqueous phase.

Now, if you tried to bring that up in contact with another one of the same sort, what you would have would be that interacting with a meniscus that looked like this. That doesn't fit. On the other hand, if you have two faces here and here with opposite chirality, those then interact perfectly to come together like this. So that by controlling the chirality of the interface of the edges of these pieces, you can control the way they come together, and here you get an AB crystalline lattice. And that is, you know, an equivalent, if you like, of sodium chloride, or something of that kind.

And I'll just show you one example of three-dimensional systems. The three-dimensional systems actually work better than two-dimensional. What one does here is to take polyhedra of this sort -- roughly this sort -- put solder in the experiments that we've done. We've put solder on the ends of selected faces. You suspend those in a roughly isodense medium, tumble them. The solder comes into contact, liquid solder, capillary forces there pull the pieces into assembly, and you end up with extended three-dimensional crystalline solids. And because these have contacts in all three dimensions, they tend to be less defective, and have more, larger extents without defects than the two-dimensional systems. So, one can make all kinds of nice things here, and that's terrific, but these are all the equivalent of crystals. I mean, they're not alive, nor is anything else we've made.

But at this point, we set out to address -- we, being Bartosz Grzybowski -- set out to address the questions I mentioned to you, of can one build energy-dissipation into these systems?

So our initial idea was the one that I showed you. We take – here is a dish. And in this dish, we float a small polymer disk – there is the polymer disk -- which is, the interior of this is just a polymer that is molded with some magnetite in it.

Now, underneath the dish we put a stirring bar, and we spin the stirring bar. This magnetic field induces a magnetic moment in that, and because this is spinning, the magnetic moment here tries to follow the magnetic moment there, so that if this is spinning at 1,000 Hz, this disk sits there, and it spins on the surface the liquid at 1,000 Hz. So it's energy-dissipating. A spinning disk, as you all know, is a pump. It takes liquid from the bottom, squirts it out the sides, and it generates a vortex. So these are small vortex generators.

Now, the average magnetic field experienced by this disk is slightly dished so that these, left to their own devices, tend to come together towards the center, but the vortex-vortex interactions are repulsive, so they tend to push things apart. So that's the system. Now, it has the characteristic that we know everything about it. It's vortices interacting. It's small particles slipping down a magnetic slope, so it's very well-defined. On the other hand, having said that, as all of you know who know something about fluid dynamics, if you don't know anything about fluids, it is really hard to calculate anything about fluids. So we know everything, but the question is what do we know? And the answer is, it's very hard to calculate these. We work with a guy named Howard Stone who does this kind of thing professionally. And I think once he gets beyond one or two particles, he feels pretty much lost.

Now, what I want to do is to show you a couple of patterns that are formed from arrays of these things, and ask the question, "At what point do these systems begin to slip out of our intuitive control? When do we begin to see things emerge that we wouldn't expect to see emerge?" And the answer is, as I indicated, distressingly soon.

Let me just take you through a set of patterns. These are, you would say, all sort of obvious. We have four of these disks. They form a square. There is the center. The system V forms something like this, VI, VII is a nice hexagon, VIII, IX, X does something a little odd, because there are two structures here which are in equilibrium. It depends a little bit on the rate of spinning.

So these two are both stable structures, but the system will flicker back and forth between them, so there are two attractors there. And that's all fine. And you can make quite large arrays here. I think the arrays that one forms are limited at the moment primarily by our ability to make a relatively smooth magnetic field over a large area. But you see, you go up to larger structures, and there are magic numbers that appear here. This is a very nice, flexible system to work with.

And here is the problem. What we do in this experiment is to take a somewhat larger disk -- here is a disk -- and we put in six smaller ones of that sort. Now, what happens here? We take this system, and we have the rotational speed what it is, and you form a structure that looks like that, very much like the seven small disks that I showed you earlier. You decrease the rotational speed slightly, and it spontaneously forms this structure – that is, it breaks symmetry – and what you get is a kind of structure of this sort. Each of these disks, I should remind you, is spinning at, in this case, around 700 or 800 RPM around its own axis, and then this set of disks here forms a kind of comet tail, which slowly precesses around that central disk.

Now, the question is, "Why does it do this?" and the answer is, "I don't know." Now, the question is, "Is that emergent behavior?" and the answer is, "I don't know." I mean, there is nothing here that's not simple physics. On the other hand, I can't predict it, and I'm not actually sure that it's predicable because we have so much trouble doing even simple problems in fluid mechanics.

So this is, in a sense, an example of a classically complex problem that has the characteristic that there is nothing hidden here. We know that this is physics. On the other hand, I am not sure that it's a predictable phenomena, at least easily predictable phenomena, based on current levels of physics.

So that raises, or let's put it this way, it blurs the boundary between complexity in the sense of true emergence, phenomena that somehow have to do with the step between DNA and the B Minor Mass -- and things that are just physics that we don't understand, or we can't calculate. It's a real cautionary example for us.

Now, let me finish off by showing you another system that we've worked with, to make a point about this kind of activity. This is work by Rustem Ismagilov. As I indicated, our approach to this problem is to try to do we call in the group "synthetic complexity." That is, we look around in the area of studies of complexity, and we find that people have tended to focus on a relatively small number of systems. There are sand piles, there are BZ reactions, there are a couple of other things – that have been very carefully studied.

The problem, the risk in science with doing this, is that you look at a phenomenon -- and of course you love your phenomenon. I mean, who could not love his phenomenon? (Laughter) So you tend to try to explain the entire world in terms of your phenomenon. This sometimes works, and it sometimes doesn't work. It's good to have a range of phenomena so that you know that there are others out there.

So our idea, since we know how to synthesize things, design and synthesize things, is to make a bunch of other systems that have the characteristic, that they also show complexity. I think these disks are going to be a very nice set of systems to work with.

Let me show you another one. And I'm showing you this to give you an indication of what the problems are in designing new complex systems, and perhaps, to ask for help. Both Rustem and I would be very interested in having your suggestions. The idea here is to look at another energy-dissipating system -- we heard about this briefly yesterday – which is the Raleigh-Benard kind of convective systems which in this case operate by taking a dish, and you heat the bottom of the dish. The fluid at the bottom is hot, the fluid at the top is cooler, and one develops spontaneous convection patterns.

There are two reasons for this. The reason that comes immediately to mind is that you see convection-driven instabilities. That is, the hot fluid is less dense than the cold fluid, and it rises. That's not what goes on here. What goes on here is that the fluid at the top, because it's cool, has a different surface tension in the regions in the center, than in the regions where the fluid is emerging from the bottom.

So that these systems, the Raleigh-Benard systems, are driven by differences in surface tension. It's a technical detail, but it's something that's worthwhile keeping in mind. Now, it's energy-dissipating, so that's fine. It's controllable, as I'll show you in a moment. And one can imagine building in feedback.

But the issue here is first, can one really make this system into a controllable pattern? And the approach that we've taken to this is to say that if we want to build patterns which would give us the potential, in principle, for building in feedback, what we want to do is to impose on the natural convection cells some boundary condition.

And what we have done is simply to pattern the bottom of our dish with posts, or things of that sort, and then watch the development of these cells. And you can see, you can lock the cells -- these are pictures taken with an infrared camera -- you can lock these cells very nicely to the posts. So left to its own devices, you would see convection cells develop spontaneously, but they would have their own pattern, whatever it might be.

Here is a square array, and there is a square array of convection cells. Here is a hexagonal array and a hexagonal array of convection cells, and that's fine. In these particular systems -- let me go back to the previous slide -- what we're looking at here is sketched here. Here are the posts, so that the posts, the fluid goes up, and then goes down in the center, so that in the systems that I've shown you, you can see the posts here. The fluid is coming up here, and going down there, so that is the pattern of fluid flow here.

Now, this system shows quite nice behavior in a couple of ways. I'm just, I don't want to go through this in detail, because of time. But here is an example. This is the same array of posts here and here. When the oil that we use is shallow, the natural size of these convection cells is of the dimension of the thickness of the fluid. So that when you increase the dimensions of the thickness of the fluid, you will tend to go to larger cells. •

With these kinds of posts with a thin, shallow layer of oil, you lock the cells to that dimension. And this will undergo a phase change as you go to a deeper layer of oil, in which you now have a different lattice on the same set of posts. So here is the lattice here, here is the lattice here, and this depends upon the thickness of the cells.

M: Do you see frustration and disorder regarding the depth between the two?

George Whitesides: No. It's a sharp transition, which is interesting. I mean, it is actually remarkably sharp. I don't have a picture here, but it does just what it's supposed to do.

Now, if you release a constraint -- and here are our not posts but lines -- here we have the same pattern this way -- What it will do as you go from shallow to thicker, it adjusts the dimensions of the convection cell laterally where it is unconstrained, but not this way where it's constrained. So it's doing all the right kinds of things.

Now, here is a system that has the characteristic then that it's energy-dissipating, pattern-forming. One can control the patterns, and we will be able to build in external feedback. Is it a good system to work with? And what we are concluding is that it probably is not a good system. It's not a good system for a couple of reasons.

One is that the cells are relatively large. They are a millimeter. We would like to be able to ultimately bring these things down to perhaps micron size, to be able to get statistical physics on them. The second is that convection is, relatively speaking, slow. And because things are slow, that means that the development of the structure is slow. And I don't know that these are essential characteristics, but I think as we as a community begin to think about developing complex systems for the future, one needs to add things that are not on my original list of what's characteristic of life -- compartmentalized, adaptive, energy-dissipating, self-replicating -- but we also need to add a few other things that have to do with convenience.

I think we want the potential, at least, for going slow. We want the systems to do what they do at high frequency, and we want them to be easily observable. That way, one can probably do a lot of science. So that is the end of the subject. And as I say, what I would really value from this audience is a discussion of some of these central issues of does complexity make sense, is there a "there" there, is emergence a useful concept? Is there a value in making new, complex systems, or is this basically a waste of time? All of those sorts of things. And my colleagues over here, and if Bartosz is here, I don't know, is Bartosz here -- Bartosz is in the very back there, that's Bartosz -- will be happy to answer questions or comment on things. Thank you very much.

M: ... do any experiments where you tilt the data so that you have shallow oil on one side, and thick oil on the other?

M: -- why don't --

M: -- an interesting example of (competition) --

M: Why don't you --

M: ... in the next session, I'll show you some of these things --

George Whitesides: But the answer is yes. We've done that, and this was the basis of the answer that I gave David. That is, you get a sharp transition from one lattice to a different lattice, really quite sharp, as the thickness of the oil changes. So it's a --

M: ... I thought you were just adding more oil.

George Whitesides: In that experiment we were, but in general, you get a nice, sharp transition. Yes?

M: Going back to [??]. It seems to me your early question about emergence, about whether emergence was the acquisition of new properties, or simply that we don't understand the physics underneath. I'm a biologist, physicists and biologists are even worse [??] --

M: ... you don't know any of that, which is true. But it seems to me that there is a question one should ask about the level underneath. How much do we know the physics under the chemistry? Even if reductionism is perfectly true, if revision -- I remember in Moscow two weeks ago -- if revisionist crypto reductionists like [??] are actually right, and all of the phenomena at the chemical level are, in principle, explained by physics, is it that we know all of the physics? Because if we only know a bit of the physics, the chemistry isn't going to restrict itself to the physics we know. And as an operating principle, it won't be any good to try and explain the chemistry about the physics, because we won't know the physics. So you don't know of your scrutiny and then do things either we don't know we should, or we don't know, and we can't. And it seems to be to be a very different question. Would you comment?

George Whitesides: I can't comment any more than in the sense you already have. The question is how much does one have to know in order to know enough?

M: Right, exactly. Thank you for that.

George Whitesides: And the problem is that in these areas -- I mean, we heard a splendid example yesterday in the discussion of water. I mean, water is -- I stand before you as a sack of water that happens to do odd things. (Laughter)

And if you say, if you're going to understand life starting from the physics up, certainly you have to understand water. People have been looking at water for years, and we don't understand water at all. I mean, we really don't understand water.

Now, we understand what a hydrogen bond is, sort of. We certainly understand what an oxygen atom is, sort of. But to say that we understand these in principle may mean that we have -- We're doing really well. We've got a tenth of a percent accuracy in the potential function. That may be orders of magnitude off from what we need. We may need to note everything --

David Campbell: There are a whole bunch of people with other questions, sorry. I apologize to those who have been waiting patiently by the microphone. Please.

M: Yes. Could you put your last transparency up, because I had a very specific question on that.

George Whitesides: My last transparency -- You know, when you ask someone of my age that, it's a risky proposition.

M: On the upper one, what I noticed was that on the lower one the squares are offset, and on the upper one, on the right hand side, what you have not emphasized, they are also offset.

George Whitesides: Right.

M: And I was just curious as to whether there was any explanation of why that would happen, particularly in the upper slide.

George Whitesides: Well, I think in both of these what you are seeing is a certain -- I mean, you're seeing the system is rigidly locked this way, and there is a certain amount of flex in the size of the cells laterally. Because there is no constraint in this direction, right? Now why there is a little variation from cell size to cell size, I don't think we would know right now.

But if you look at a Raleigh-Benard system, the cells in that are not absolutely regular, and those may be slight heterogeneities in surface tension, they might be slight differences in rates of heat transfer into the bottom. These are things that are very small.

M: -- variable when they go through time, or not?

George Whitesides: ... are the cell sizes here laterally variable with time, or do they rigidly lock once they form?

M: They will. Once they form, they will gain more footage. It will take a long time for them to come to the surface structure. And if there is any [??] start moving, and that could take hours to settle down. This might help [??] structure after an hour, but usually [??] several days [??].

M: My background is in bifurcation theory with symmetry, and nothing that you've shown me here is the least bit surprising. I think that absolutely every single one of those things can be -- The existence at least, if not the stability of all of these solutions, can be understood in the context of representative theory. And I think that's something which I haven't heard too much. I'm new to this subject, so I don't know that much about it. We may not know all the physics underneath there, but we know that it's SO_2 symmetric --

M: -- symmetry --

M: That's right. And I think that gives us a handle here. It might give us some sort of -- Or it helps us classify those things which might be possible. And even if you've been living the good life, you can even sometimes establish stability results by virtue of getting some Taylor coefficients. If it's a local bifurcation, you can do quite a lot using that center manifold reduction, or a Lyapunov-Schmidt reduction. Have you done any of them? It is really a pretty good industry of people like Golubitsky and Stuart have done a lot of this stuff in Raleigh-Benard and other systems. And I think it could be applied quite directly to some of the --

[?? overlapping dialogue, inaudible]

George Whitesides: -- In the Raleigh-Benard systems I don't find these things surprising per se. What is more surprising, the surprising result in all of this was symmetry-breaking in that set of seven disks. Now, is that obvious to you?

M: Well, you still have -- It doesn't surprise me. None of the patterns that you showed me surprise me there, including the ones where the comets go around, because that's a spatio-temporal symmetry, where you essentially combine the rotation with the time translation. So yes, I would say again, the existence of those solutions is not surprising whatsoever.

George Whitesides: Well, but the existence of physical reality is no longer surprising.

M: If you gave me a couple of hours, I could have listed that as one of the possibilities a priori of the behavior. So I think the idea is without knowing --

M: We're running into philosophy --

M: Okay, without knowing -- The last thing then is when a system has symmetry, if you have that taken into account, it's very difficult to understand the generic properties of it.

M: We know that is [??].

M: Yes. So I think this may be a case where those sorts of considerations may be called for.

M: Let's go to this side, since [??] apologize.

David Campbell: You think you've got it tough? We're in the social science business. (Laughter)

M: But, this question may sound silly, and maybe it is, but I'm not sure. You said the little bits and pieces you were making play with each other and form structures, they have the analytical solutions to these problems, and they do it all the time.

M: Right.

M: Now, whether they're just super mathematicians or what, we don't know. But if you could interview them and ask them about why they're behaving as they are, would it help you understand the dynamics of that system? Because we can interview our bits, you can't interview your bits in chemistry and biology.

George Whitesides: Well, we can interview them in the sense that in principle we can ask them what forces they feel at any given time. We can ask them what gradients they're moving along. The trouble with these systems is -- One of the characteristics of these systems is that they are at zero degrees, a temperature of zero. And what one means by that is at least in the absence of stirring, the Brownian motion which keeps particles going in solution, the thermal bath, doesn't exist here. So that left to their own devices, the agitation that pulls them apart is weak compared to the strength of the things that holds them together.

And that introduces some complexities in these systems in terms of locking things in in meta-stable states. This doesn't, strictly speaking, answer your question. Your question is you have much more complicated behaviors that you observe, and much more information that you can get out of them. We have simple pieces, and very limited information we can get out.

M: But if you could interview those people, in those simple cases, would that help you or not?

George Whitesides: Yes, it would help, absolutely.

M: What questions would you want to ask them?

George Whitesides: The question I think you'd like to ask is what do you see out there? What is there? What is your universe, and why are you responding to it the way you are? What causes you to form the structures that you form, rather than some other structure?

In the case of most of these little hexagons, I think the answer to that is it's the most stable structure. But in the case of these spinning disks, the issue there is decidedly unclear. I mean, despite the fact that one can, in principle, talk about symmetries, the fact of the matter is that we would not have predicted that symmetry broken structure. So I'd like to know why they do that.

And talking to good fluid mechanisms, they can't tell me why they do that, even though they can also see the symmetry issues. So, I mean, I think we really are open to questions as to any way that one can get information as to what motivates these things, if I can put it that way.

M: I'll be really brief, okay? There is a very interesting problem in microbiology which is how does a cell position its site of division? We don't know how it positions it, nor why it does it when it does. And we have speculated, perhaps foolishly, that maybe this has got something to do with the formation of an interface between different proteolipid domains within the plane of the membrane.

Now, it's very difficult to investigate that. There are hundreds of different types of lipids, and thousands of different types of proteins. Could one of your systems -- and you've shown several -- be adapted to address the fundamental principles involved in the formation of domains, the formation of interfaces, and so on?

George Whitesides: Yes. The answer to that is I am sure it could be done. Don Ingber and I -- Don talked on Sunday -- have done a lot of work in patterning cells on surfaces. And in that kind of thing you can take, by various techniques, you can make a square block of something, grow a cell so that it's square. If the block is big enough it will divide. And you could, in principle then, make that square such that half of it was A and the other half is B. That is, impose, regulate the size of the cell and shape of the cell, and then do the chemistry in such a fashion that it would reflect that.

M: I'm sorry. I did not ask you the question properly. There you're talking about a real cell.

George Whitesides: Right.

M: I'm talking about your model system. Were you looking at the classical things?

George Whitesides: Absolutely. We can certainly do that.

David Campbell: Let's thank George again for a very stimulating talk.

(Applause)

Irv Epstein
From Nonlinear Chemistry to Biology

Irv Epstein: I think there are some common themes with the previous talk.

I'm going to talk about chemistry. I'm actually going to talk about synthetic complexity. Since I only have 35 minutes, I won't bash physicists.

But what I'm going to try to do is to give you --

M: That's because you are one. (Laughter)

Irv Epstein: Different people think I'm different things. What I'm going to try to do is to give you an overview of an area called "nonlinear chemical dynamics." What we synthesize are basically systems that show complex dynamics, rather than complex structure. And then I'm going to talk about a specific example that introduces feedback, a theme that was raised at the end of the previous talk. Along the way, I'm going to commit the fallacy of suggesting that maybe some of the things our phenomena that we see in chemical systems, are relevant to biological systems. But at least I know it's a fallacy.

Now, unfortunately I was working with another complex system, yesterday and I wasn't here. Did people actually see the BZ reaction?

M: Yes. I'm doing it again this afternoon.

Irv Epstein: Okay. I have a different reaction, different color scheme. Shall I --

M: Yes --

Irv Epstein: All right. Now, the question is where can I put it so it will be visible?

M: On the outside.

Irv Epstein: Yes, here we go. I don't need the --

David Campbell: While Irv is setting this up, I will announce again the [??] Prize which has been unclaimed for 20 years, which is to make a drinkable chemical [??].

David Campbell: The prize is ten bucks, and a handsomely engraved certificate, and I get the right to [??] your drink to the entire [??] bar.

M: Excuse me, mine is drinkable.

M: Is it delicious?

M: Drinkable, fine, but you die. (Laughter)

M: That's a serious [??].

Irv Epstein: All right. This is meant as motivational more than any thing else, but it does illustrate some important points.

When I got into this business, I was actually hired to do quantum mechanics, and to worry about those bonds that George was talking about. But I got lured into this largely by watching reactions like this one. I hope you can see it from wherever you are. In these days of environmental protection --

M: Light.

Irv Epstein: This? I don't think this is going to help me. You've got to watch what's going on in here --

M: The light helped.

Irv Epstein: This helped?

M: Yes.

Irv Epstein: All right. Let that go for a little bit. When I started in this business which was in the seventies -- there we go --

M: -- ooh --

Irv Epstein: -- there were only two known reactions. One of them, the infamous BZ, and the other reaction that had been discovered in the 1920s called the Bray reaction. This one is actually a hybrid of the two that is a little more colorful. They had both been discovered accidentally. And what we did was to say, well, if you really understood this stuff, and it's a different level of understanding than physicists are looking for -- sorry -- you should be able to make reactions that behave this way. So what we did was to think about what was going on, and reach a certain level of understanding, and then build, develop a systematic algorithm for designing oscillating chemical reactions.

Actually, I'm going to show you a movie at the end, but since I do only have 35 minutes, let me suggest if you want to know the details, you should read the book. So there is the book.

M: It just proves [??].

Irv Epstein: We may see life forming then.

M: Right.

Irv Epstein: All right. One of the tricks to making chemical oscillators was to steal from chemical engineers. That is, to recognize that a key to making a system behave in this way is to keep it far from equilibrium.

For 40-50 years before the BZ reaction, chemists, for the most part, didn't believe that this kind of stuff could happen. And in fact, journals tended not to publish reports of this kind of behavior, even when it contained recipes, because it clearly violated the second law of thermodynamics. How could the free energy be increasing monotonically if the color is going back and forth?

Well in fact as you'll see in a bit, the second law will win out in this system. But if you adapt a tool of the chemical engineers, the continuous flow stirred tank reactor, which is basically a flask like this, but it has some tubes going through it. And we can bump fresh reactants in, and then there is an overflow tube to maintain constant volume. Now, you can keep the system far from equilibrium, and it will oscillate indefinitely, or at least until the reservoirs run dry.

In a sense, this is an analog of a living system. It's bringing in high free energy reactants, food, nutrients, and it's emitting low free energy products, and we don't have to discuss that in any detail. But as long as you keep the flows going, the system keeps oscillating, or living.

And by utilizing this relatively simple technology, and a little bit of knowledge of chemistry, we were able to develop a whole set of new chemical oscillators. And here is a sort of periodic table of chemical oscillators. There are a number of families, and there are some family relationships. What I would point out is that here and here were the only two known oscillators, again, both discovered accidentally before this method came along, and here is life.

So, now if you run a reaction like this, not in a stirred flask, but in an unstirred system like a petri dish, then something else can happen. You start to get pattern formation. There we are. Okay here, and unfortunately the top won't fit on there, but this starts off as a homogeneous dish of red -- [?? overlapping dialogue, inaudible]

All right. (Laughter) Here is what started off as a homogeneous dish of red liquid. A little dot of blue appeared, and then it grew into a ring. Then we get concentric rings. And then when the rings come together, they annihilate each other, and you get these very complex patterns. And here is a close up of one of these patterns.

Now, what we see here, depending on the number of defects or dust particles in the system, and whether we shake it a little bit, we either get so-called target patterns of concentric rings, or we get spiral patterns. In the next slide, you can see -- Well, can you see the one on the left? The one on the right -- This is too multi-media for me. (Laughter)

Okay, this one is the BZ reaction, the one I showed you on the previous two slides. This one, which takes a little more imagination or younger eyes, is a slime mold. This is Dictyostelium discoideum which when it's deprived of food, aggregates ultimately to form spores. And if you stain the medium -- and I guess they didn't stain it all that well -- you can see these spiral structures which look remarkably like these, although as one of the questioners suggested, I think the symmetry of the system may have as much to do with the actual structures as the biochemistry, but it's still rather suggestive.

Now, let me -- let's get rid of this slide now. Let me show you the level of complexity from a chemist's point of view that we're dealing with. This is a mechanism that is a list of the individual chemical molecular level reactions that contribute to a particular chemical oscillator. This one is a purely inorganic reaction. This is rather typical, and we've done studies to try to unveil the individual molecular details of these reactions, but I don't want to spend any time on that today. I just wanted to give you a sense of the complexity.

Let me focus for the rest of the talk. The pattern formation phenomena that I showed you just now are traveling waves. I'll show you a movie in a little while, but those patterns move around in the system. Now, another sort of pattern formation phenomena that's been of considerable interest are stationary patterns. In 1952 Alan Turing published a seminal paper on morphogenesis, and suggested a mechanism by which the combination of reaction and diffusion could lead to pattern formation.

His ideas stimulated theoretical biologists, and people in fields from ecology to astrophysics for almost 40 years. But there was no definitive experimental demonstration of Turing patterns until the early 1990's when Patrick De Kepper in Bordeaux, using one of the new chemical oscillators that he had actually designed when he was a postdoc in our lab, showed very clearly Turing patterns in a chemical reaction, the chloride iodide malonic acid reaction.

And here are some examples of Turing patterns in this chemical reaction. We've given them clever little animal names. But basically, they consist of different arrangements of stripes and spots, and here is another with some computer-enhanced color.

Now again, focus on stripes and spots, and one gets bifurcations among these various patterns by changing certain parameters, for example, the temperature, or the composition, the chemical composition of the system. Here is a fish. This is -- and I never remember whether it goes this way or that way --

(Laughter) I'm sure the fish remembers.

This is a tropical fish called -- well, I only know the French, but it's either the box fish or the chest fish in English. One of these -- and again, I can't remember which is which -- but one of these is the male, and one of these is the female. Again, presumably the fish remembers.

But the point I want to make here is that one of them shows this pattern of spots, and one shows this pattern of stripes -- on here it's not just symmetry. And basically, the difference between the male and the female to a physicist's approximation is a control parameter which is presumably some hormone level analogous to the change in chemical composition going from here to here. So --

M: May I say just one little thing here?

Irv Epstein: Yes.

M: It's ever so dangerous to talk about diffusion here. I'm absolutely fine with these Turing things, but the long period of time when you said nobody was doing anything, what was happening was lots of us were showing that many biological systems did not follow the Turing pattern. And the fish aren't diffusion because there are blood vessels in their skin. It's something comparable to diffusion, but don't say diffusion.

Irv Epstein: Okay. I will not use the word "diffusion" in the next 15-20 minutes. No, you have a very legitimate point, I agree. Okay, so let me move on. This is actually in deference to David. Among the phenomena that these systems can exhibit if you push them in various ways, is chaos. So I just wanted to show you an example of chemical chaos. This is the chloride thiosulfate reaction. If you count, you'll see that there is an irregular alternation of large and small amplitude oscillations here. One also has other systems where there is more classic period doubling. These are something we call a period-adding sequence.

I just wanted to again show a suggestive, and I think even less accurate, analogy here. These are signals from a neuron in the stomatogastric ganglion of a crab. Well, just leave it at that, a suggestive analogy.

M: I like that one.

Irv Epstein: You like that one better?

M: It's got a nice [??].

Irv Epstein: All right. Now, let me get into some more recent work, and this involves feedback. And what I want to talk about here is global feedback, that is, a system in which the feedback is governed by the behavior of the whole system, rather than just the behavior at a single point.

How can you do that? One example comes in electrochemistry where the electrochemical potential serves as the transmitter of the feedback. Here you take a piece of nickel and you put on some kind of a potential. The nickel is one of the electrodes in the system, and it's sitting in some kind of acid. And the potential is controlled, and depends upon the behavior of the whole system.

Here we have heterogenous calyces, an example that was referred to earlier, where the information is transmitted by the composition of the gas phase above the metal which is well-mixed, and the biological example involved neural networks where in a theoretical neural network you can couple every element to all of the other elements in the system. In a brain, you might have each element with a coordination number of something like 10^6, which if not absolutely global, is pretty close.

The experimental system that we look at -- being mere chemists -- is actually the BZ reaction. This is a simplified model, where X is the autocatalytic species -- bromous acid -- Z is the oxidized form of the metal catalyst, and the details of these model equations are not critical, and I won't say the D-word ["diffusion"] about these terms.

But what's critical is the feedback here, and I'll show you experimentally how we create it. We monitor the concentration of the species Z, and we have a feedback on X which is proportional to a parameter GZ, the feedback coefficient, and depends upon the spatially average concentration of Z in the system. This is a target value, so the feedback is proportional to the difference between the instantaneous average of Z and this target concentration.

And the way we do this is we run the reaction in a disk of silica gel. Again, this is a photosensitive variant of the BZ reaction. Basically we have a light source that illuminates the reaction and effects the reaction. When light shines through it, bromide, which is an inhibitor of the reaction, is generated in a series of photochemical stems.

There is another light source to monitor the concentration of the oxidized, the red, form of the catalyst. There is a third light source we can use to produce an initial pattern on the system. And here are a pair of polarizers. The angle between them is controlled by the amount of light that passes through the system, and that gives us our feedback. So we monitor how much light passes through, which is a monitor of the integrated concentration, and in effect, we are dealing with a set of integral differential equations.

Now, the behavior we get -- What we're going to be particularly interested in is a phenomenon called "cluster formation." Clusters are analogous to standing waves. Basically, what you have is a small number of domains in the system which oscillate with the same phase -- so picture a sort of square standing wave -- except that there is no intrinsic wavelength. Because of the global nature of the feedback, there is no intrinsic spatial scale. And therefore, you can impose essentially any pattern you like on the system initially, by shining light through a mass, and then that pattern, under the right conditions, can persist and oscillate.

And typically you'll see two-phase clusters or three-phase clusters. That is, think of it as domains that are alternating black and white, and white and black, and I'll show you some pictures.

This is a diagram of what happens as we increase the feedback parameter. At zero feedback -- that is light off -- there is bulk oscillation of the system, the sort of thing I showed you in the flask, but now we're in a disk of silica gel.

As you increase the feedback, pattern-formation begins to occur and you start to see traveling waves like the spirals that I showed you on the slides. Then, at a certain feedback level, those traveling waves break up, and you get what we call standing clusters, which I'll show you in a minute. Then those evolve into other kinds of clusters that I'll show you.

And then finally, at very high levels of feedback, the pattern formation again disappears, and you get small amplitude bulk oscillation. These are large amplitude to start with. There are hysteresis phenomena along the way that I don't want to worry about at the moment.

Here I show you some standing clusters. These evolved out of initial spirals, and I'll show you the movie in a minute. But basically during each -- This is a periodic phenomenon. This is what it looks like, and then half a period later black has turned to white, and white has turned to black. This evolved out of uniform initial conditions, so these are the standing clusters.

Now, if you go a little higher with the feedback, you get irregular clusters, where these flash black and white, but the boundaries move. The domain boundaries move with time, and again, I'll show you a movie in a minute. These are all experimental data. At somewhat higher feedback you'd get localized clusters, and what that means is that in most of the system there is essentially nothing going on, very low amplitude, bulk oscillations, but you'd see these standing clusters in a part of the system.

Now, we're interested in cluster behavior, in part because neurobiologists have suggested that this occurs in neural systems, particularly in reticular thalamic neurons which are responsible for -- or a certain kind of oscillation called the spindle oscillation in these neurons is thought to be responsible for -- drowsiness, and what happens in anesthesia.

We are actually working with some people who are interested in memory. And the notion of localized clusters is a very attractive one. If you wanted to store information in a system, having localized clusters rather than global clusters seems like an attractive way to store lots of information.

Here I show some of these phenomena in a different way. This is a standing cluster, and now this is a space-time plot on a diagonal cut here. So what you can see is that it's temporally periodic, and we go white-black, white-black with somewhat different sizes of the stripe widths here. This is an irregular cluster, and what you see is temporally and spatially irregular behavior. This is a localized cluster, again, this is a diagonal cut, so we just get things happening on this end of things.

Here -- Okay. This shows the evolution of one of these clusters. There are long transients before we get a standing cluster. And what tends to happen, there are curvature effects. The system -- a little bit analogous to surface tension effect. The system tends to try to minimize curvature. And so what tends to happen is that small pieces tend to link up with big pieces, and you get a minimal number of these domains.

There is also something which happens that may or may not be evident from this, that we call phase balance. There is a tendency for the system to equalize the areas of the domains of different phases. Here is another. Here what we've done is to sit at points A and B, and you can see that they're oscillating 180 degrees out of phase, whereas if you monitor the total concentration in the system, the solid line, you can see that it's oscillating periodically at twice the frequency.

Here is one of these irregular clusters. If you sit at a single point, you see a signal like this. Here is the power spectrum. But if you average over the whole system, then you get something that looks nearly periodic, and there clearly is a fundamental frequency. Maybe what I should do at this point is to turn to the video. Then we'll stop.

[Pause in dialogue]

[Video] These are the traveling waves. They start off as spirals. Now we're increasing the feedback, and so the traveling waves are starting to break up. I'll fast forward if I can --

Irv Epstein: ... it's also a sort of chemical Rorschach test.

(Laughter)

M: ... at what level --

Irv Epstein: Yes. The feedback shows at the beginning of each one. That's the one that evolves over random initial conditions. This one is sort of boring. What happened was that we had a [??] so that one half of the medium was illuminated and the other half wasn't, and it just persists. And really, if I could find a fast forward button, I would do it.

Okay. This one had a lot of spots, patterns, a little bit like the one that we saw in the last one. This one you want to go slower. This is three phase clusters, and there is a black, gray, and white -- black goes to white, white goes to gray, gray goes to black. All right. These are our regular clusters --

[?? overlapping dialogue, inaudible]

M: Can you explain what the mechanism is for feedback [??] pattern [??] waves --

Irv Epstein: Yes. Basically, you have this chemical oscillator -- these are localized clusters. Things only happen here, and this area is sort of dead. Basically what happens is you have a chemical oscillator that is inhibited by bromide ions. When you shine light on the system, the ruthenium gets pumped up to an excited state, which reacts with a bromine compound to produce bromide ions, thereby inhibiting the reaction.

We have set it up so that when the light -- When the ruthenium, the oxidized ruthenium is at a certain level, the angle between cross polarizers changes, and changes the level of light coming into the system, thereby modulating the bromide production in the system, and that's the feedback mechanism.

M: What about the global [??] is that correct?

Irv Epstein: Yes. It depends on the integrated -- By the way, the colored ones are simulations, you know, experiments with black and white, simulations

[??]. But these are irregular clusters that grew out of spiral waves –

M: Why do you say [??] dead?

Irv Epstein: Excuse me?

M: [??] what do you think about that? Something that is [??].

Irv Epstein: Yes. When I said that, I meant that there are no large-scale oscillations. Actually, if you monitor carefully, you can see low-level bulk oscillation, but they're not part of the clusters where you have clumps of medium that are at one phase or the other. This I believe is growing into a localized cluster. This transient period, you'll end up with relatively few of the active spots, and just fall [??].

This is real computer time. Yes? And you can see the three colors more clearly here than in the other. The trick is to sort of focus on one area, and then you can see that these patterns are essentially [??]. It turns out that in the three-phase cluster behavior, the boundaries tend to drift, and there is some theoretical work that suggests that three-phase clusters shouldn't be stable, or two-phase clusters are. I think I'll stop there. Thank you for your attention. (Applause)

David Campbell: For questions, please use the mike.

M: The previous two speakers, I have questions. Maybe both of you can contribute. The theme of this morning is self-organization. And I think what both of you talked about is organization, and any organization requires participation of free energy to satisfy certain dynamics.

So you have both dealt with dissipation of energy, but I'm just wondering whether there are two kinds of dissipation, or two kinds of organization. One is self-organization, which you demonstrate beautifully, if you would, which is driven by the internal dissipation, by chemical reaction.

There is another organization which may be called the other organizations. [??] two self-organizations. The other organization means that that information is driven by energy supplied from outside, just like [??] as Professor Whitesides illustrated.

So I'm just wondering whether we have to introduce other organizations, the concept of other organizations, as well as self-organization. One advantage of this is to explain artificial life that we'll hear about. That may be a beautiful example of other organizations as well. And I want to know whether this is the right way to think about it.

Irv Epstein: I'm not sure I would distinguish so rigidly between systems that dissipate energy from within, and systems that have energy sources from without. Because in fact, the second law is okay. And if you only have the energy from within to deal with, you're not going to last very long.

Life evolves on this planet, presumably with help of energy from the sun. And all of us are – we continue to dissipate energy, because we continue to take in fresh reactants. So in that sense, we're just as much other organization, or self-organization, in that terminology.

M: I think you're right. In your case two chemicals have to be supplied from outside. But one distinction may be time delay. In the case of self-organization, the organization and energy-dissipation is synchronous, whereas in other organizations, what I call other organizations as the example of Dr. Whitesides, there is –

I'm sorry, in self-organization there is a time delay. Supplying of the energy source and the time it takes to [??] chemical reaction, there is a time gap, a time delay. Whereas in other organizations they are synchronous. Now [??] and that information goes together in time.

Irv Epstein: Again, I mean, I think it's useful to think about those as maybe ends of a spectrum, but I'm not sure that one can always make a bifurcation between them. I think, well certainly, any kind of delayed feedback promotes at least temporal self-organization. And the system I showed you with the global feedback, that feedback is almost, but not quite, instantaneous. And in fact, one of the things we're thinking of doing is deliberately increasing the delay time, and seeing whether we get different kinds of phenomena.

M: I want to mention another system, a very different system, that produces very similar patterns to what were on the [??]. If you take a video camera and point it at its own signal, turn it at 60 degrees, and do it so that the image is the same size as the picture, you'll get squiggles and dots, they look almost insane. Do you speculate on whether there is an underlying similarity in the mechanism? (Laughter)

Irv Epstein: The only thing I would say is that one does have to be very careful about reasoning from similarity of patterns to similarity of mechanisms.

M: Of course.

Irv Epstein: And that symmetry is hugely important in determining the kinds of patterns one sees. Beyond that, I certainly don't understand how the fish formed this pattern, and I think there might be a chance I could understand how the camera thing works, but I don't --

M: I have a question about -- In all [??] reactions that [??] this role reaction to diffusion equations.

Irv Epstein: Yes. You're not supposed to say –

M: Well, only with respect to fish, not with respect to chemical reaction. I think then it's okay. The question I have is in all of this reaction, in all of the -- With all the diffusion coefficients, were the diffusion coefficients different between the different chemicals? In other words, there is this idea in [??] that you need to have different diffusion coefficients in order to be able to get patterns, and that was the case as well with reactions that you got?

Irv Epstein: In the BZ reaction, the diffusion coefficients are either the same, or within 10% of one another. I mean, the metal catalyst is actually a [??] complex, and it's sort of big, so it's a little slower to move around. And that works because it's the global feedback that is driving the pattern-formation.

In Turing patterns, we have demonstrated that in fact the reason you get Turing patterns in this chemical system is exactly because the diffusion constant of the activator which is essentially iodine is much less than the fusion constant of the inhibitor, because the iodine is essentially complex to stat molecules which are trapped in the gel in which the reaction is run.

So yes, in order to get Turing patterns, you definitely need a difference of a factor -- depending on conditions -- of something like 6, 8, 10 at a minimum to get pattern-formation. That's a different phenomenon.

M: Thank you.

M: My question relates to George Whitesides' apparently conflicting definitions of emergent behavior, one being something totally new and unexpected, and one [??] physics.

If you consider that the phenomena you're studying have emergent behavior, I would like to suggest that those aren't necessarily conflicting definitions, and in fact by doing the simulations, you have also demonstrated the whole hierarchy of levels among the standing [??] going, "Wow, look at this experiment. Let's [??] a model which generates similar behavior." But that does not necessarily get [??] understanding, understanding the physics, it still could be emergent as a new phenomenon.

Until you actually come up with a mechanism, back [??] hand-waving description. And I'd like to just end, do you have a hand-waving explanation and understanding --

Irv Epstein: You mean an intuitive as opposed to a mathematical --

M: Well, simulation doesn't necessarily endow you with an understanding. It's a tool on the way to understanding. And I would ask you just one step further, just to illustrate the full hierarchy to, "Wow, isn't that amazing in nature?" to "Let's build a physics model and simulate it" to "Let's try to extract the essence of this" [??] extent, through all of the other applications and it may be [??] a new phenomenon.

Irv Epstein: No. I agree with your entire characterization, and I think it's fair to say that we are groping towards a more intuitive physical explanation for this. We're not at the point where I'd want to say it in front of an audience of this size, but I like the [??] because I think that's a very important part of all of this.

M: I just want to make an observation that the visual patterns that you showed reminded me of emergent behavior of a much more complex, biological system than the fish, mainly the abstract painters of the mid-20th century such as Motherwell, Mondrian, Rothko and Klein. Do you want to comment on that?

Irv Epstein: Well, one could speculate -- I guess there was more of this kind of speculation going on in the Sixties about whether there are patterns inherent in the brain that somehow find their way onto the [??].

David Campbell: If that's [??] nothing happens to these people. [?? overlapping dialogue, inaudible]

M: A comment again. I'm happy about [??] requests for understanding. I don't understand understanding. You want --- and [??] saying, "We can explain this like this. We explain that like that." Now explanations we can see [??]. I can explain this to you [??] what would happen. Being [??] is all

196

symmetry-breaking, and that's fine. This [??]. So the understanding somehow fits beneath the quantum level are never going to affect it.

David Campbell: Let's thank the speaker.

(Applause)

Chris Adami
Artificial Life

Chris Adami: I think it is my duty to provide light entertainment. At least part of my talk today will be light entertainment at least the part that will be –

When I'm going to tell you about what there is, and actually it could be a large deal about artificial life. I'm going to be a bit more serious when I talk about foundation, and you know, why we think we can do what we are actually doing.

And then, you know, I'm going to talk to you a little bit about what we do at my lab at CalTech, but that's going to be maybe like the last third of the talk. So I am mainly going to talk about stuff that's [??]. But I think the beginning of this whole thing of artificial life is really that in order to investigate life what we usually do in the sciences, in the hard sciences, is very difficult.

Usually, what you do is you start by analysis. And if you actually have taken apart the system into its components and studied every component you think, "Okay, maybe I know enough now to actually put things together. Maybe that leads me to the path to understand things." And of course for life that's a bit difficult. Because if you take life apart, then you know, each piece of it is most likely not alive, and the thing that you're after is gone.

So instead what you do in order to -- or at least, what people in the [??] community do -- to investigate life, is to construct, in a sense, simple living systems, or, as we shall see, parts of simple living systems, so that you can actually conduct important experiments, analyze results, and maybe even predict some results, which, you know, is obviously one of the extraordinarily hard things in biology.

So one of the questions that you might ask about what goes on in artificial life, and one obvious one, which I get asked at almost every talk is, "Why is it at all possible that we can do this?" That we can actually study artificial life, or construct artificial life systems. I'm going to try to answer why is it plausible that we can. I'm going to, you know, maybe try to convince you why we should care about the results of things as artificial rather than real, or by chemical life, I should have said.

I'm probably going to get questions like, "Isn't artificial life sort of like artificial intelligence?" and I'm probably just going to gloss over it and say yes and no. Another question is, "Has this field produced any results? Have we learned anything just by studying things like that, where I see there are some results?"

So here comes the serious part. And it's serious enough that I want you to all try to concentrate on what I'm saying, because it's actually non-trivial. I actually may have usually troubled to convince people that what I'm saying is actually extraordinarily fundamental. It's not anything that I've said, it's what people that have worked in the theory of computation, let's say -- And the important thing is how that applies to, you know, fields of study like artificial life. And the general idea is that there is, in fact, a duality between physical dynamics and computation. And you'll see that I have a pointed arrow that goes both ways, and I'm going to try to explain to you what I mean with these both ways.

First of all, it is true that almost any physical dynamics -- that means degrees of freedom interacting according to the laws that these degrees of freedom usually abide by -- can be interpreted as a computation. We see in fact one example today in some talks where it was obvious that, you know, certain particles that were interacting in certain ways were actually solving an equation. And in a sense, and that actually, there are stringent results. Almost any physical dynamics, in a sense, solves a calculation. Okay. So that's actually this direction. Any physical dynamics, in a sense, leads to computation, or can be understood in terms of computation.

Now, what about the other direction? Well, the other direction, just think of your usual computer. What goes on in there? What goes on in there is physical dynamics of degrees of freedom. There is actually on your hard disk, on the RAM, there are little memory wells so to speak where bits are stored. And these bits are interacting physically according to, in this case, the laws in which these things have to interact.

However, it is physical dynamics which is going on inside of the computer -- that's this arrow. Any computation actually has to be a physical dynamic, because in the end, there is nothing else but physical dynamics. This thing, the computer, is not interested in [??] thinking, even though thinking actually alters physical dynamics. It goes on in neurons in your brain.

To make it even clearer, this direction is a rock falling down on a cliff can be thought of as accomplishing some rather uninteresting confrontation. More interesting, maybe [??] on the table can actually be made to function as a computer.

I can actually, or at least my students can, program any computers to actually have, let's say, a standing wave, a sine wave, inside of the memory of the computer, just by making sure that all of the right memory bits at the right time interact the way they should. So you could, actually, you could take a microscope, see a wave going through this piece of [??] just to convince you that you can actually have physical dynamics inside of your computer.

And if this is so -- and I'm going to get back to this point -- and in a sense, it's the theory of computation, and [??] which I'm going to get back to often enough -- then in fact, any physical dynamics can be viewed as a computation, and any computation involves physical dynamics. Then there is no reason why I can think that I can implement physical dynamics in a computer that would actually give rise to the kind of dynamics that we're used to [??].

And we've seen a lot of people trying to do that in biochemistry, and it's just extraordinarily difficult in biochemistry. We would have to wait a long time on Earth, or anywhere maybe in this galaxy, for this to happen. We're just lucky now that it happened on our planet, so to speak. But it is difficult.

What I'm going to try to convince you is not so difficult in the computer. In fact, it's almost the ideal system for life to emerge. I think that's essentially what I just said. So the reason why artificial life is not crazy is maybe that dynamics of learning systems can be implemented within a computation of chemistry. And that's a metaphor that I'm going to [??] back all of the time. You can't actually implement a computation of chemistry inside of a computer. This chemistry is necessary and sufficient to actually support all those phenomena that we are familiar with as far as last [??].

There is actually one more component to this story, why in fact if I have created such a system, why I should believe it has anything to do with a system that, you know, let's say we have [??] biochemical living systems on earth. And the reason is that what Turing has proven in 1931 is that you can have a chemistry in a sense -- You know, I'm going to go back between computation and biochemical chemistry.

If you have a chemistry that's complex enough, then it can, in a sense, simulate any other chemistry. And when I say "simulate," it means it can look like any chemistry, you wouldn't know the difference. One computer can simulate another computer, and you won't know the difference. And, if in fact you have this duality between physical dynamics and computation, that means if my computation is complex enough, I ought to be able to implement any type of dynamics [??] simulates, just like a Turing machine simulates another Turing machine -- these dynamics -- And it means that they are from a fundamental point of view indistinguishable. That means if I study one, I actually study the other. There is no reason to believe that there is a fundamental, metaphysical difference between them. And this underlying assumption is that of [??].

Just to reiterate what [??] means here, it means that there are general principles, and if you implement these principles in any substrates, then the dynamics ought to be comparable. And there is no a priori difference between them. That's what I mean with why can we think that we can, you know, study things artificial.

Now let me go to the definition, the first third of the talk of the century over, the serious part. Can we or should we define artificial life? No, let's not. Let's just say, you know, a sentence, and see if that sort of lets us think about what all the different types of artificial life could be.

Usually I like to say if I emulate, simulate, or construct living systems or a part of living systems in an unnatural medium -- which means essentially you're not going to do what life already has done. I mean, obviously that's been done. (Laughter) And done very well. And on top of that, we don't actually think that we can reproduce it. So we actually are out there reproducing, or producing something much more simple; something that hasn't evolved for four billion years. And Chris Langton coined it life as it could be, rather than life as it is.

So there are a few milestones in what is known as artificial life. There are, in a sense, two main eras. The first era is that of Turing and von Neumann. There is no possible way I could under- or overemphasize the role that Turing has played in this. Because if it wasn't for Turing who has essentially brought home this point of universality; that if I understand computation, I understand absolutely everything, in a sense, because, you know, computation underlies everything, then John Von Neumann would not have thought about, you know, the universal neural odometer. McCulloch and Pitts wouldn't have thought about universal neural odometers. The whole field of neural networks goes back to that.

But after John von Neumann died, essentially the field did not really exist anymore, because it was John von Neumann who was the one who thought about the idea that, you know, if I have universality, then I ought to be able to build a universal ?? odometers, and I ought to be able to observe the phenomena of living systems in an artificial medium. In a sense, you know, that created this whole field.

It was revived in '86 -- Actually, I shouldn't say it was revived then. It was revived in 1989, roughly, with Chris Langton and confidence that all came out of that. But there is more stuff going on, and what I'm going to talk about -- I'm going to talk a little bit about simulating, then I'm going to talk a little bit about robotics. I'm not going to talk at all about wet [??] life, even though that's an extraordinarily interesting subject, and I want to prime you for --

There are people like Gerry Joyce and Jack Szostak who are doing such amazing things with biochemistry; trying to recreate a biochemical living system that within 10 or 15 years Gerry will have succeeded, and everybody will just stand there with their mouths open, because he will have living systems in a little test tube, which has absolutely no ancestry with the biochemical living system that we are, that we know of here. So I think that's extraordinarily exciting, but I can't possibly do it justice here.

And then there is this feel of digital life which the main proponent, or early proponent was Tom Ray, when he created Tierra in 1991, and I am going to -- When I talk about the research I do at my lab at CalTech, I'm going to talk to you about Avida. So that's history. Now I'm going to tell you about the different forms of artificial life that there are, because it's actually an umbrella term for many, many different things.

And I tried to sort of put them into three main categories: emulation, simulation and construction. And even though the difference of emulation and simulation is so much semantic, let me just tell you what I mean by that. What I mean is with emulation you don't actually look at population, you look at either one unit of a population, or in fact a piece of a unit in a population. I'm going to give you an example where this object is emulated in hardware, and that's crickets -- What's important is that I have emulating here, simulation, and construction here. There is hardware/software, hardware/software, biochemistry and computation. So that's the main ways in which I tried to make you understand the breadth of this undertaking.

So I'm going to talk about crickets as far as hardware emulation of living objects is concerned. I'm going to talk about Karl Sims' work, and the illusion of creatures in software.

Then I'm going to talk very briefly about what can be done with simulation. You can actually simulate populations of robots by – populations of animals, by basically building robots as we've seen, but I won't talk much about that. I'm going to talk a little bit about swarms so to speak, about large populations of adaptive agents, and what you can learn from them, and I'm going to talk about [??].

Then I'm going to go to construction. As I said, I'm not going to talk about biochemistry because I couldn't possibly do it justice, but I'm going to talk about computational artificial life in a particular [??].

All right. First, crickets. So there is, in fact, a problem in, I guess you would call it ethology but it's really the problem of how crickets find their mates. How do you do this? How do you even begin asking that question? Well, you can do lots of experiments with crickets, but it's very hard to do it sometimes in order to actually make sure that some boundary conditions are met, and that they aren't doing something that you don't think they should be doing. So the way this was done in the study that I'm going to show you is by building an artificial cricket.

What I want to say is that the objective is to understand cricket behavior, in particular in the [??] we're going to briefly focus on, and the solution, that I've said, is with a robotic cricket with neural controls, and a peripheral auditory morphology – And that is supposed to tell you that it is a highly complex robot, built out of the most complex materials that there are.

And let me show you where these -- I guess let me see -- The receivers are on there because in fact, on the cricket they are at the end of the antenna. And the question is when the cricket goes forward, how is it actually processing the signals that come from the source?

And what they have done is they've put in a hypothesis, they have an idea on how that might possibly work. And the idea is the following. Each of these antennae actually receives the signal, and then sort of integrates the signal until the integrated neural signal reaches the threshold, at which point this neuron fires and one of the wheels stops turning.

Well, the equivalent of the wheel is the cricket, obviously. And in a sense, whoever is – I mean, it depends on how this cricket is rotated towards the source. Whichever antenna is closer will actually fire first. And then you ask yourself, "Well, what about after it's fired? How actually do you reload this system and make sure that like the other neurons, it actually did not fire? How do you make sure that that is not going to fire with the next piece of [??]?"

And the idea is well, that's where the fact that crickets are an intermittent [??]. Because for a certain amount of time, there is no sight, sound, and in fact, the activation level drops down again, and it reloads the same. So that was their idea, and they implemented that. And I don't actually have the graphs there, but essentially what happened when they implemented this algorithm, they saw this cricket move in first and fifths towards the source, and they could vary, for example, the lengths of the chirps, and see how their movement depends on the length of the chirps, and could reproduce precisely what the cricket – you know, which they were using a parallel experiment also, in fact the shapes of these approach curves looked the same.

In other words, they have created a model in a completely artificial system, that carries, most likely, the necessary ingredients. And even though you can't prove that this is exactly what happens in crickets, it's pretty likely that in fact it's not a bad explanation. So this is the case where I wanted to emphasize that this is, you know, real research done on artificial life, published in *Science*, where actually people have answered a question that they have.

Let's quickly, let's actually stay in hardware. I just want to point out with this slide that there are actually people who are building small robots out of little pieces of electronics which are very cheap. In fact, one of these robots is called "The Walkman." Which one is it? Probably this one -- I'm not quite sure. You can imagine out of what kind of pieces it was built.

What this guy can do -- this was built by Mark Tilden at Los Alamos -- he can actually put them together. They actually can learn how to work by a very simple neural controller, and they interact with each other. They have light sensors, and they can go towards food, and things like that. And he can actually put five, or ten, or 15 of them together, and see, you know, how they attack each other, and how they actually form some sort of ecology.

They can actually do – even though it's a quite primitive way -- interesting experiments about populations of robots. In fact, there is one other group that does something like that, even thought it's quite different. Here that's the Immobile Robotics Group at now USC, Maja Mataric'. What she does is she programs robots to interface in certain ways, and to actually, you know, behave socially with each other.

And robots usually, if you just put them somewhere and make them go, they're usually not very nice to each other. They like to bump into each other and things like that. It is actually not easy to make them behave socially. And what she has done is she has, you know, she implemented certain algorithms so that these robots essentially can do interesting tasks. Here she is following the behavior -- It's not a very good movie, and I'm not going to certainly show a lot of it -- [?? overlapping dialogue, inaudible]

So you know, this is simple, following behavior, but you know, if you talk to Maja, you'll find out that it's very, very difficult to implement, you know [??]. One guy goes to the right, and the other guy goes right to it. You know, that's a form of behavior, basically like a pet. And then there are all kinds of things like flocking behaviors or I'm not going to bore you with that. (Laughter) But here you have several flocks of robots.

It's actually extraordinarily hard, but you can actually learn something about the general principles of the types of algorithms that you have to implement. Now, let me talk about the emulation of — I'm going back to emulation. Remember, I was at simulation in the population before, now I will go back to the individual level because I am jumping from hardware to software, and I'm going to talk about Karl Sims' work. And here is what Karl Sims' has done.

He says, "I am willing to understand the illusion of morphology." Okay. "And I am willing to understand how the world in which the organism evolves, shapes their morphology and the structure, and the interactions of the objects that are in it."

And he does it in quite an interesting way. I think it's one of the most beautiful examples of the implications of artificial life.

His organisms are quite simple. They're basically made out of blocks. Here you see a central block with two slide blocks. Here is the representation. This circle is a central block, and this circle here is hiding one of the side blocks, and these two arrows tells you that there are two of them. So in fact the morphology, the external morphology represented by this diagram -- two circles and these two arrows.

Now, on top of that, there are actually neural controllers, sensors and activators in these blocks. In particular, there are many in the side block, and in fact, they are controlled through a master controller here. And in that sense, they can sense pressure from the ground, and they can do certain movements, and in fact, this entire wiring diagram is defining the entire creature, not only its morphology, but also its neural network, its brain.

And you can actually mutate them. You can add circles, you can add arrows, and you can add and delete connections between those. And by mutating these things, you can create new types of creatures, and he actually will test their fitness in an artificial world. Now I cannot, again, overemphasize how important this artificial world is. And I think I can make you understand how important it is, because it is what shapes entirely and completely the morphology and the behavior of these animals.

Here, for example, are a bunch of creatures that adapted to a particular task. Now, he is actually very good at implementing physics in a computer. In other words, what happens here is entirely according to the laws of Newtonian physics: there is friction, there is gravity, there are all kinds of things. And these creatures, they have to compete for this little black box in the middle, and whoever wins it, gets replicated a multitude in the next generation, and essentially is a GA that creates new objects in the next generation, and they again compete. And the best ones are taken to the next generation, and the bad ones are left to [??] I suppose. And you see here very different approaches to actually winning the fight for the cube. This guy has developed [??] things. And this guy actually, I guess, hoards it and tries to push away the other.

Then he discovers that there were different ways of doing this experiment. And one of the things he has to do, he has to make sure that the distance of a longer and higher creature would be different from another one. Because when he evolved this at the same distance, he found the solution where essentially there would be a long plank that would develop and just fall over, and cover this brook which is apparently a very effective, but inelegant, solution.

So it turns out that he has to play around with these a little bit. But in fact, you find very interesting -- this is very uncanny. There are actually movies about these guys doing this, but I don't have that here. What I have is evolved virtual creatures when he tried to evolve organisms that could either swim fast -- and you had to basically program all of the rules of hydrodynamics, or walk fast, and had to basically implement Newtonian dynamics.

So let me actually play you a little bit of that video, because in fact it tells you quite a bit about how it really worked. So here we see, for example, sort of typical morphology, and it swims pretty well. This one sort of looks strange, and you see immediately that well, we're not quite used to that. We are not quite so used to this type of swimming behavior either, but it apparently works. This also evolves -- and that was obviously the one that we are actually very used to -- This we're not. But again this one, obviously very, very familiar to us.

And what you see here is that there is stuff that evolves that we do expect, and there is stuff that evolves that we don't expect at all. None of this was pre-programmed. All of these things are creatures the essentially evolved from scratch from these blocks just by selecting those that, you know, did well, into the next generation. And obviously the success of this is actually getting the laws of physics right.

Now here, this was selecting for speed of movement. And again, there is no telling how they are going to use whatever their mutation morphology is to go forward. But whatever works, works -- in the next generation, that's what's going to be the answer. That's obviously also a crazy way of trying to move forward. (Laughter) One story that he told me, that is, at the beginning of these experiments, okay, he actually made a mistake in programming the world. What he did in fact was he incorrectly implemented momentum conservation. And then ran it. He didn't know he incorrectly implemented momentum conservation. But when he found these creatures that moved forwards by hitting themselves. (Laughter) Then it was obvious. Then he made this little change in the computer to say, "No more violation of momentum conservation." You just had these creatures hitting themselves, and they just look idiotic, and they don't go into the next generation. So let me show you a few other of these.

Obviously, that's something that we're familiar with, a mode of locomotion. And here is an interesting one. Let me just pause this for a second. This is relatively familiar, and also it has a structure that we previously we -- Look at this. What the heck is this?

It just happened. It doesn't matter. It doesn't disturb you going forward, and it's probably going to go on in several generations. It's a neutral mutation. It doesn't actually have any selective value. But, you know, it's piggybacked on something that worked, so it's just going to get replicated into all kinds of next

204

generations. And you can see this type of thing in all kinds of creatures that look odd and weird. Because the only criteria in this move forward -- whatever works, works. That's obviously not something that you'd like to meet on the street. (Laughter)

This again is actually a well-known way, the previous one, of going forward -- this also, this is a snake-like motion, a sideways snake-like motion which is very well-known in snake dynamics, let's say, even though that is obviously a very strange type of board. That's actually also not totally, you know, weird. That is some sort of [??] ; we've seen that.

So what I really want to show you here is that evolution does these things. It gives rise to things we know, and it gives rise to things we don't know. And you have to think of this type of evolved phyla as the type of things that would have involved very early in the Cambrian explosion, where everything goes, and where evolution hasn't had time to hone the really good solutions. Because here, you know, it was just a matter of getting something. All right. Enough of these little wigglers. (Laughter)

There is actually another project by Maciej Komosinski in Poland, who has done a similar job, but he has implemented the type of morphology differently. In fact, his objects are much more complicated, in the sense that they have all types of different joints that they can, you know, choose from. But each one of them, also, is determined by a genotype. He also does experiments in trying to evolve swimming. Here is, for example, one of his creatures swimming.

Which again, I mean obviously this type of swimming motion would always evolve. It's obviously not different -- difficult. Here is one where he selects also for forward motion. And I'm just showing you an early genotype that's not good at all at doing what it does. And you see here, it's sort of like, just sort of a weird vibrating thing. But it moves forward, and that's why it was selected into the next generation. (Laughter)

Here is a much later offspring. I don't know even if it's of the same line. But which, you know, has developed actually quite an interesting way to move forward in this environment. And what I like about his simulations is there's is actually much more detail as far as shape, shades, and all kinds of things are concerned. But in any case, what is important is that they get the physics right, because otherwise, your things are going to adapt to a physics that has nothing to do with the type of physics you are used to.

Here is actually an organism which is not involved, but which in fact was constructed. And this guy is supposed to just -- these things are food sources, and if there is a food source it goes towards it, and then eats them, and so on. That's enough about that. So this is a type of research which really shows you some of the general principles of evolution, and how the world around you shapes the kind of organisms that you obtain.

So now I'm jumping to simulation in software of populations. The general idea -- and it's sort of like the paradigm -- these things are like anthills or ant colonies; and trying to, from understanding the interactions between the ants, understand how the superorganism of the colony behaves as a whole. I think this is even more interesting, because this is the simulation of cooperative behavior, in particular, distributed building of wasp nests.

What happens here is that you have wasps that do very simple tasks. For example, you have a wasp that only does one thing. Like if it finds a block on top of a bunch of flat blocks, it just puts the block on top of it. It has no idea what this thing is, it just knows, in a sense, this is a stimulus, maybe this single block on top of the surface is a stimulus for the agent, and whenever it sees something like that, it puts one on top of that. If there is no such thing, it just sits there. And there are a bunch of other wasps that have different programs, of different rules, which are stimulated by different other shapes.

The interesting thing here is that you can start out with one block or no block – if you have one wasp that just basically always puts a block somewhere where there is nothing. Then the structure that is already there, stimulates the rest of the population to take the next step. So in fact, they're changing their environment all of the time while they're building this structure. They have no clue what they're doing. But at the end, you get very complex and interesting structures.

Here is, for example one, where these researchers -- Eric Bonabeau and Guy Theraulaz basically just implemented a few of these wasp agents with certain local rules, and they would build this very complex structure, which actually is known to be the equivalent to a particular type of wasp's nest. In fact, they played around a bit more with the rules, and they found that they could actually get all of these very different structures of well-known species of wasps.

So the idea to take home from this is that again here, just by implementing the idea of stigmergic buildings -- mainly the idea that you can have populations which are queued by the structure that they are forming to take the next step. You can actually do very complex things from very, very simple rules. Because after all, they, Eric and Guy actually programmed these rules. And in a sense, these rules themselves, they define already what this is.

But it is not exactly easy to, in a sense, write a computer program that will just take a look at this and says, "Oh yes, I know what kind of structure these things will give rise to," because this is an iterative process. And, in a sense again, this solves, or at least gives a hint, to a particular problem. I mean, it's not just pretty. I mean if you're asking, "How do wasps build their nests?" this is a pretty good clue as to the fact that they have these local rules in order to build them.

So let me go towards the last point, namely, that's computational artificial life. And let me reiterate what I said at the beginning, to in a sense recall into your memory the reason why we think we can do that. Universality promises that any computational chemistry which is pure and complete, and supports self-replication, can give rise to life. Period.

Now, once you've said that, and it says, "Okay. So let me actually find a good computational chemistry." Some people have tried cellular automata. And it turns out, and it is my opinion, that even though it is in principal possible to do it, a cellular automaton is extraordinarily difficult. And there are other media which are many orders of magnitude more amenable to this kind of thing. And it's sort of odd, or maybe inspirational to realize that the medium in which this works best is, in fact, von Neumann computers, and not von Neumann cellular autometa.

So you can ask the questions, "Are computer programs self-replicating in computer memory?" because that's what I'm going to use, "Are they alive?" So let me show you something.

This is a phylogenetic tree that Dave Hull gave me, of computer viruses of the Stoned family. And what this guy did is he just did the usual genetic distance analysis of computer viruses of this family to create this phylogenetic tree. And you see that there are all kinds of mutations. These mutations are, of course, done by computer hackers, who try to evade the efforts of the people in computer security. So you see, there is, in a sense, a certain arms race going on between the two and gives rise to this type of phylogeny.

You can see here that every one of these viruses is compared to the DOS boot sector in order to see the absolute distance. It's not supposed to indicate that the DOS boot sector is also a computer virus -- (laughter) -- even though many people have suggested that. I will not comment about that. But what you see here is that obviously there is some sort of immediate connection, or at least intuitive connections between computer viruses, self-replicating computer programs, and at least simple living systems. And that's where the general idea of digital life comes from. And let me give you a very short rundown of the history of digital life.

In 1990, Steen Rasmussen of Los Alamos came up with a VENUS simulator where he just took basically the computer language of the Core Wars game, Redcode, to implement self-replicating computer programs in a confined environment, and that didn't work very well, because in fact, he did not compartmentalize these programs. And as a consequence, they were writhing all over each other, and killing any population of self-replicators that [??] the Core Wars.

And Tom Ray was actually looking over his shoulders then. And he saw that and didn't tell them that he knew how to do it better. He was like, "Oh, oh, I know where you went wrong," but he went home and programmed it himself, and then he came up with the ubiquitous Tierra.

Now, I saw Tierra some time in 1992, and because I had this job where they told me to sit around and think and you can do whatever you want, I said, "Okay, I'm going to think about this." And with a bunch of students we made a different system, which was obviously inspired by Avida, but which actually -- by Tierra -- but which actually was amenable to real scientific experimentation. So we made the system in such a way that we could control any parameter, that would take a lot of data, and then we could have reproducible results. And also we had, in a sense, more physical type of environment. As you'll see; I'll show that. There's an offspring of this by Andy Pargellis that I'm not going to talk about. We're going to talk about this system, Avida.

So I've already primed you for what it is. It's a software the implements a population of programs that self-replicate in the memory of a computer. They are subject to mutation and survival of the fittest. In other words, we are dealing with very pure Darwinism. And I want to also remind you that because of the universality and the duality that I talked about, when I'm showing you runs with this software, these are not simulations. I'm not simulating anything. These are experiments.

The dynamics of self replication actually takes place on the RAM, in the RAM of the computer. So in a sense there are patterns of zeroes and ones which do self-replicate because of the laws of the computer are such that they can. Not everybody can. You actually have to write a program that self-replicates, and see the memory of the computer with that. If you don't do that, and you think oh, let me just have random programs and I'll mutate them, and see if one just jumps out.

I did a little calculation. If you're lucky, you only have to wait between ten and one hundred thousand years, with one hundred processors of your biggest Crays for that to happen. Because even though this is computational chemistry, which is simple, the simplest self-replicator is still about eleven instructions long, and one instruction is taken out of the set of anywhere between twenty-five and whatever you want.

And there are just too many programs to try. Just out of the question. So we don't talk at all about the origin of life. We write our own self-replicator. And Gerry Joyce always tells me how jealous he is that I can do that. Whereas he, he has the advantage of, he has a population of 10^{15}. But he can't write his self-replicator.

So let me show you what happens here. This is an artificial patronage in a sense. These computer viruses, in a sense, these self-replicating programs, have been put into an artificial environment inside of the computer, and they can't really get out of there, okay. Because if they did, in whatever other environment, they're just data. But in here, they thrive.

And let me just show you an accelerated run or, in this case, experiment. This is very accelerated. Different colors denote different genotypes of programs. So I should tell you that − I shouldn't have done that, and restart this. At each lattice point here is one organism. The organ consists of, as you will see, a genome and a virtual processor that sits there. And each color denotes a particular genotype.

And there's obviously tons of mutations going on, and there's tons of diversity. And there's all kinds of evolutionary transitions which happen in there. And what you'll also be able to see is that at the beginning there's tons of activity. And then you know it's sort of, it's going to settle down into a metastable state, a plateau. Nothing much is happening. They're just basically, just go on with their merry lives.

And at some point you'll see a very strong transition that is like a punctuated event. Actually, this is the plateau I was talking about, so basically it's pretty much equilibrated. And you'll see that there's a new guy on the block that is going to wipe out everybody in almost a few seconds at least in this simulation. That's just -- there it is.

M: What does the location represent?

Chris Adami: The location is a physical location. Let me show you the world. Okay? Here is a grid. And in this case it's a grid of ten thousand of such programs. And at each location is one such program. Okay? And this program -- that's a genome -- is circular. You shouldn't be too surprised about that, because most genomes of simple organisms are circular; DNA in the mitochondrion is, too.

And what that means is that I have information coded like this, and I have an instruction pointer that can basically just run over this. And that's the chemistry of this information. And obviously you don't want the chemistry to stop. So we have made this circular. In Tierra this was not like that. And so obviously this program is being interpreted by this very simple CPU, three input buffers -- three registers, three input buffers, two stacks and so on. And essentially that's all there is to the chemistry.

Now I want to show you briefly, this is what happens, you know, when you measure fitness for these organisms, which, you know, I sort of laughingly call digitalia sometimes. And just to make sure that in the end this is a form of life that we're investigating, and it's just one of them. And here's compared to the evolution of E. coli. This is the actual chemistry.

So you see, there's about twenty different instructions and you can sort of understand why we sort of select twenty. But we can play around with instructions and chemistry of all kinds of sizes. Cell division in a sense goes on like this, so it's nearest neighbor. And here's just a list of the kind of stuff that we have done -- actually it's a very partial list of the kind of stuff we've done in the last few years. We just have an organism here. We can investigate anything we want.

So we investigated taxonomic abundance patterns, the propagation of information, the evolution of complexity. And I don't think I have time to talk about this, but I had actually two more slides. I'll tell you about it. This was just published in *PNAS*. We did a bit of investigating the complexity of genomes in epistasis in a recent *Nature* article. We looked at differentiation of genomes and organization, roles of chance in history and evolution.

And we have a bunch of other things that we want to do. In fact, there's no limit really of what we can do. It's like, have organism, will travel, or will do experiments. And we're working together with the microbial ecology lab at Michigan State where they do actually the E. coli experiments. And we on our side do the same type of experiments with digitalia, and we try to compare results, in order to understand what is it that is actually in common to all forms of life? What is universal about life, and what is actually due to simply the particular type of chemistry that you're dealing with?

So we are able to make or try to generate statements about things that are common to all forms of life, and not hindered by the fact that all forms of life on

Earth actually have one ancestor. So we're looking at molecular evolution, selective pressures, Muller's ratchet, robustness of computer languages. We're just about ready to actually grow ecologies of organisms and study the dynamics of ecological webs, and many, many things more. So I'll leave you with that. (Applause)

M: Can you show the slides on universality, about a half a dozen slides back?

Chris Adami: Yes.

M: While you're searching for it, I want to thank you for the informative and entertaining presentation.

Chris Adami: Thank you. I actually had a bunch of slides on universality.

M: Universality – the last one.

Chris Adami: Here's one. Is that the one you meant?

M: I wrote it down here. You said, any computational chemistry which is [??] complete, etc., and support self-replication, can give rise to life, you said.

Chris Adami: Yes.

M: Now I think in your enthusiasm, you I think committed confusion between life and artificial life. And that definition there, the last word, life, should be artificial life. You're dealing with artificial life, not life.

Chris Adami: The beginning of my talk insisted upon that there is no fundamental difference between biochemical and artificial life.

M: How do you know that?

Chris Adami: This is what I spent the first third --

M: Have you studied life?

David Campbell: I think that this is going to quickly go into philosophy that's beyond the scope. Dave?

Dave Meyers: So I'm very sympathetic with that assumption, so I'm not going to question it. But my concern is that, as far as we know, physics is probably reversible at the bottom. Really, it's unitary. [??] some questions about quantum gravity.

Chris Adami: I [??] agree with that.

Dave Meyers: So you sort of hinted at the problem I'm worried about when you said that it was easier to not start at the CA level but start at a higher level and get the kind of interesting results you're talking about. One of the things I've worried about a lot is, can you possibly get these kind of results starting from unitary or reversible physics? I'm curious what your thoughts on that.

Chris Adami: Energy-dissipation plays a very big role here. Any mutation in fact is irreversible.

Dave Meyers: That's right.

Chris Adami: And, for obvious reasons. And you know, I can't give you a hundred percent answer. But my gut feeling is that the answer is no. That what you would get is periodic behavior, rather than this type of increase of complexity that you have.

David Campbell: [??] analysis, the answer must be yes. Because we have a real world where we've evolved and from that.

Chris Adami: I just hate to make general statements.

David Campbell: You just made the first general statement; you might as well [??].

Chris Adami: Without thinking about them. (Laughter)

M: In your descriptions of emulation experiments you've described microbes and sort of pre-Cambrian organisms, such as those down in British shale. Is anyone doing emulation experiments where they put in a realistic representation of a known organism to see whether they can emulate the way it responds to things like changes in food availability or predation [??]?

Chris Adami: The answer to that question -- and I'm not sounding like a politician -- is yes. There is a paper in the Artificial Life VI Proceedings where a guy has actually constructed an extinct pre-Cambrian actual organism. If anybody here can help me out – it's a guy that sort of like, you know, you didn't really know how it swims, because all you did, is you had fossils.

But he basically put these fossils or this structure into a computer and then essentially put in an evolutionary algorithm that would find the best swimming behavior – Anomalocaris, that's what it is called -- the best swimming behavior of this object. And in fact he found that it sort of like swam like a flounder.

M: Pre-Cambrian.

Chris Adami: And that was, at least in this constricted view of things, the most likely way in which this thing went towards food and so on.

M: Okay. Well, what I had in mind was something that could be tested experimentally where you actually put in an animal like a rotifer and see if it responds like a rotifer when it's threatened with predation or something like that. I gather that has not been done yet?

Chris Adami: Not to my knowledge. But obviously the cricket work is sort of going in this direction, where you try to look at interaction.

M: I'm very sympathetic to this. I enjoyed it enormously. I think you are talking about life. I think if we, people say that mitochondria are life, then we would then -- these are alive. But I think you misuse the word genotype. Because you've haven't got phenotypes.

Chris Adami: Of course I have phenotypes. I hadn't talked about this.

M: But they're the same as –

Chris Adami: I hadn't talked about this, and I hadn't had time. But actually what these things do, they live in a world in which computations reward you. They develop genes for computation. And we can actually look at each organism -- what kind of computation does it do? It actually has a phenotype. And the genotype-phenotype mapping is extraordinarily complex. Many different genotypes can give rise to the same phenotype and vice versa.

M: Right, that helps me a lot. But you still don't have a two-phase construction like you have in metazoa. You work beautifully for E. coli, you'll work beautifully for rotifers and protozoa. But I think that developing creatures you're still missing.

Chris Adami: Well, I will tell you something. This is a very simple form of life; a form of life that maybe can be compared to like an RNA world where you have chemicals that basically act as templates for their own reproduction.

M: Yes, yes.

Chris Adami: And obviously I cannot tell you that this form of life is anything more complex than that.

M: There's just your use of the word "universal" for all life in the middle of your –

Chris Adami: What I wanted to say, that in principle, if I have a world complex enough, this form of life can develop into anything. There's a big if.

M: When we say artificial life, I think we're getting to artificial life the level of biology. I don't know whether we have artificial social life, in the sense that we have entities that could come into the classroom and argue with you about your interpretation of their behavior. Is our mind or mental understanding of the world in which we exist -- is there any – does that emerge in your world in any form? Is it necessary that we reflect upon? I mean, that's the [??] --

Chris Adami: If I understand your question correctly, are you asking about any type of social behavior in this?

M: Not social. If you have a mind, which I'm not sure – essentially, it is self-reflective. You have a theory about the way you behave, like we're having a conversation. How does that play a role in artificial social life? Is that coming up?

Chris Adami: I really can't answer this question, for a very good reason. I do not study sociology here. I do not – I mean, these communities are entirely on a level of symbolic sequences interacting with each other. And we have made sure that they actually do not interact with each other except for a trivial way. Why? Because we'd first like to understand the simple system before you try to make them mate and do all kinds of things, which is only going to complicate them.

M: So there's no survival value for mind?

Chris Adami: No, they have no minds. They are sequence only. Think of RNA molecules, polypeptides if you will. But nothing more than that. But the interesting thing is that that's already enough to show actually quite interesting dynamics.

M: I want to know if you think that brain functions can be mapped or reproduced or simulated by a Turing machine.

Chris Adami: If you ask me whether the brain is computational, it's emphatically yes. And that there is no reason to believe that it is anything else but that.

M: Well, if it's computational it can be mapped into a Turing machine.

Chris Adami: Yes.

M: And brain function, one of the brain functions is producing Turing machines. So you are saying that a Turing machine is able to produce a Turing machine?

Chris Adami: How does a brain produce a Turing machine? A brain can emulate a Turing machine but not produce one.

M: Oh, yeah, I mean to --

Chris Adami: Well, you can have machines that emulate machines that emulate machines.

M: Yes. And also I mean, how could you think that would be possible? I mean, if it would be possible, it should already have been done. I mean, if it's possible, it's not easy.

Chris Adami: It's certainly not easy. It's just like the origin of life. It's possible. But it's extraordinarily improbable. And that's why it's so hard to get a handle on it.

David Campbell: Let's thank Chris again. (Applause)

Duncan Watts
Small World Networks

Duncan Watts: What I want to talk to you today, is a problem that's really about science, but it starts off in an interesting place, which is in sociology. And I think that's interesting. Because I think it's fair to say that historically the relationship between the social and natural sciences has been something of a one-way track with ideas being appropriated by the social sciences and not so much the reverse.

And I think one of the interesting things about this, this subject of the small world problem and the small world networks in general, is that it's a problem that was initially studied in sociology and has been thought about for a good thirty-odd years in that field, and has recently appeared to be a lot more general than that, and is now having some, at least a potential for some interesting applications in some other disciplines.

So the idea of the small world problem was studied initially in the 1950s, but really became well-known in the 1960s through the work of Stanley Milgram who was a social psychologist. And it is a very interesting experiment, where he was investigating the idea of, the question of how distant people were in the social world. In a sense, how many degrees of separation separate individuals over a large geographical area? And in his experiment he was talking about the continental United States.

At the time, he was at Harvard. And naturally that was the center of the universe. And what could be further from the center of the universe than the Midwest? And so he started several hundred chain letters in Nebraska and Kansas. And these letters were of a very specific format. They were to be sent to one of two target people. One was a stockbroker in Sharon, Massachusetts. And the initial recipient of the letter was given this person's name and their rough location and some information about them. For instance, their profession. And was asked to send it to them if they knew them on a first-name basis. They probably would not, and so their instructions on that circumstance were to send it to somebody else whom they did know on a first-name basis who they thought were more likely to know this person than they were.

And so you can see this sets up a chain/search procedure, which we'll talk a little bit more about at the end. And the next person who gets the letter has the same set of instructions. And this procedure continues until either the target is reached or somebody can't be bothered to continue the process. And it says something about the era, that people actually did this. I mean, if you were to receive a letter like this today, you probably wouldn't get past the first sentence before tossing it in the garbage. But back then, of the three or four hundred letters that he started, several dozen arrived at their destinations. And Milgram found that the median path link of these chains was about six.

And that's what led to this famous phrase, six degrees of separation, which has since been embodied in a play by John Guare in 1990, which really doesn't have much to do with six degrees of separation, but it's a catchy title; and recently has achieved the ultimate level of social acceptance by appearing in *The New Yorker*. And now Malcolm Gladwell is making a killing out of his theories of tipping and six degrees with his new book, *The Tipping Point*.

Well, you might think this is all very trivial and needs no explanation. And in fact, there's a very obvious calculation that you might do. Which is to say well, hang on a second, here is me or here's the person who gets the letter, and here's their friends. And let's say for the purpose of argument you have a hundred friends, and in fact, the work that sociologists have done indicates that in fact the number is much greater than that, but of course it depends on how you define a friend. And then each of those friends has another hundred friends.

And so you say well, you know, at one degree of separation I have a hundred friends, and at two degrees there's ten thousand. And very quickly you can see that this number, it's growing exponentially. And so within only a very few degrees of separation you've already exceeded the Earth's population. And so of course everybody is separated by only a few degrees.

This is a trivial fact about what we might think of as random graphs. And this is a well-known property of random graphs, that the typical length scale or the diameter even of a graph is logarithmic in the size, in the number of elements. Of course, if you think about it for another second, you realize this is a ridiculous calculation, because if we go back to this picture, and you ask yourself well who are my hundred best friends, and then you say who are their hundred best friends, you'll probably come up with a lot of the same people.

And so in fact, the mystery of the small world -- inasmuch as there is a mystery there -- is not, "How can everybody be connected in some big social network in such a way that we're all only a few degrees of separation apart?" but, "How can this be true despite the fact that the world is nothing like a random graph?" If we were all to be selecting our friends at random from the pool of six billion people in the world and this were a uniform selection, well of course, nobody here would know anybody else in this room. You wouldn't know any of the people you grew up with, you wouldn't know any of the people you work with.

In fact, we have a very few organizing principles that determine almost all the people that we know, or at least a large proportion of them. And as a result, there's a lot of what we might call local order or local clustering in these large networks. And so the question is, how is it possible for the world to be small despite this?

Now there are many ways you could go about answering this question, and certainly one of them is empirically by looking at large networks and asking about their features. This is something which is starting to take off now in the age of the Internet where we can actually collect a lot of very accurate data about large collaboration graphs for instance, or large social networks, people emailing each other in online chat services. Of course getting it out of the companies that own this data is a problem that I've been running into.

But a few years ago even, this was an exceedingly difficult exercise to actually get network data. And sociologists had been battling against this problem for a long time. Where if you have to use survey instruments to actually figure out who people's friends are, people are notoriously bad at telling you accurate and consistent information. And so sizes of more than a couple of hundred were rarely possible.

So the approach that we chose to use was to, rather than looking at real networks, was to construct very simple models of networks that embody the features that we're after. And the features that we're after can be briefly summarized in the sense that we want networks which are large -- many, many components; networks which are sparse, in the sense that no one element is connected to any more than a tiny fraction of the total. Because again, obviously, if you have a hub and spoke-like network it's trivial that everybody should be closely connected.

So these are very decentralized sparse networks. And they have some mixture of order and randomness. And because we're relying on models rather than the real thing, we don't actually know what that mixture is going to be. So you'd like to look at a kind of family of networks that interpolate between something which we think of as being very ordered and something which is very random and see if there's any interesting structures in the middle.

So to start with the very simplest possible kind of network, we look at a structure which is just commonly called a graph. And I use the words interchangeably. And it's a very simple kind of graph in the sense that there are no self-loops here -- there are no connections to oneself. There are no weights on the edges. Every edge is equally important. And they're undirected. And these are obviously oversights from, you know, the Internet has directed edges, brains have directed edges. But this is again the obvious place to start.

And the only parameters that I'll really be talking about are N and K, where N is just the number of elements, and K is the typical number of friends, if you like. And the assumption is that N is much larger than K. Okay, so again, we want to construct some simple model which interpolates between a structure which we think of as very ordered, and something we think of as very random. And there are in fact many ways to do this. And this is just a very simple way, which captures some of the interesting phenomena.

And the idea is as follows. You start with a one-dimensional periodic lattice with coordination number K. And so in this case it's N = 20 and K = 4, and each element is connected to its nearest neighbors on the ring and its next nearest neighbors on the ring, although typically we use much larger numbers than this. And having constructed that, then you define some parameter P between zero and one, and you go through every single edge in this lattice and randomly rewire it with probability P.

Now what does that mean? Simply that you now treat this -- you roll your die, and if it's less than P, you treat this edge as an infinitely stretchy bungee cord. You leave one end fixed; you pick the other end up and move it somewhere uniformly at random on the lattice. So obviously if P is zero, nothing happens. If P is one, every edge gets tested and every edge gets randomly rewired. And you see, you end up with something which is close to that of a random graph. If you think about it carefully you'll see that it's actually not the same thing as a genuine Erdös random graph. But the differences are not important. Well, not for this exercise. Obviously if P is somewhere in the middle, then some things get rewired and many things do not. And so this is for some small value of P.

And what we're interested in is what happens in the middle. And so we need some statistics to measure. There are again many, many bearings on statistics -- the director, the computer scientist and social network people. We're just going to think about two of the simplest here. The first is simply a measure of distance, in the characteristic path link or might as well just say average point-to-point link between a pair of vertices measured purely along the edges. And you're interested in the shortest path link, which may or may not be unique. That's not important.

And then you average that over all pairs of vertices. And if you have the entire network, you can do that exhaustively. And that's what we call a global statistic, in the sense that in order to measure the path link, the shortest path link between a single pair of vertices, you have to know, in essence, everything about the network. The other thing that we're interested in is a local statistic where we only need to know information about the neighborhood of a given vertex.

And the idea is to pick your vertex V, look at its neighborhood, which is defined simply by the vertices it's connected to, and then simply look at that as a subgraph in itself, and the edges which connect the friends to each other. And the question is how dense is that neighborhood? What's the average probability that two of your friends know each other? And so we can calculate that -- it's strictly between zero and one for every vertex -- and then we can average that over all vertices. And we can come up with another measure of the clustering. And again, this is a local statistic.

So before looking at the intermediate case, we can actually make some statements about what happens at the extremes. Because these things are known or trivial to figure out. When the lattice is completely ordered, L is large, in a sense that it depends linearly on -- you double the number of nodes in the network; you double the typical separation. And so for large end this is going to be a big number. And the clustering is likewise large. And, in fact, it tends to three-quarters for a regular periodic lattice when K, the coordination number, is sufficiently large. And so you know, that's on the scale of zero to one. That's close to one. It's a big number.

For the opposite extreme, if we look at purely random networks, we see the opposite thing. We have a small L, which is, as I said before, logarithmic, and N. And so note that as N gets very big, the difference between these two things is growing roughly linearly in L. So you're going to see very large differences between the big world limit and the small world limit. But the clustering is also small. And it in fact, vanishingly so, as N goes to infinity.

So you might expect and you might predict, and it would be quite reasonable to do so, that you can have one of two cases in one of these large sparse networks. Either the thing can be large and highly clustered, or it can be small and poorly clustered in the sense of -- small in the sense of path length. So maybe it's surprising to you, maybe it's not. It depends on your intuition. It certainly was surprising to us when we first saw this that in fact a large regime in the middle -- note this is on a logarithmic scale, so if this were on a linear scale the length would just dip almost like an L shape -- that for a large interval of P in fact we see this combination of very small characteristic path length close to its asymptotic limit. But very large clustering. And note that this things are scaled to put them on the same axis.

And so this intermediate range of high clustering and small characteristic path length is what we call small world networks, simply for the reason that it corresponds to this notion that we have in sociology that this is something about the world that we find surprising. A lot of work has been done recently on this class of networks. And in fact, a nice review has been written recently by Mark Newman at Santa Fe, and that's available on line, I can tell you, at the Los Alamos arXiv, along with a bunch of other papers that have sought to find solutions essentially for the length problem.

So there's a lot of technical details that are now available. But the intuition is actually pretty clear. Which is just that a small fraction of long-range random edges has a dramatic influence on the global properties of the network in this sense of characteristic path length. But negligible effect on the local properties.

And so to see this, just imagine the world consists of people standing in a ring, and you're all holding hands, and you know the fifty people on your right and the fifty people on your left. And let's say there's a million people in the ring. So the idea is to pass a message to somebody who's on the opposite side of the ring. And you can only go around obviously in steps of fifty. And so that's about ten thousand steps around the ring. And thus the characteristic, the typical length is about that. So you're talking about a network with five thousand degrees of separation.

Now imagine that you and the person on the opposite side of the ring are given a cell phone each, and you can communicate directly. Well, the powerful effect is not just that you can now communicate with them directly and thus cut your path length down from ten thousand to one, but that your friends can communicate with their friends.

Instead of just under ten thousand steps, they can now do it in three. And your friends of friends and their friends of friends can now be connected through that single shortcut. And so one connection can have a vast impact. And in fact, you can show that in an infinite limit, about between five and six shortcuts will have the typical path length of a ring lattice in this sense, regardless of the size. But obviously at the local level most of your friends still know most of your friends, and nothing much has changed.

So there are a number of other issues to move on to. The first of which is, this is obviously, you know, a very simplistic model. And so there's an obvious question of whether it can actually tell us anything about real networks and about the real world, if we're to get anything really interesting out of it.

Secondly, and related to that question, is how general is it? I mean, is this something which is only true for social networks? It's certainly something that we only started thinking about in social networks. But can it be shown to be applicable to other kinds of networks as well? And does it actually have any effect? I mean, is it simply an intellectual curiosity, that we know this about the world now and we can all go on with our lives? Or is this something which is important in the sense that the structure of these networks actually has an influence on the sorts of things that can happen?

And finally, if we get to this, we often say with awe, oh my God, six degrees of separation, this is wonderful. We're all connected. But in fact, six is a big number. And obviously if everybody is six degrees of separation away from you, well, you're kind of back to where you started from. And so there's been some recent progress made on this question of how do people actually find things in these sorts of networks.

So the first thing to observe is that social networks are just one kind of network, and in fact, depending on your background, when you say the word "network," you have a completely different idea in your head about what we're talking about. And we were careful in the description of the model itself not to put any information in about the actual constituents of the network. And so in principle this sort of thing could apply to any of these, any particular application, as long as its topological description is correct.

And so we actually went looking for some examples of real networks. And as I said, even a few years ago, this was kind of hard to do. But we found some in some unusual places, the first of which is a parlor game that was popular at the time known as the Kevin Bacon game. And the idea is this, that you -- I'm sure you've heard of Kevin Bacon. He was in *Footloose*, right? And in fact, Mark just got to meet him recently, so that was interesting. But so the idea is that there are three fraternity brothers at William & Mary, were in some altered state of mind when they realized that they could actually connect people to Kevin Bacon in a remarkably short number of steps. And how they do that is through a collaboration in movies. If you've acted in a movie with Kevin Bacon, then you have a Bacon number of one. If you haven't, but you've acted in a movie with someone who has, you have a Bacon number of two.

And some of you are already thinking this is like Erdös numbers, and indeed it is. And what they found is that they couldn't think of anybody who had a Bacon number greater than say four. And so they came up with this hypothesis that Kevin Bacon was the center of the Hollywood universe. And his career has indeed since enjoyed a revival. And you'll see the irony of that in just a bit.

So what we can do is we can calculate not just Bacon's number but Sean Connery's number and everybody else's number. And then we can average all those and see what the average actor number is. In fact, what is the average path length for this universe of movie actors? And feel free to guess amongst yourselves what you think that is.

Meanwhile, the second network that we looked at is a completely different structure altogether. It's actually the power transmission grid of the Western United States. And we were very fortunate to get this done. I should say we got the Bacon information from Brett Tjaden who was a computer scientist at University of Virginia, but better known to the world as the Oracle of Bacon. Which was for a while the most popular web site in the world. Which obviously means some people have too much time. (Laughter)

But this was data that was given to us by some colleagues of ours -- Jim Thorp and Koeunyi Bae at Cornell who are power engineers who are actually interested in things like large power failures. But they also have the complete network of power stations and sub stations. There's roughly five thousand of them in the states West of the Rocky Mountains, and all the high-voltage transmission lines that connect them.

And finally, the third data set that we have -- any biologists in the audience will know what this is -- is the worm, C. elegans, which has had its genome sequenced and many other remarkable things discovered about it in the last thirty odd years. But one of the things that is known about C. elegans is the complete structure or almost the complete structure of its neural network. And that has been mapped out in a book which comes with a, back then actually came with a floppy disk, which has the adjacency matrix of the neural network.

And so we were able to do these calculations of length and clustering with these three networks, and here are the various parameters. And you can see that the -- here's the comparison. You have the length, the characteristic path length of the real network, compared with the characteristic path length of a hypothetical random network with the same parameters. And here are the clustering coefficient comparisons.

And you can see, let's look at the film actor, that the typical number of steps from any actor to any other actor is about 3.5. And it could only be about three. So, this is a surprising number, right? If we take these, if we take -- imagine it were a ring instead of the graph that it is, and we had N of two hundred and twenty-five thousand and K of sixty-one -- then the average path length would be roughly about two thousand.

So in principle, given any topology at all that this graph could have, this number L could be anywhere between two thousand and three, and it's 3.65. So it's close to a random network. And yet the clustering coefficient indicates that it's nothing like a random network in terms of its local structure. It also indicates that there's nothing special about Kevin Bacon. (Laughter)

They could have picked anybody, anybody you've ever heard of, and almost anybody that you haven't heard of, and they would have found exactly the same phenomenon that they -- maybe it would have been 3.8 instead of 3.5. But no one would have noticed that. It still would have been this remarkable phenomenon that everybody could be connected to them in only a few steps. And it just happened to be Kevin Bacon, of all people.

So anyway, since the last couple of years, some more recent work has looked at other networks. Again, just completely different kinds of networks with completely different coordination numbers and different size, you know, from the World Wide Web to biochemical reaction networks to ownership networks of German firms. And even more recently than that, there's -- I've been looking at a network of interlocking boards of directors in Fortune 1000 companies in the U.S. They all show this same combination of statistics, which is small lengths and large clustering.

And so, this is really just saying something about the world, first of all that real networks are not random networks because they have, for various reasons, either social or physical or biological, utilized local ordering. Local order, local clustering is a useful thing or simply a cheap thing to do in real networks. But that even a small amount of randomness will suffice to make the global path length comparable to a random graph.

And so you could probably tell elaborate stories about evolution and convergence to an optimal solution, and maybe that's true. But it is also just a very simple reason, which is just it's very likely that that sort of thing is going to happen. So okay. So this doesn't answer the question of whether any of this matters, in terms of whether behavior is affected by the structure of the network.

And in fact, to me this is a much more -- you know, we started on this project thinking about dynamical systems, in fact, thinking about crickets and how they, the very same crickets that you heard about earlier, except the question here was how do they synchronize to make more noise so that however they find them, they find them more easily.

And so we started wondering about the topology of coupled oscillator networks, and whether that was influential in whether things could synchronize. And then we started to think about, you know, prisoners' dilemma games and cellular automata solving density classification problems. And disease-spreading dynamics. And it turns out that in all these examples the structure of the network has a dramatic effect on the global dynamics of the system.

What is unclear is that it seems to be a different effect in every case. So unifying that particular set of phenomena is a very interesting and open question. But let me just describe very briefly a problem of disease-spreading. It's just a very simple, the classical SIR model of disease spreading. But you can just think of it as rumor-spreading if you like.

And the idea is that you take a graph like this, and you fix P somewhere between zero and one. And then you run the -- you have an uninfected population. A single node gets infected. And then they have one time step to affect each of their neighbors independently with some other probability, which is the infectiousness.

And then you just run the dynamics until one of two things has happened. Either the disease has died out and infected some fraction of the population in the meantime. Or else it takes over the whole network in some characteristic time. And so you can see that in the random, the solid triangles or the random limit, you see the classical tipping point here where no one is getting infected, and then all of a sudden when more than one person gets infected you see it per, on average, when each person infects more than one on average, you see exponential growth of the disease.

And a large chunk of the population is getting infected. You see a much less rapid transition in the case of a one-dimensional lattice. And then the small world graph, predictably enough, is somewhere in between. And the question is how rapid is this transition? So we can look at that by asking, well, for what value of the infectiousness here do we see some fraction of the population infected? So at what point do we see a real epidemic?

So we can take a cross-section of this and ask when is it crossed, for what value of the infection is it crossed? For each value of P. So here we have this aha, the epidemic point, if you like, for a disease as a function of P. And so you can see that initially there's this very rapid drop. In fact, it's just kind of a straight line on a log scale. Although I don't -- it's not actually an exponential. It's more complicated. And we understand this a lot better now.

The point is really just that a very small fraction of random edges has an initially dramatic effect, and then a lot more random edges have very little additional effect. And finally, that -- and this is a slightly more obvious point -- that in the world of a very highly infectious disease where everybody has been infected, the question is, how long does it take to happen? And you can see here that again there's a very rapid drop in the characteristic time as a function of the randomness. And in fact, it's very closely related to the characteristic path length.

So I just want to briefly address the other issue that I mentioned. This is a model of broadcast communication. You tell everybody and they tell everybody, and they tell everybody, and how fast does this message spread throughout the population? It's not a terribly realistic method of communicating for, you know, for personal messages. And particularly, it's not a very realistic method of searching.

So let's go back to Milgram's experiment, and say well what if we just send the message to one person and they had to send it on to one person, how do we find people in a small world network? Rather than, how do we just tell everybody? And it turns out, this is actually a very hard thing to do. And I know it's hard because the predecessor of the Bacon game was the Erdös game where Paul Erdös was the founder, you know, of probabilistic graph theory, and published fifteen hundred papers in his lifetime and a few after.

And it was considered very prestigious to have published a paper with Erdös, and thus have an Erdös number of one. But if you didn't, and there's about four hundred and sixty people who did, then maybe you published with one of them and you had an Erdös number of two. And in fact, if you have an Erdös number of one, it's obviously, you know that it's one, and if you have an Erdös number of two, you can just kind of look at the list and see if you've published with anybody there, you probably know that.

And so two is easy as well. Beyond two it starts to get difficult. And in fact, I know this because when I published my first paper with my advisor, Steve Strogatz, he spent about two days trying to figure out his Erdös number. And he couldn't think about anything else in the meantime. (Laughter) And his is four so mine is five. So my calculation was trivial.

But anyway, so this search problem is a hard problem, and it's been shown for instance for graph coloring that in fact in these small world networks, in fact it's much harder to solve the problem on a small world graph than either on a regular graph or on a random graph. And some recent progress have been made – oh, which raises the question of how people actually do it.

So it's hard, but it should be extremely hard. In fact it should be no more difficult, it should be no easier to pass a message in this kind of small world graph than in this kind of graph where you simply have to – in the sense that it takes order and time. And the reason is simply that these shortcuts, although they are large in the sense of typically spanning great distances, they're all over the shop. And so the chances are that one will take you close to your destination, assuming you have a destination. But then the next one you reach is just as likely to take you back far away again. And so the question you might ask is, well how do people actually do it, and what does that tell us about real networks as opposed to these toy models?

And Jon Kleinberg at Cornell has recently proposed a solution to that problem, which is a very nice idea. Simply to say that the uniform distribution that I used here was not realistic for real-world networks. In fact what you need at the simplest level is a distribution that decays like a power law where the exponent of the power law is the same as the dimension of the underlying lattice. So what that does is it still involves creation of randomly rewired edges. But now their distribution is correlated with a dimension of the lattice.

And so you actually have information about the lattice encoded in this distribution of random edges. And it turns out that with that stipulation, not only are there short path lengths, but you can actually find them. If the power law is too steep, then there aren't enough short path lengths. So there's this compromise between needing enough random edges to have short path lengths, but having enough information about the lattice to actually be able to use them to construct these path lengths.

And so really the point I want to raise by that is to say that there is a lot that we still need to know about these kinds of networks in order to say useful things about the useful world, and that it's going to involve contributions from, not just from sociology and economics but also from applied math, from computer science, from statistical physics. So it's a very interesting problem in that it overlaps so many of these fields. And we're really just at the beginning. Thanks very much.

(Applause)

M: Two quick questions. Is there a strong K dependence to these results -- two, three, four, five, six?

M: I mean, in the sense that the numbers change with K in a predictable fashion. But that the -- I mean, there's a universal scaling law that just scales out K.

M: And has anybody studied inhomogeneous K, if you randomly vary K across the network?

Duncan Watts: Actually there has been some very nice work done by -- I mean, in fact, all these models have a distribution in K, once you introduce randomness, in the sense that a random graph has a Poisson distribution in K. But in all these models that I've been talking about, the distribution is tightly coupled -- you know, it's sharply peaked around the average.

But László Barabási and his group at Notre Dame have been looking at -- and also Bernardo Huberman and his student at Xerox Park, have been looking at classes of random networks where there is in fact a power law dependence in K. And this is much more representative of things like the Internet where you have some very highly connected sites like Yahoo! and so on.

And in fact, it turns out for those things that they can be even smaller than small world graphs, because they're more hub-like. And that's a different -- in fact, one of the big questions is, what classes of networks are there, and what parameters do we need to measure to be able to put a given network in a class? And certainly this idea of say exponential tail versus the power law tail in the distribution of K is one distinguishing feature that we need to be thinking about.

Lionel Sacks: Hi. Seems to be a factor missing from your presentation which is also largely missing from everybody else this morning, which is the budget. To create pretty patterns takes energy. And most organisms or systems will select which pretty patterns are most effective, most cost-effective somehow. Because that budget is finite, whether it's energy or whatever resources are the cost function. So okay, fine, as you said in your graph, one graph will take you further away and nearer and further away and stuff. But then there's no economics of which of the connections you actually keep and then take [??] the research --

Duncan Watts: Oh, yeah, I wasn't very explicit about that. I think that actually that is the explanation for the existence of clustering. It's precisely for that reason. You know, why -- so we can state that most of our friends are friends of each other. And in fact, most of the people that you become friends with you were introduced to by people that you already know.

Why do you do that? Well, because it's easier. Right? For that reason alone. It's energetically less costly to know people who are socially close to you than people who are far away. We could in principle introduce ourselves to people on the street, but you know, depending on where you live, that may be a costly exercise. (Laughter) So you cut down on a lot of that by simply relying on introductions through your current social network. Nevertheless we have different groups of friends who don't know each other. Sometimes we know a single person who doesn't know any of our other friends. And so there is this -- and we just model that -- that there's presumably complicated reasons for that -- we just model that with the introduction of a small amount of randomness. But I think that those energetic considerations are absolutely what leads to clustering in all these different kinds of networks. And it needn't necessarily be social costs.

M: It's a question which you might just have answered, but it concerns the insides of a cell. Cells, biological cells, are full of networks, genes, controlling enzymes, and you mentioned metabolic pathways. Can you explain to me why a cell would evolve small-world networks, or why it might evolve small-world networks in its regulatory systems? And secondly, if you can explain why one should expect to evolve them, why it should continue to maintain them when it could adopt perhaps other structures?

Duncan Watts: It's hard for me to answer that question in terms of metabolic networks. But certainly neural networks, the answer is more obvious. You have -- in a sense, it depends what your connection is. Like if you have a physical connection, then you can obviously say that it's less costly to build a short one than a long one. In terms of metabolic networks, it's not quite so clear to me.

M: Okay. Perhaps it was an unfair question. But the question I would be answering if I were to look for small-world networks [??] on one side robustness, and on the other side efficiency.

Duncan Watts: Right.

M: Now presumably robustness and efficiency are the sorts of parameters which evolved from the small world networks inside of neural networks world.

Duncan Watts: Right, if you're talking -- so there's an argument in terms of dynamics as well. You could say that these things happen simply because they're easy to build. But it also may be true that they do actually perform functions better than either random -- and there's actually a significant amount of evidence that suggests that that's true.

That, and precisely along those lines of either flexibility or speed of adaptation to an external signal versus coherence. And that these things actually tried off against each other, and that this is some kind of compromise in the middle. In social networks the example is fast diffusion of information, but maintenance of cooperation.

M: So they could be the equivalent of an edge of chaos? This would be a broadband [??].

Duncan Watts: Well, I don't know about that. (Laughter)

M: Sorry, actually.

Duncan Watts: But it's an optimization problem, you know, between trading off between different constraints. And it makes a good deal of intuitive sense. Pinning it down is the tough part.

M: You mentioned epidemics in your talk, and I assume that there's lots of movies and things like that which talk about this issue -- you know, Twelve Monkeys and the Andromeda Strain. And the question is, I think it's probably fair to say that small networks are how you would imagine disease factors would actually get propagated. And the question is, from your studies, what do we have to be concerned about? I mean, we haven't all died of AIDS or some --

Duncan Watts: I think there's something you can say, in that in fact, you can construct essentially a correlation length for these networks, in the sense that you have this low-dimensional lattice and it's superimposed, and now you have this small fraction of random edges. And so you can measure the typical distance between ends of random edges.

And you can define that as the correlation length for this network. And how that is useful, in that if you have a cluster growing -- so if you have this infectious disease spreading -- initially it spreads like the dimension of the lattice. So it spreads in a low-dimensional fashion. But then once it exceeds the correlation length, then it starts to hit shortcuts which then of course lead it to hit other shortcuts, and so you see that the growth of these things is exponential.

And so the dimension of these small world lattices is actually size-dependent. And how that relates I think, if you were to wave your hands for a little longer to real disease-spreading is that, in fact, what you want to do is make the correlation length as big as possible. So you want to -- if the diseases are going to spread from person to person, you want that geographical spread to continue for as long as possible before they get to you. So abolishing airports would be a good way to slow down the spread of disease.

But also needle exchange programs do the same thing, right? So you know, preventing the transport of dirty needles from one place to another -- which is essentially like introducing a shortcut, a random shortcut between anonymous strangers -- and so in that sense you can, by reducing the density of random edges you increase the correlation length and you slow down not just the spread of the disease, but also you increase the likelihood that it will die out before it hits the correlation length.

So I mean, these are very simple intuitions that we can get from these very simple models. But you have to be really careful about it, because disease dynamics are much more complicated than these SIR models, and the networks are more complicated as well. But there's the hand-waving explanation.

David Campbell: If you're actually interested in the AIDS dynamic, that was looked at at Los Alamos some years ago by Mac Hyman and Stirling Colgate with similar analysis to what Duncan said. And if it had been around a network we will [??].

M: It's pretty bad in Africa right now.

David Campbell: Yeah, it's a combination. Well, thank you Duncan.

Chapter 6

Complex Engineered Systems

Dan Braha
Session Chair
Nam Suh
Complexity and Design Engineering
Steven Eppinger
Product Development Complexity
Michael Caramanis
Scale Decomposition of Production
Dan Frey
Complex Systems Integration

Introduction

Dan Braha: Good morning. Welcome to the complex engineered systems section. You know, it is often believed that engineering is fundamentally different from the natural sciences. This belief is based on the contention that engineering is a prescriptive science as opposed to the natural sciences, which are descriptive in nature. Consequently, it's said that engineering is placed at a lower rank as far as complexity is concerned.

Personally, I believe that this contention is somewhat artificial. But even if we accept it partially, the complexity of engineered systems increasingly grows exponentially, over time. This is reflected both in terms of the artifacts as well as the processes that lead to the creation of those artifacts. It is

228

compounded by a multitude of dynamic aspects that affect the engineering process, such as the social and human factors.

In addition, engineered systems affect the micro- and macro-economics environment, which in turn affect customer needs that actually trigger the development of new technologies. Interestingly, customer needs are a product of natural evolution, and this creates, in my view, an interesting link between engineering and biology.

The role of complexity in engineering is twofold. First, we'd like to develop description languages for capturing the information that are related both to the producers as well as to the dynamic aspects of the product development itself.

So, just to illustrate, we may want to develop languages for capturing the intrinsic structures of bridges, of architectural design. We may want to develop dynamical models that will capture the part information of product development teams.

The second role is prescriptive in nature, and it is related to identifying the main sources of complexity that are inherent in engineering systems. These will enable us to eliminate or to minimize those complexity elements, thus increasing more traditional performance criteria such as cost and performance.

Just to illustrate this point, we may think of two examples. In engineering design, a common strategy is to divide the problem into loosely coupled components, solving the different problems of each component, and then gluing together all the solutions.

In manufacturing control we may adopt several strategies that will lower the complexity. We can mention hierarchical control, or we may use decentralized control, thus distributing the information among the different agents in the system. Our speakers will address some of these points today. Overall, we have to remember that, while some scientists indeed describe the inherent complexity of nature, engineers have to deal with it.

Nam Suh
Complexity and Design Engineering

Dan Braha: Without further ado, I would like to proceed with our speakers. Our first speaker today is Professor Nam Suh. Professor Suh is the head of the Mechanical Engineering Department at MIT. He was the Assistant Director for Engineering at the National Science Foundation. He's the recipient of several honorary doctorates. He's a foreign member of the Royal Swedish Academy of Engineering Science. And he is the developer of a well-known design theory, called axiomatic design, and today he will talk about this theory.

Nam Suh: Good morning. It's my pleasure to be here. Today I'd like to talk about sort of my view of complexity, which may be quite different from what my colleague, Ed Crawley, I understand talked about. And that's very natural at MIT. At MIT, we have a thousand professors and a thousand opinions, maybe more sometimes. So I'll just talk about how I view complexity. Usually I talk mostly about hardware-related things, and then near the end, I'll talk about complex-associated software, how you measure complexity in software, and so forth. And now I'm beginning to do some work with a colleague at Mt. Sinai Hospital who is dealing with biology, and hopefully some day I can talk about the biological systems, too.

So, as all of you know, engineers design many different things: hardware, software, and systems, and you can define all of these in your own way, but we roughly have some idea as to what we mean by hardware, what we mean by software, and so forth.

So, when we deal with engineering design, always there is the question of this complexity, because we like to know whether or not what you design is more complicated than it should be, or if it can be made less complex, and how you measure and how you reduce complexity, if you knew how to measure it.

So let me show you a few pictures here. Some of you who are not mechanical engineers may not know what this is. This is a throttle body, and Qi Dong is one of our Ph.D. students, and she did a term paper on this thing, so that's the reason I have this thing. First of all, I didn't have to do very much work; I just copied what she sent in.

And what this thing does is, obviously, mix air, try to mix the right ratio of air with the fuel, so this is the thing that lets a certain amount of air in. But the functional requirements, what the car companies want to do with this, are not only do you have to control the ratio of air with the fuel, but it also has to give the driver the right feeling, by making sure that your acceleration panel, when you depress it a little bit, you've got to have the right response from your vehicle.

So what they did with this is they fine-tuned this thing, and this is the cross-sectional shape of it. Actually the way many of these things are made and controlled hasn't changed very much. They put it in the car, and they fine-tune it. And then the people in industry go through I guess it's called DSM, design system matrix, something like that, to see how certain parts are related to other parts, and then try to organize their companies around those issues, and then be able to control it, and so on and so forth.

But really the reason I'm showing this to you is, is this complex? Is there an absolute measure for the complexity of this thing? Can you reduce the complexity of this thing, if I give you a lot of money and a lot of time? Depends how much I guess, huh? I'm sure there would be a lot of takers if I had said you can have an infinite amount of money and an infinite amount of time to improve this.

It turns out the car companies are spending a lot of money on this, year after year, all the time, without making an awful lot of progress. The question is why. So I want you to think about that. And in the end I can talk about some solutions to that, how you can make it very simple.

This is another one of these problems that we deal with in industry, and this machine is called a track machine. Essentially what this machine does is take a semi-conductor wafer—now it's about eight inches in diameter, and we are going to twelve inches in diameter—and we have to spread on top of the semi-conductor wafer a thin layer of photo resist. This is a chemical that reacts to light, so we can print a circuit on it.

It turns out that the photo resist you put on this thing has to be very thin, less than about half a micron, and the thinness of the whole thing has to be within few atomic distances, because the wavelength that we are using now is 19 nanometers, and the depth of focus is a fraction of that, and therefore you have to have the whole surface reflect.

So what this machine does is the following. These are what I call modules; the modules consist of various different kinds. Some modules simply clean the surface of wafers; some, they spray the surface with solvent, and then the photo resist comes in, and then it spins the photo resist and tries to make a thin layer.

And when you do this—this is a very simplified picture of this—there could be 48 or so modules, and wafers come in at this end, and wafers leave at the other end, and then this robot has to come and pick it up and put it into the right stack. The question is, what's the best way of getting the maximum productivity out of this machine? It turns out the productivity of this machine is very limited by the speed with which this robot can transport the part. Because we can put in as many modules as we need to be able to do that.

So in this machine, wafers may come in one every thirty seconds, and then this wafer is picked up and put into one of these modules, and after that operation is done, you take it out and put it into the next module, and so on and so forth.

The problem is, when you try to schedule this robot, you find that sometimes the wafers you have to pick up from one of these modules are all waiting for the robot to come and pick them up nearly at the same time, within the time for the robot to make one motion.

So if you're in a maximized productivity, how do you schedule this thing? And indeed, lots of people went in and tried to optimize it. But, interestingly enough, it's not a problem that you can optimize, until you design it correctly. And the reason is, when you design a complex system, when you have a system that -- I'll talk more about what I mean by complexity, but for the time being, if you can accept the fact that if you have a wrong design, you cannot optimize it.

Part of the problem I think the community has had over the years is the fact that they think mathematics will solve some problem. In a lot of companies, like Ford and Chrysler, what have you, there are a bunch of these statisticians who think they can improve the quality of their product. Now they are going through the Six Sigma cycle; they are fascinated by this thing. But the point is, they are not going to improve the product by simply applying statistical techniques, until you get the design right.

And why that point is so difficult for these guys to understand, I don't know, but that's really a very important point in terms of product improvement.

This is another problem. I'm trying to tell you what the problem is, and I'll quickly go over what the solutions are, so you understand where engineers come from and how they look at it. This is a machine developed by one of the largest, in fact, the largest printing machine company in the United States, maybe the world's largest. And what they wanted to do was come out with a printing machine that prints commercial labels, so they can print the commercial labels very quickly. And then, if different customers come, all they have to do is put the sample in there and, just like any other copier, just keep copying it. But since it has to be done at a very high speed and high quality, they had to develop a new machine.

This machine sells for about a million dollars, so it's not exactly a desktop copier. This is about 3 feet in diameter, an aluminum drum, coated with selenium, and what you do is put the image there and when light hits the selenium then certain parts retain charges, some places do not retain charges, and then it comes around here, picks up toner, which is charged, comes around here, and then this wiper roll removes excess ink, and then it goes and prints on the paper again. So this is done at a very high speed. The toner in this case happens to be liquid, because of the speed required.

The problem they had with this was the following. When they came to see me and talk to me about this, they came to see me because they had scratch marks on the surface. They found that once in a while the selenium coating would be scratched off, so they said there must be abrasive wear. And I happen to dabble in tribology, and so they came to see me about that. And I said, gee, so abrasive wear is a very simple problem. They went away, and they called me, saying that you've got to come see what's happening here, because still it's there, after trying many different things.

So I went there. I found that the company added a lot of people, a lot of Ph.D.s and so forth, and put in fancy filters and all sorts of things, because they thought the scratch marks occurred because of abrasive wear, and that by putting filters in so that the toner material would not have any hard particles at all, they could solve the problem. And they couldn't solve it, and it turned out that the problem was a design problem rather than anything that had to do with particles or anything like that.

So when you look at a problem like this, the question is, how do you know whether it's complex or not? Why was it so difficult for 30-some top-notch engineers to solve this problem? Was it really complex, or was it just in their heads that it was complex? Whenever people cannot solve something, they think it's very complex.

Once they solve it, they think it's very simple. Why is that? And these are the questions that one has to address, in my opinion, when we deal with the complexity issue. So that's what I'm going to talk about.

So that's the topic that I'll be addressing. And I understand that not only scientists, but a lot of engineers, all think that there's an absolute measure of complexity, and I tend to believe that's not the case, and I'd like to sort of go over why I don't think that's the case. And I'll use some of those things I talked about.

In my opinion, complexity is a relative quantity; you can only compare complexity relative to something else; you cannot come out with a single number that says the complexity of this is so much. And I'll try to make my case as to why I look at it that way.

In my opinion, complexity can only be measured relative to what it is you want to know or you want to achieve. Depending on what it is that you are trying to achieve, the complexity's going to be different. And it turns out that lots of times when people are, say, "complexed," they are really not "complexed," but rather they do not really know what they are doing, and therefore things appear complex.

And also I'll first give you my conclusion of this, and then I'll just go through the details. I think there are two kinds of complexity. One is called time-independent complexity, and the other kind is time-dependent complexity, and they have entirely different characteristics. And within the time-independent complexity we have two kinds; one is real complexity, and the other is imaginary complexity. The one that I talked about in terms of the printing machine problem, that was imaginary complexity, namely, that it wasn't really complex, but people thought it was complex. In other words, they had a perfectly good design, but they didn't know that they had a good design, and they just fiddled around with all sorts of knobs, and they couldn't make the machine run correctly.

In the case of time-dependent complexity, I said there are two kinds. One is combinatorial complexity, and the other is periodic complexity. And I'll show you that, in the case of the track machine, most people tried to program the robot simply treating it as a combinatorial complexity problem for which there is no solution, no deterministic solution. And I'll show you how you can change, sometimes, not in all cases, but sometimes you can change a combinatorial complexity problem into a periodic complexity problem, and you can solve it, and you can get a solution very easily.

So that is, to go from this to that, you have to redesign the thing slightly, not very much. And then you can get a very simple solution to a problem that appears to be very complex.

So we'll define what we mean by complexity, and this is my definition. You can either accept it or leave it, but all my talk will be based on this definition of complexity. I define complexity as a measure of uncertainty in achieving the specified functional requirements. It's very important to realize that I'm only concerned with the complexity associated with what I'm trying to get done, and engineers have to get things done. Translating that into a situation with biology, it's the same thing: biologists want to know how certain biological systems work, but the problem that I see in biology is that biologists typically deal with the interaction of one DNA with something else, and they keep studying this, and each time they study the pair, they publish a paper, and so on and so forth.

But in real biological systems, one has to satisfy a large number of these requirements at the same time and be able to have a living cell, living body, a living human being, and that is a different issue than simply studying one at a time.

So before I can go into details, we have to know a little bit about axiomatic design, because it's all based on this idea of axiomatic design. In axiomatic design, we say there are domains, and in order to come out with the design, you have to map from domain to domain, and in order to come out with the complete design, you have to decompose it, and so on. In the case of the throttle body that I showed you at the beginning, there are a lot of parts, and you have to assemble all of those parts, but at the highest level the functional requirements of that are only limited to 5 or so. In order to achieve these 5 highest requirements, you have to decompose.

And so we say, in terms of the customer's requirements, you want to be able to push down on your paddle, and car, you're going to go. That's really what the customer wants. In order to do that, you translate it into functional requirements for that, and then you say, well, the valve has turned so much, the butterfly type of valve has to turn so much, if you want to go at a certain rate, and then it has to match the fuel rate, fuel coming in, and so on and so forth.

And then, once you decide those are the functional requirements, then you have to come out with the design parameters to satisfy this. At the highest level, design parameters are roughly the kind of thing you want to achieve. And then you have to figure out how to make it, and then you decompose, back and forth, to come down to what we call the leaf level, to come out with a final design.

It turns out that when you map, we say you have to satisfy two things. One is independence action that says all the functional requirements (FRs) in the functional domain must always remain independent from each other if you're going to come up with a system design. When you try to come out with a very complex design, as far as I can see that's the only way you can do it.

One of our graduate students, a naval officer, just finished a thesis on how to design a ship, a naval ship that has to carry weapons and so on. He had to decompose ten layers and several hundred functional requirements, and so forth. And at the highest level, if you violate this thing, it becomes a hopeless task later on, because everything becomes coupled, and you have to deal with it.

The second action says minimize information content, and I'll talk about that more in relation to software, because that's sort of an interesting question that one can look at.

So one of the things that we say is, in engineering—this may apply to in science as well—we are trying to satisfy this FR. FR is one of those, not all of them, one of those things that we're going to satisfy, and this is density, and this, here, is what I call engineering range—that is, as an engineer, we'll be happy if your functional requirement, not your physical things but rather the functional things, can be satisfied within that range.

So you go ahead and design and make a throttle body and then put it together. And when you put all of that together, and then when you try it, the real system behaves this way, rather than being inside. In that case, when you're in here, some of the parts you made, some of the throttle body you made, if it's in here, you are very happy. You go home happy and be nice to your kids.

But if you find that one of your products that happened to be here, you think you wasted your time, and so on and so forth. So, in terms of this relationship between this engineering, or this design range and the system range, you think it's very complex, because some of the time it works, sometimes it doesn't work. On the other hand, if you came out with a design that lets this system range be right inside the design range, then you'd think, oh, it's nothing, that's a really simple problem. I didn't get my MIT degree to work for this company. They gave me such a simple problem.

But you can see that the difference between that is simply shifting this curve sideways, right? Sometimes you cannot shift it, because the way you designed the throttle body is such that you cannot shift it. Sometimes, you may. The question is, how do you design it such that you can shift every one of these curves for each one of these FRs such that you can make the thing work all the time, every time? If you can do that, you can say that it is necessarily very complex.

So we can figure out probability associated with being successful and achieving the functional requirements, and then we can translate into algorithmic functions, and we call that information, and then that's the information you have to supply to be able satisfy functional requirements within specified range.

So I define real complexity simply as uncertainty associated with achieving that FR. So if I didn't know any better about the throttle body, for that particular design I can indeed come up with a measure of information content, and I can sort of figure out, for each one of those FRs, how much information is associated with it. And I define the real complexity as being nothing but equal to uncertainty, as measured by the information content. And then you can optimize it, and so on and so forth.

But there is also a thing called imaginary complexity, and let's think about that. What is imaginary complexity? I showed you that printing machine case, and this machine here, right? I said that the problem really has to do with not really the real problem but an imaginary problem.

Suppose you have a design that consists of so many functional requirements that you must satisfy at the same time. And suppose that your design is such that when you choose these design parameters, you find that what relates these functional requirements to design parameters gives you a triangular matrix. In this case, if you come out with a triangular matrix, it turns out that it's consistent with maintaining independent functional requirements if I change these DPs in the sequence given here.

But if you don't follow the sequence, you may think the machine's no good. It turns out that, over all the possibilities, the only successful combination is the one that gives you the triangular matrix. Anything else will not make the system work. So there's the probability of finding that one combination out of so many is given this way. And you can see that if you had 7, say, functional requirements, which is on the high side, for most engineering design—most engineering designs may have around five or less—there are more than 5,000 combinations. Out of 5,000 combinations, you have to choose that particular one that will work all the time.

Now what that does is that, even if you work 200, I guess maybe in Germany 180, days a year, every day, and perform each experiment, you'd never get to the solutions. So what do you do? You produce something and then sell it, compromising, maybe, the performance.

So, in this case, you can see that you had a good solution. In this case, it turns out that it is a problem with the 7 FRs. And then, they had all the things done correctly. It was digitally controlled, but they couldn't make it work.

So when you look at it this way, it turns out the solution is just like that. We were able to solve the problem after discussing this for three hours one evening, two hours the next morning; we solved it, and that was the end of that.

Now I'll go into the details of how we solved it. But it's the kind of argument I just gave you. If you're interested, we can then talk about how you decompose it, how you figure out what to do.

The solution to this problem was very simple. The solution was to turn the wiper roll, which rotates in the opposite direction from this—this is the same direction, counterclockwise—but you have to turn this thing on first, and you have to make sure the surface speed of this roll is greater than the surface speed of that. It comes down to that simple control. It works all the time. You can put the chips into the tanks; it does not scratch the surface anymore.

Coming back to this problem, you can now appreciate what the problem is of this thing. The problem with this thing is just like airline scheduling problems. If you have bad weather in Detroit, and therefore airplanes cannot land and cannot take off, it will affect northwest flights everywhere else, right, Boston, and so on and so forth. And so it disrupts entire flight schedule.

236

Same problem here. If something goes wrong with any one of these things, then it will disrupt. Not only that. In this case, depending on which one you pick up first, when you have two or more wafers waiting to be picked up at the same time, depending on which one you pick up first, it will affect all subsequent operations. It's a common engineering problem, right?

The question is, how do you figure out which one to pick up? So the question is, how do you solve this common engineering complexity problem? So we start out—engineers always think about making money, because our employers are in the business of making money, and then we go from there, and then we decompose. So one of the things we've got to do is process wafers, and then transfer the wafers. And in order to process wafers we use modules. To transfer wafers we use a robot; those are the design parameters, at the very high level. Obviously, you have to go down to the nitty-gritty. And there are constraints, such as the throughput rate. You have to process so many per hour. Manufacturing costs must be less than something, quality must be, and so on and so forth.

It turns out that the scheduling was like so, such that processing wafers, transferring wafers, using the modules, using the robot, created a full matrix, and there's a coupled design. That means, if I want to only change the process, that affects my transport. If I want to control transport, that affects the processing of the wafer. And that's what I call coupled design.

So, the solution is, somehow come out with a design that will decouple it. When we have a triangular matrix, I use the term decoupling. You decouple it; that means to make this thing zero. That's a design intent. I'm showing you, this is the design intent, because if I do that then I can control my process; once I control my process I can independently control my robot motion.

So the question then as a designer is, how do I make this zero? For a long time, I had a hard time teaching students this, because as soon as they realized this, at the high level what we write down is a design intent, such that all subsequent decomposition or subsequent decisions have to be consistent with it.

Typically what people try to do, they try to figure out whether this is zero or not by going to the nitty-gritty level even before they design it. So typically what they do is they look at an existing machine and see why it's not zero. The idea is, at the highest level, it's design intent.

So one way of doing that is you can put in what we call decoupling, and you can solve this thing. I'll give you sort of an example also of what a solution is like.

Suppose you have five different modules that process this in five different ways. There's a temperature which is set, and then, at that temperature, the wafer has to sit in each module for so long. In this case, the tolerance is zero, because as soon as it's done you have to take it out, otherwise you ruin the wafer. In other cases, you have some tolerance associated with it.

So what we do essentially is we introduce Q, or what I call a decoupler, such that the decoupler will separate the need to have these modules picked up at the same time. So this was previous to the original scheduling, gives you all its wafers, four of them to be picked up at the same time, and then, by putting decouplers in, you can now stagger them so you can deterministically go in and pick up each one of them separate from each other, and just simply repeat this thing.

So what we have done is we've changed a combinatorial problem into a periodic problem. So now, every period, this repeats itself. So that's the solution.

Let me just say a few words on software systems and then conclude. What is complex in software? Sometimes software either works or doesn't work, right? And also the question is, if I give you two pieces of software that perform, that fulfill the functions equally well—say information content is zero—are they the same? Are they different? Those are the issues.

So there are a number of things you can do. You can measure the software. The main point is, you cannot talk about complexity of software by number of lines of code—it doesn't mean a thing. And again, you can determine the complexity in terms of real complexity. But also, another way of doing this is to count the number of times a program that performs certain functions fails, and to reboot it so you can measure information content this way. And then you can, again, define complexity in a similar manner.

And when information content is exactly the same between two pieces of software but one is ten times longer than the other one, is one that is ten times longer more complex? Now, according to this way of looking at it, the one that's ten times longer is less efficient, right? So you can define what we mean by efficiency, which is different from complexity. You can have two things—two designs can do the same thing, but one happens to be much more efficient than another.

The reason we cannot come up with an absolute measure is because we cannot say whether or not the design we have come up with is indeed the simplest design, because I cannot prove to you that although the thing works fine and what-not, that someone else cannot come up with a better design than that. And I don't think anybody can come out with the proof of that, and therefore I say complexity has to be measured in a relative sense rather than in an absolute sense. Thank you very much.

M: I think I have a rather strange design problem for you. The design problem is actually on my browser, and I've called it building a bridge, designing a cosmological bridge between science and consciousness from the viewpoint of complex systems.

I feel that this problem fits many of your criteria very, very beautifully. It is certainly a primary problem in complexity issues. And it is also a time-independent problem in that it is made up of both real and imaginary aspects.

When I look at the criteria that you have for informational content, there's neurology, biology, evolution, psychology of mind. And all of this, I think, builds into a very complex set of relationships that I would like to see sorted out and related to one another in such a way that we can make sense out of this issue of what it is about consciousness that has some kind of a root, in the body, in science, in issues that are very real and functional in terms that we already know.

So I would like to pose that to you, and see what you think about that as a design issue.

Nam Suh: That's sort of like the problem that Ravi Iyengar and I have been looking at. It's a biology problem, in the sense that in biology, really the ultimate answer they want to get at is not how one DNA molecule interacts with something else, and so on and so forth, but they, rather, want to know how the living system really works.

And so when you have all this data generated, and there are lots of papers coming out in biology showing the interrelationship between one to one, they do not ask the very question that you raised. Namely, if you want to have, let's say, a biological system, and translating your question to biological systems, one of the things a biological system has to do is you have to supply energy, it turns out, right? And once you supply energy, you have to supply reactant. If you supply reactant, you have to initiate the reaction. And third, it's not very different from engines, it turns out.

And another thing you have to do is you have to regulate, control the rate at which it goes, and then you have to replicate. In a biological system, unlike engineering systems, you have to replicate all cells. And in a situation like that, then it turns out, if you look at that, then I come to the conclusion—and people may say, that's foolish or it's too speculative, and someday, I hope, I can prove it—is that unless the design of a biological system is such that it is also a triangular matrix, living beings cannot live. That's the conclusion I come to.

So in going back to the very profound questions you raised there, I think the first thing you have to be able to answer is what it is you want to know. If you cannot state what it is you want to know clearly, then I say you cannot get very far, because in the absence of what it is that you want to know, it could be very difficult to talk about what it is you've got to do. So that's where the problem should begin.

Steven Eppinger
Product Development Complexity

Dan Braha: Our next speaker holds the General Motors Leaders for Manufacturing Chair at the MIT Sloan School of Management. Steven is co-author of a widely used textbook on product design and development, and besides academic research, he has consulted for more than 50 firms and international corporations.

Steven Eppinger: Good morning. I'm going to speak about complexity in my own perspective, which is, I think, related to several of the perspectives that are themes of this conference. So one of the conference themes is about emergence, structure and function, and I'm going to talk about a perspective which I call architecture of products, systems and organizations. So I'll talk about architectures and how we look at those.

And second, the theme of complexity that is defined about information content, I'll show you how we look at information content from what I call the architecture perspective. In particular, I'm looking at product development, that is, developing complex products, or systems, or complex systems. And what's interesting about fairly complex products is that these are both interesting technologically—that is, they're technically complex—but also, the process and the organization that create these very complex products are also a very interesting social system, and I'll show you how we blend this perspective of looking at both the technical complexity or the technical structure with this social structure and complexity.

So I'll start by showing you a quick perspective on information density in product development. So, this is a product; this is an office copier produced by Xerox. And as Prof. Suh said, there are many ways we can measure the complexity, and one of those might be with statistics such as these, in terms of how many people, or how many parts, or how many decisions are made in that complex organization process and product to create it.

And so these are useful perspectives. But it's really not clear which of these things we should be looking at. And so what I'm going to share with you today are three particular perspectives that I think are helpful. One is what I call the product or system level perspective. So, how do we particularly look at the product and ask about its complexity and its difficulty?

Second is the process that is creating that product. We go from beginning to middle to end. How simple or straightforward is that product? Is it iterative? Does it loop? Or is it a more structured, straightforward process?

And third is the organization that is creating this complex system. Is it a simple organization? Are there many, many complex interactions that we could not possibly understand it all, that need to be maintained, or is it a pretty simple structure?

And so I'll show you how we'll look at each of those. The perspective that we use to look at those starts from how we start the composition, as Dan just defined in the introduction today. We decompose things. Why do we decompose them? Because it helps us understand them, and it helps us develop them. So we decompose a complex system into subsystems; we do that all the time. We decompose a complex process, because we cannot understand the complex process, into simpler things which we might call subprocesses. Those might be made of many, many tasks, and then further, maybe, down into decisions.

We decompose a large organization, certainly, into also manageable units, and those might be called teams, or departments, or divisions. Eventually, we get down into individuals. So we need to look at these as problems, and we decompose them. We decompose them to simplify them, to manage them, to structure them. But more important than just the decomposition, I believe, is how they're related. So that's what I call the architectures that relate to the decomposition. So the pattern of the interactions among all these decomposed elements, which are now at the bottom of this tree, the pattern, which are these little blue lines, down there, shows us the architecture.

And we can look at the architecture of those systems, of those processes, and of those organizations—they may be the same, they may be different. This particular architecture shows a very simple structure, where we've decomposed it neatly, as Dan said—or modular is the word that's used these days—we've decomposed it in a modular structure, such that each of these, say, subsystems, which might be, say, the red boxes, only within them, interact inside and not across. And that's a very sensible way to decompose. If you read books about system engineering, they tell us basically to do just that, maybe processes in organizations that have similar things.

On the other hand, they're not always quite so simple. A more complex pattern is this one. So here we have interactions across those subsystems, or across those teams, or across those subprocesses. It would probably be much more difficult to develop that product or to live in that organization or to work in that process.

And so I believe that there's something about that pattern of interactions which is important. And this relates, I think, pretty strongly to what Prof. Suh was talking about in terms of the first axiom.

So, that brings up what some potential complexity metrics are. The number of elements is certainly one potential complexity metric. I think we've already rejected that one today. That's like looking at lines of code, or number of people, or number of parts in a product. So I think there's something about the number of elements that makes problems difficult, but I don't think it's what truly makes them complex.

So the second is this pattern of interaction among the elements—that's what I call the architecture. I think that's a more reasonable metric of complexity. So I want to look carefully at those patterns.

And the third is the alignment of the patterns from one perspective to another; that is, the alignment of the system architecture to the process architecture, or the process to the organization. I think there's something about that matching, or mapping, as Prof. Suh called it. And so I want to look particularly at that.

So what I'd like to do is to show you how we actually look at these patterns, and I'll show you how we look at them for each the system and the process and the organization. And then I'll show you the newer work that we're doing: how we look at those patterns across them, how we look at how one maps to another. And so I'll show you a bit of each of those.

So, as I said, the approach is that we study these patterns of interactions, these architectures, at these three levels; and I'll show you three examples first; and then I'll show you how we look at them across.

The first example is an aerospace example that comes from Pratt & Whitney, where we studied the development of what is actually the largest jet engine in the world, the Pratt & Whitney 4098 engine. It's what goes on the Boeing 777, of course, because they have a very, very big plane, with only two engines. The second is an automotive example. It comes from Fiat auto, where it will just show you an example of a development process that we've studied in some detail, where you can see the complexity of the process. And the third example is a second automotive example. It comes from General Motors Powertrain. I'll show you how we look at the organization which develops a product, an engine. And then, fourth, I'll show you how we look at a comparison between the two, and I'll return to the Pratt & Whitney, the aerospace example, for that.

So, first I'll show you how we look at system architecture. So this is, as I said, just one example of how we look at the system architecture, and you have to point to things now. So, this is the product; this is this engine developed by Pratt & Whitney. This engine includes 54 components. This isn't actually 54 parts. It probably includes tens of thousands of little parts, but 54 components, because there are many fan blades, many turbo blades, and there are many smaller parts as well.

And these are grouped into nine systems. The first one's called the fan—that's this front end of the engine. There's a low pressure compressor, a high pressure compressor, a high pressure turbine, low pressure turbine, and this burner diffuser, sometimes called combustor. Those are the six main systems.

And then there are systems which are a bit more what I would call distributed across, and they're the mechanical components that are distributed. And that includes, for example, the main shaft; so these are not trivial components, but these are not focused in this particular place.

And then, finally, externals and controls are two other systems, which include the piping, the wiring, the controls, across the whole engine.

So we can define the decomposition. I can just list the 54 components and that would be the decomposition. I don't think that would tell much of the story; so, as I said, we define the architecture as a set of the interfaces. So there are 569 design interfaces across these 54 components which we documented. And this matrix just shows you a sort of a graphical picture of all those interfaces. So, simply, these are the 54 components, listed down the rows. Same across the columns. And these marks indicate the interactions or the interfaces across all those components.

In particular, we've been working on ways to document design interfaces, and we do it a little bit differently for each type of product. For this aerospace product, it seemed appropriate to look at spatial interactions—that is, constraints, spatial/geometric constraints between the structural interactions, energy and materials interactions. And then data controls other kinds of information interactions, which tend to be simpler or more flexible.

So there are lots of these types of interactions, and the pink and the red represent the strength of those interactions and how many there are. So there's more detail in how we create all that, but let's just say it's possible to document those all. And I believe that this pattern of interactions tells us much more about the complexity of the system than simply, say, the number of components. And as Prof. Suh says, this pattern of interactions may be coupled; they may be decoupled; they may be uncoupled; and that all relates to this pattern. So that's the system perspective.

I want to define one more thing, which actually relates to the way I'll use this later. These first six systems are what I call modular, that is, they're local, or focused. So the fan, the compressor, the turbine and so forth, they're in a single place. So this is the fan, this is low pressure compressor, high pressure compressor, and so forth.

So these boxes represent those systems, or subsystems. So these are the first six. And the last three are these. And the interesting thing is—and I'm going to differentiate these when I show you what we do with these data—the modular systems are those that are focused, and the distributed systems are these last few at the bottom, which are not just distributed, because they're not focused in a single place, but, more importantly, they're integrative. They create the integrated function of the product. And they actually behave differently in the social context, which I'll show you later.

The second perspective is, this is an automotive example of a development process. And a development process for an automobile, as you probably can imagine, involves thousands of people, millions of activities going on, a billion dollars or so, a long period of time. So this is a big activity. And we don't usually document entire development processes in much detail, but rather --

Q: (inaudible)

Steven Eppinger: Yes. There's a different reason for this in the next slide, so I'll do them both. This one is actually largely symmetric. It's largely symmetric because if my component has an interface with yours, then you ought to tell me that yours has an interface with mine, and so it's largely symmetric. It's a fairly symmetric question we ask when we document the design interfaces. Whereas the development processes, they may be much more asymmetric because, in particular, what we're mapping are the information flows. An activity provides information which is required to execute those other activities. And those information flows may be more uni-directional, so these tend to be less symmetric matrices.

So this is a documentation of a development process, and in particular, this is a subprocess, this isn't the whole process for developing an automotive, this is just a piece of it. This is a particular piece, which Fiat calls digital layout, or digital mock-up, and this is particularly for the engine compartment. In the US auto industry we would call this engine packaging.

So we've got a few vehicle styles fairly early on in the development process, and we want to make sure the engine's going to fit in there—the engine's the big thing you fit underneath the hood.

But then you fit a lot of other things—the fan and the air-conditioning system and lots of other accessories, they call them, and other things—and they may not all fit. So then you may have to iterate the style, or other features or components. And so that process used to be just built up on what they call a buck or a mock-up, which is about this big, about a vehicle size, and they just literally build it up and see how it works.

Well, today they do that digitally, using what they call a DMU, a digital mock-up, so the DMU is the digital mock-up. So this is a subprocess. This is just that process for what they call the DMU, the digital mock-up. And in particular, this is a new process, which is why it is important to document it and teach it to people.

So I'll just run through real quickly how the process works, so you just get a sense how to read one of these process perspective diagrams. These are the responsible organizations, which I'm sure you can't read very well, but it has names like Styling Center and Core Layout Team and Extended Layout Team, and so forth. These are the names of the activities in the process, so each row and column of this matrix is an activity in the development process.

And the marks in the matrix represent interaction between them, which in this case are information flows, so information largely flows from early to late, which is why they're down here, below the diagonal. So the upstream activities feed information to later, downstream activity.

The process, like most or I would say all development processes, is phased. We do a certain phase of activities, and then we move onto the next phase. And these phases are represented by these boxes, which capture the interactions within a phase, and then, below the diagonal are interactions across the phases.

This first phase is called planning or project planning. The second phase is called data collection, where while they're building up this digital model of the whole engine compartment, they actually go and collect all the CAD data, the CAD models that are made by the engine people and the climate control people and all the other folks.

So they collect the data, then they pass it onto this next phase, which is called DMU prep, that is, they actually build up the model, so they collect the data. It's iterative. They usually find the missing data, so they go back and get more data, so that's an iterative process.

The next phase is called digital mockup verification—now you've got the mockup, what are you going to do with it? You're going to make sure it looks right. You start to analyze it. So the verification is a large number of analytical activities. They verify a style compatibility; they verify overall DMU with something. So there are a variety of analytical activities to make sure it's right. And this is very iterative, because it's not usually right.

But then they have to do what they call release it. Release means this is it; this is the model. Maybe we've narrowed it down to one or two, and now everybody can use this data, so all the rest of this very large development organization can use these data. And they have what they call three releases—in Fiat they call them freezes. Freeze DMU, step one; freeze DMU, step two; here's the freeze, step three. So they release it once, twice, a third time. And that says, this is preliminary, but you guys can use it and do the rest of the planning of body and interior and everything else. And then they do more analysis, they get more data, do more analysis, do a second freeze, more analysis verification, they call it, and a third freeze.

So this is a development process. We've done dozens of these development process models. We learn things about how to improve development processes. So this is just one I picked to show you, to say that it is possible to capture development processes. It's not the same as capturing the product architecture. This is the process which creates the product; this is not the product. And it's not the same as capturing the organization architecture. This is not the organization; this is the process they go through. So that's this second perspective.

The third perspective is an organization perspective, and I'll show you this in the context of a second automotive example, and this comes from some work we've done with General Motors. And it's the development of a new engine; it's a small block V-8 engine, which was just released recently; it's on the new Corvette and also a small pick-up truck.

And the interesting thing about this is, you may not have thought much about what it takes to develop a power train for an automobile, but it's not just a big team, it's 22 teams, in this case. There were 22 product development teams involved in developing this product. And why 22 teams? It's because they decomposed the product into 22 systems, or subsystems— they actually called them modules. So they decomposed the engine into 22 modules, and for each module they assigned a product development team, what they call the PDT.

And by the way, that's a pretty typical structure. You decompose the system into smaller things, subsystems or modules or components, and then you assign a team to each. You can imagine other possible assignments, but the simplest possible one is the one that almost everybody uses, so that was the case in this case. So they decomposed the product, or the engine, into engine blocks, cylinder heads, cam shaft, pistons, connecting rods, etc. And so each of those are what they call the modules, not because they're modular in any sort of mix-and-match way. It's for completely other reasons they call them modules. Actually, it has to do with the production system. In the factory, each of these teams is responsible for developing an entire module of the factory. That just means a big piece of machinery that makes one engine block every 25 seconds, and so forth.

And then, each of these teams has a structure. Each of these teams has what we would call a concurrent engineering structure. That is, they've got a cross-disciplinary flavor. They've got the right people to develop their pieces and do a good job of it. So this is the PDT, the product development team. The composition for the crank shaft team is what's shown, exploded, here, on the right.

And they each have that cross-disciplinary, very capable set of people, flavor—release engineers, designers, manufacturing engineers, and so on and so forth.

Well, that's just the decomposition. What the theme here is, I'm going to show you interactions, right? You can assume for the moment that this crank shaft team could design the crank shaft, and do a great job of it. And that the pistons team can do the pistons and do a great job of it. The question is, how do all those teams have to interact to make sure that, not just that we have great pistons and crank shaft, but we have a great engine, right? And this is the set of interactions, across all those teams, that they have to maintain.

And these data are also largely symmetric, because we do a survey, and we simply ask each team how much they need to interact with each other team. So there were 22 teams. We needed 22 surveys, surveys of the type where you actually need 100% response rate, so, you can imagine, there's a lot of bugging the folks to get these data.

Nevertheless, each team tells us how often and how importantly they interact with the others. And the units for this particular model basically are daily, weekly, or monthly interactions between the teams. So let's say the crank shaft team, which we looked at on the previous slide, told us that they work daily with a) engine blocks, weekly with b) cylinder heads, and monthly with c) cam shaft, and so forth.

And you can see that the engine block team said, yeah, we work daily with those guys. The cylinder heads team said, we work monthly with crank shaft, while crank shaft said they work weekly with cylinder heads, so there's not perfect symmetry, there's asymmetry, and that's due to differences of opinions. But you can just kind of scan the pattern, and you can see that it's largely symmetric, as you would expect.

So this is the set of interactions, and, rather than stopping right there, let me show you what they were doing about it. As you can imagine, this is a lot of interactions to manage, and you wouldn't just say, do you work and talk to each other as you see necessary? They actually have a plan for managing these interactions. The plan at General Motors, the plan is what they call system engineering, and system engineering actually means teams of teams. So these are PDTs, these are the teams, and then they group the PDTs, they cluster them, into what they call system teams. And this is the assignment of system teams. Basically each PDT is assigned to one of these system teams. So, the short block system team, on the left, is defined by these six PDTs— engine block, crank shaft, and so forth.

And they're essentially looking at what I would call the major mechanical side of the engine, the big stuff, where the valve (inaudible) team is looking at the upper side of the engine, or the minor mechanical side of the engine. And then there's an induction system team and emissions and electrical system team. And they would literally meet every other Thursday afternoon. It would be system team time, and all the people would go to sort of the four corners of the building, so to speak, and the short block system team folks would get together, and the valve train system team folks would get together.

And that was the plan for what they called system engineering. And if I mapped that into this matrix, you can actually see what they're talking about. Well, this is the short block system team, the valve train system team, and so forth. Now I've just moved the rows and columns and the matrix around, so you can cluster the team so you can see all of the interactions inside the short block system team, all the interactions inside the valve train system team, and so forth.

And you can see, there's a lot to talk about. There's a lot of interactions inside there. Unfortunately, while they probably do a great job of facilitating the interactions inside one of these boxes, they probably don't handle all the interactions, right? That is, in essence, they certainly don't handle all the interactions out here. And so I actually asked the program managers in this organization, how do we handle those? They said, hmm, it's not being handled by the PDTs, and it's not being handled by the system teams. Must be our job, right? The guys at the top.

And I scratched my head, and I thought, you know, we're talking about hundreds of interactions per week between engineering level people. These couldn't all be going up, over, down, that is, from engineer to PDT leader to system team leader to program manager and back down the other side. And they said, you're right. We spend our time in budget meetings, in scheduled meetings, in program reviews, etc. And I believed that. So I said, well, I've got a graduate student. Maybe he can handle those. No. Maybe we can analyze this structure and come up with maybe a better structure of the teams.

And what we did—and this is kind of a reasonable system engineering practice—we clustered as many interactions as we could inside the systems, in this case the system teams, and as few interactions as possible would be outside those. But that, of course, requires, as you can see, us to have cross-membership across those teams. That is, we wouldn't have mutually exclusive team assignments any longer; now the engine block team is assigned to team 1 and also team 2, because, well, they interact with all these folks who I assigned to team 1 and all these folks on team 2.

There are some teams that needed to be in three places at once, for example cylinder heads and intake manifold are B-1, K-1, over here. They're also B-2, K-2, over here. So you can see I put them on team 2 and 3 and 4. I just put their interactions in two different places.

But, see these guys at the bottom? They basically told us, these five PDTs said, look, we need to interact with everybody, right? And you can see what they are: the accessory drive, ignition, this is control module, this is electrical, and this is assembly—they basically are integrating the whole thing. So there's no one or even two or three system teams to assign them to. They said, we need to talk to everybody. And, by the way, they're not delusional. Everyone else says, yes, we need to talk to these guys, too. That's why the columns are almost as full as the rows.

So there's no obvious system team to assign these to. In fact, just calling these guys the important system engineering integration team doesn't handle all their interactions, it just handles them, maybe, working together, which is this corner, and that's obviously not the solution.

So I worked with this team and tried to help them figure out maybe a way to have these guys working with all the other teams, and it just sounded to me like more and more meetings. Like maybe they'd go to Monday meetings with team 1 and Tuesday meetings with team 2, and so forth. Maybe they had Friday meetings; maybe they go to meetings all week. And they said, how about, we have these program meetings. Really? What's that? Well, program meetings is where once a month, Tuesday mornings, the entire organization gets together and they talk about status of the whole program and big problems we're having, and so forth.

They said, why don't we just let these guys, which I call the integration team, run the program meetings and kind of handle the integration problems as they come up across there, or at least coordinate it as best they can. That was actually what they implemented in this particular organization.

By the way, it's interesting, while I think this diagram is very insightful, they said, look, Professor, just tell us who needs to work with whom. And so this is the diagram they actually put up on the wall. It's got the five little teams. It's got the team 1 assignment, the team 2 assignments, 3, 4, and so forth. This group at the bottom, the integration team, which, you see, I call it. They call it the platter: it holds everything else up. I wonder why they called it that.

But it's interesting. If you want to know why cylinder heads has to be on three teams, you've got to look back at the matrix. Or you can ask the cylinder heads guys, and they say, well, we just interact with these eight teams, so why don't they just all be on our team? So that's a local view. You need to have kind of the overall view.

Just kind of backing up for a moment, those are our three perspectives I've showed you. What is it like to look at products, their architectures and all the interfaces within; what is it like to look at complex processes that create those things and all the information flows within; and third, what is it like to look at these organizations?

And the perspective I take on all three is that it's not about decomposition, it's about interactions. So that's kind of the theme. The question I now raise is, well, can we look across them? I think there's something more than just that set of interactions. I think those perspectives, while I think they're interesting, they're a little narrow. We need to look at them together, and if there's any pair of those perspectives I can take, for the moment. I haven't figured out how to look at all three simultaneously.

So the first idea I'm going to show you—I should say, the last idea I'm going to show you—is that we can look at the product decomposition and document that as a set of technical interactions. We can look at the organization decomposition, if you will, and look at that as a set of team interactions. And then we can ask several questions.

The first question I'm going to ask is, does this drive this? That is, does the architecture determine who talks to whom? And, as a engineer, that kind of makes sense to me, right? If we're developing these systems, I'm developing one, you're developing another, and mine interacts with yours, I probably ought to talk to you about it, and to make sure that they're going to work together. So I would hope that this documentation of all the interfaces ought to predict quite well who interacts with one another in this large organization. And if it doesn't, maybe we're to learn something from that, too.

So let me show you how we do that. We actually use two of these together. We document the technical interactions, the design interfaces, in the first matrix. Then we document the organization interactions, the team interactions, in the second matrixes. And we use this thing I learned in kindergarten, that red and blue makes purple, so the idea of the graphics is, if there's a design interfaces are red here, if there's a team interactions that's blue in this matrix, and if there's both, there's purple.

And I would hope it's largely purple, right? Because that means there is a design interface, and they know it. The teams actually implement it by talking to each other. That's good. So the way I would read that matrix is, okay, if there's a design interface, that's red, or purple if there's also a team interaction. If there's a team interaction only, it remains blue.

So this is the way I might read that matrix. Let me show you actually how we study it, and I'll do this, just to remind you, in the context of this jet engine. So the complex system we're looking at here is this engine by Pratt, and I described already the decomposition of the engine and the systems, and then components. And we documented all the design interfaces, so that's the reds; that's the red matrix. So now red means design interfaces.

The second matrix we'll look at is the blue—that's the organization. So for that same organization we did the other trick, that is, we asked everybody who talks to whom. And in particular for each of those components there is a team, so that's 54 teams. There are actually six more system integration teams similar to the type that I showed for G.M., except these guys actually have no hardware responsibility. These guys, their job is explicitly integration. So there's a little separate job.

And, for the moment, let me not look at them. Let's just look at the 54, because I can compare it directly to the 54 components and all their interactions. So this is the intensity of interaction of the teams, which took place during the detail design period for this product development process. And just as with architecture, where you had high, medium and low, we had strengths of interactions on a multi-point scale. We also have frequency and importance of the interactions, here. I won't get into that into that much detail today, however.

So now I want to compare them. And the overall results of this comparison—as I said I was looking for a lot of purples—the answer is, there are a lot of purples but there are also a lot of whites, and I'll take credit for the whites, too. The white just means there's no design interface, and they don't talk to each other.

Well, okay, so the bottom line is that there's 90% purple and white, that is, for the most part the team interactions are somewhat predictable, although, honestly, I don't take too much credit for that. And actually, I think the more interesting story is in the reds and the blues, okay?

And by that I mean, here, there's a design interface, there are 569 of them, and the majority, 341 of them, which is about 60% of all those, predict team interactions. The teams talk to each other when their components interact. That's good, right? But there's 40% of them, that is, 228, where the teams don't interact. Even though there's a direct interface between my component and yours, we're not talking to each other. Why not?

And so we actually studied these very carefully. We developed hypotheses about why not. We can study that, and I think it teaches us something. And, similarly, I think these are interesting. That is, there is a team interaction; the teams are talking to each other; yet their design interfaces are none. That is, their components don't interact with each other, and yet the teams are talking to each other.

What are they talking about? Why? Are they talking about football games, or are they talking about interaction of their components that aren't there? Are they imagining them? So what is that about? Actually, they're not talking about football, because we only documented technical interactions about work. So we didn't just talk about who's going out with whom.

Which is interesting, in other social contexts. So we actually focussed our study on both of these two. But actually the overall result is that yes, the interactions are largely predictable.

Actually, there are a lot of things we tried to study. Let me show you an example of how we studied just this first issue, that is, the remaining red ones. And there are several hypotheses.

Let me just show you how we test #1. That is that, why are only 60% of all team interactions predicted by technical things by the design experts who give us these? Only 60%. Forty percent are not, so that's very interesting.

And one hypothesis that I'll show you how we test is that maybe these design interfaces aren't matched, that is, they're not implemented, because they're across system boundaries. Which kind of illustrates the hierarchy of the composition and why it's important.

Let me show you how we tested that. So, as I said, 40% of these design interfaces were not matched—that's what we're looking at. 60 percent are. And so, of these 60%, what we looked at is, well, within these boundaries—now, these are both organization and system boundaries, which I'll return to in a moment. So, let's say within the boundaries they're more purple, that is, they're 79% purple in the boundaries. They're only 48% purple outside. So that says that one explanation for why they do not implement known interactions or interactions that are really there—somebody knows about them—is that they're outside their organizational boundaries, or they're outside their system boundaries. And in this particular case their organizational boundaries are these groups, and the system boundaries are the same. So I can't tell you whether it's an organization or a system boundary issue. I just say they're outside.

And let me show you that it turns out that these particular systems that are more distributed are actually better at handling that. They do better than that 48%, which is just one (inaudible). Of that 48% of interactions across boundaries, it turns out these teams do worse at it, they only do 36% of them, and these integrated teams that have a very distributed sort of nature of what they're designing do 53% of them. And what I believe is going on is that these guys are designing things that are very integrated, like control systems, so they're kind of focussed on everybody, that is, on controlling everybody and talking to everybody. So they're actually better at handling the interactions that are across the system than these guys whose job is very focussed, you know, compressor and fan and turbine, and so forth.

So let me just say some conclusions. I think we've learned a lot by studying architectures and organizations together. We can predict a lot of the interactions, which is nice. There are a lot of interesting reasons why teams that share design interfaces don't communicate. I've showed you just the first couple of hypotheses and how we started to test those, but let me just say that we tested several. Some we can support and some we cannot.

But the interesting thing is that design experts know a lot about what they're doing, but the teams don't necessarily implement all the design interfaces, even though they're known. And also I guess what we learned is that the design experts don't know about all the interfaces, but eventually, to get a high quality product you have to handle all these bugs or interactions and things, and so they actually do interact to create them.

So I think we learned a lot about development of complex products or systems by studying things in this way. The first is probably the most important lesson, that product complexity has to be considered in the context of the process and the organization. I think as an engineer I believe this second point, that if I know the architecture I can design an organization around that. But that's a very narrow perspective, right? That is, it says, look, I know what the architecture is, and all you people organize around that. Of course, that forgets about whom you like to talk to and who your existing organizations are and your existing incentives, instructions, and all that kind of stuff.

So these perspectives are in opposition, and I think this helps us look at that contrast

Dan Braha: So, the obvious question here, if I understood things, is that the first speaker said things should be lower diagonal, and the second speaker showed us that they aren't at all. Can you two reconcile that for me?

Steven Eppinger: Well, maybe we'll each comment on that, if we can. Yeah. I certainly agree that if things are less coupled they would be easier, which is I think what Prof. Suh is saying, that they should be lower diagonal, or actually, better yet, there should be no interactions at all off the diagonal. For each FR there's a DP.

And what I'm showing you is that, in reality, there are many, many of them. Now, these may be largely lower diagonal, I didn't try to make it as low a triangular as possible, I just showed you the pattern. But, sure, I'm showing you the interactions as we know them. Prof. Suh may comment that we don't actually know all of them; it's worse. Do you want to say a word?

Nam Suh: Well, I think I agree with everything you said. If you take the throttle body, that first picture I showed you at the beginning, (inaudible) ask the question, it turns out that initially when the student analyzes this very coupled system, one of her jobs as a student in a class was to come up with another design. It turns out that there is a design that completely uncouples the thing. That makes it very simple to make, and so forth.

So it turns out that her solution turns out to be the way I think industry's going. Kind of interesting, a student who doesn't know anything about anything, just about.

Steven Eppinger: I thought she was pretty bright.

Nam Suh: She's a very bright young lady, but one night, when she first showed me this interaction chart, I said, so, what do you do with it? And then she came out with a design solution in two weeks, which is amazing. Until the end of the term, she couldn't handle it. And I think she's a very bright young lady.

M: I have one question. The system that you show us is a sort of very interesting complex system where you have complex interactions, but people basically break the rules of the system, where they shouldn't interact, and they interact among themselves anyway, which is what you analyzed to try to find what you should be doing.

And then you go ahead and try to reorganize that to put all the interactions that (inaudible) already into boxes where they should interact, trying to make this self-organized system do something that is sort of in a top-down organization. Does it actually improve the performance of the companies?

Steven Eppinger: Probably the best example of what you're referring to is the G.M. power train example, where we essentially did and proposed and they implemented that reorganization.

Well, unfortunately, they wouldn't run the experiment both ways for me, right? That is, they didn't split the team and say, okay, you guys develop an engine the old way, you guys develop the engine the new way, with the help of this MIT solution.

So they didn't. And for a complex system, I'm not sure we'll ever be able to run that experiment. For simple systems, I can, that is, I can do those experiments for simple systems, and yes, it works. But I don't believe in the scaling from simple to complex, so I'm not sure that tells us a whole lot.

The fact is, they implemented it, and they only implemented it because it made sense to them, and they were glad they did, and then the next time, they implemented it again. So that says that they believed it could help, even though I don't think they can prove it to you or me convincingly.

M: I think in this discussion it's always important to distinguish between complicated products and complex systems, where you always have adaptation and learning. I mean, I see in your network a lot of parallels to how the brain solves comparable problems, where we have something like (inaudible) learning, where you have continuous emergence of cell assemblies that spontaneously interact and change their interactions when there is no need for communication.

Do you have any ideas how you could mimic or learn from the brain how the brain is doing this, by implementing something like (inaudible) learning, where if they have a lot of meetings but they really don't have a lot of efficiency in the interactions, that you just reduce this, or have you implemented other types of interactions, like phone or e-mail, that is based on how efficiently they interacted in the past?

Steven Eppinger: Well, obviously, I haven't studied it in that way, but it's a very interesting comment, and so I'll just give you my top-of-the-head response to it. I think that we do see patterns here of learning, or, I should say, changes in behavior. That is, if we document the personal interactions one week, and then the interactions a year later, and that says, look, they've learned a lot about who needs to interact with whom. They didn't know it earlier, but they know it now. And so there is that learning.

And that learning comes on, comes up, comes about for maybe the reasons that wouldn't surprise you at all: we had bugs, we had to learn, we had to talk to each other, and I would have to go and talk to so-and-so and so-and-so and that's the guy who really has the answers, so next time I'll know to go to them more directly.

But the same happens in processes. If we look at these process diagrams of information flows, for a product that's new, that is, they've never designed it before, they don't know what they're going to do. Yet then we look at more mature ones, where they've done it many, many times, they actually have a lot more interactions that they can document, that is, that they know about. So I'm sure that learning goes on.

I don't think I have yet figured out how to apply the very nice idea that you had, which is, how to bring some cognitive science into it, but it's a nice thing to think about. I appreciate that.

M: You've described the organization charts in terms of functional requirements of the projects. You haven't said anything about keeping complexity at a manageable level. But I kept thinking of Ashby's Principle of Requisite Variety all through this, which seems to work quite well in determining the degree of organization that one should have in putting together teams, committees, and so on.

Does that play a role? It seems to me that some of these components are going to generate a lot more information, or variety, if you will, than others. And I'm just wondering where that fits in.

Steven Eppinger: I have to admit ignorance. I'm not familiar with that principle, but I think the things you said in describing it are very much true. There are some components, or elements, depending on which type of map we're looking at, that require a lot more information from others and provide a lot more. And some provide a lot more, and sometimes they're the same, and others don't. Sometimes, we look at the complexity, the absolute, as terms of, how many interactions are there in one of these rows or columns? And the answer is, on the order of five to ten. That is, on average, in these architectures, each component interacts with a handful of others, not 100.

And this doesn't grow with size. When we go from a system of 100 components to 1000 components, to an entire airplane, which has millions, it's still on the order of six or eight. And that is, I think, relates to this sort of limited complexity idea. On the other hand, there's a huge range. There are some that are very integrative—we call them integrative—and then there are some that are more modular, that is, that they're focussed.

Michael Caramanis
Scale Decomposition of Production

Dan Braha: Our next speaker is Prof. Michael Caramanis. He's the Associate Chair of the Manufacturing Engineering Department at Boston University. Prior to joining Boston University, Michael served as a consultant to the World Bank and directed the Scientific Secretariat of the Greek National Energy Council. He also served as Deputy Director of the Utility Systems Program at MIT.

He's the editor of the prestigious journal IIE Transactions, and his talk today is based on two consecutive $1.2 million awards that he received from NSF.

Michael Caramanis: Thank you very much. Today I would like to talk about time scale decomposition and other types of decomposition of complex systems which allow us to deal with the complexity. So I'm not going to, per se, try to characterize complexity, but I'll try to say a few things about how we can handle complexity.

In a typical manufacturing supply chain environment, one is confronted with decisions that are made at different levels of an enterprise that have to coordinate investment across plants, that have to coordinate and plan and schedule production within a plant, and, finally, within focussed factories within a plant, within a factory, controlling processes and so on and so forth.

And this is, as I'm going to argue, a very complex and difficult problem which traditionally—as you've heard in the previous talks, Steve made this point—we've been handling by decomposing the decisions into various levels and groups and decision making nodes.

Fortunately, in at least a (inaudible) manufacturing environment, cellular manufacturing suggests a natural decomposition framework where at least the scope of decisions suggest that you can apportion decision-making into various levels of decisions that have different scope, and then somehow deal with the problem.

Of course, one has to handle the interaction between those decisions and try to achieve something that I call vertical integration, which, however, is not sufficient. In addition to vertical integration, one needs to succeed in horizontal integration, namely, handling the interactions of various decisions that have practically the same scope, the same time scale, and so on.

So let's see how we can structure the problem a little bit more carefully. Clearly, there are computational challenges in complex systems like this because of the size of the problem, presence of uncertainty we've all heard about, the need to act in dynamic settings—we need to be making decisions dynamically. The performance objectives are changing, at least in the manufacturing world. We are not just interested in scale and cost and so on, but we're interested in quality, we're interested in time, in reliability, and so on. And finally, the hybrid nature of manufacturing dynamic systems, and by hybrid I mean the co-existence of different structures and different time scales.

If you could excuse me the use of some equations, they're only sort of stylistic here—the overall problem of centralized decision-making would be to somehow look at the state of the system and the dynamics of the state of our system. But the state of our system, in a big enterprise, is a huge vector that spans everything from the state of the economy, the price of the stock, all the way down to what is happening at a particular process, how far we have advanced in processing a particular part, and so on.

And, in theory, one would like to be able to come up with this mapping of the state of the system into the action space. That, again, involves a huge vector of decisions, spanning investment decisions all the way down to controlling a machine. And obviously, that's something that is very difficult to do in a centralized fashion; so, traditionally, we've been doing the only sensible thing, which is to decompose. So now, rather than asking the same team or the same person to make all of these decisions, you assign these decisions to various groups, and each one of those groups tries to map not the whole state space, which is unmanageable and huge, but some sort of a subset of the state space, some sort of a mapping of this huge information of the condition of the system to information that matters. That sort of projection I indicate there, by the phi function, into the subset of decisions that they're responsible for.

So this is what we've been doing. The question is, have we been doing it efficiently? It works somehow; that's how we take care of things; but my conjecture is that we can improve a lot by increasing the interaction between those decision-making teams and by essentially designing a better architecture of this decomposition, and also designing a better information exchange between those various decision teams.

So I'm showing here just two decision-making levels. Perhaps you can think of the first one as being at the factory level, and I'm going to be focussing on a specific example. So let's think of, let's say, the plant level, where you do production-planning, and the cell level, where you do sort of shift-by-shift or day-to-day decisions on what to produce, how to produce things, and you determine the capabilities of the cell.

So, let's say at the plant level, you need to decide what production requirements you should impose on the various focus factories or cells, over time, and that time is sort of a broader time scale. You can call it a time bucket; you can think of it as a week; that's t_1. And whereas this decision is typically made on the basis of some subset of information—let's say the average capacity of the cells and what the demand requirements are out in the future—you will be able to make a much better decision if you have information on the performance capabilities of the various cells.

So this phi-bar is somehow average performance information, first and second order, and that's important, which can come up to this decision-making team by the cell level. So that means that you can improve, perhaps, by designing the right amount of additional information that is made available to the decision-makers at the plant level. And that, of course, imposes an additional burden or requirement to the decision-makers at the cell level or the focus factory level who have to generate this information.

So essentially what I'm going to argue for is that it is possible to come up with a systematic way of assigning decisions and a systematic way of assigning additional information exchange and additional information generation at the various decision nodes, which allows better, still decentralized, but coordinated now, decision-making, through an augmentation of the state, or an augmentation of the information that's available to each decision-making.

And I'm going to show you an example of how this can be done, and how much better one can perform if one engages in this additional information exchange and better improved architecture over existing, not existing, but over general practice or state-of-the-art in at least production-planning and scheduling.

Of course, there's a cost of doing that. That's the cost of information, which has to be balanced against the value of information, the improvements and the additional efficiency that we can get by exchanging this information.

So here are some guidelines, with some characteristics, which can help us structure the problem and come up with a better architectural design. And these characteristics of production system decisions can be thought of as falling in three main categories, or three main characteristics, that help us structure the problem and create this architecture and information exchange.

One is frequency, and frequency of course is the characteristic time scale, the frequency with which we make decisions. There are some decisions that we make very rarely. For example, investment decisions regarding the building of a new plant are not made very frequently. The characteristic time scale is, let's say, years, and you go all the way down to controlling a manufacturing process, where the characteristic time scale is almost continuous. So you make decisions in an almost continuous, sort of second-by-second basis.

Then you have the scope, which is something that you saw very nicely in previous talks, and that has to do with the portion of the enterprise that the decision affects directly. So, obviously decisions at the enterprise level affect everything; decisions at the plant level directly affect the plant, and decisions at the cell level affect the cell, decisions at the process level affect the process. So the scope kind of narrows, and as the scope becomes narrower, the time scale associated with those decisions also becomes smaller, so there's some correlation.

And finally, you have the nature and the functionality of the decision, which again maps or correlates to time scale. You have resource allocation, and resource allocation takes place at various scope levels: you do resource allocation at the enterprise level, at the plant level, at the cell level, at the process level. As the scope decreases, the associated time scale of resource allocation becomes narrower.

However, in addition to resource allocation, you have uncertainty analysis contingency planning, which, again, is done at various scopes, and sequencing or scheduling decisions that have to be done at various scope levels. Within the same scope level, as you go down this functionality, the time scale again decreases. So you make uncertainty analysis contingency plan decisions more often; the characteristic time scale is shorter than the resource allocation. And sequencing, of course, is even shorter.

Let me give you an example. I'm only giving three broad functionality groups. There are more that you can come up with. But if I were to give an example of how scope changes on the right, and how functionality changes going down, and by scope I broadly use here the enterprise, the plant, the cell, the process -- And the cell, you can think of the cell as a focussed factory or a subset of the factory that is specialized in performing either similar types of operations that would be a function of the cell, or is specialized in processing similar products—that would be the product cellular design of a factory.

And then, here, you have the various functionalities, resource allocation, contingency planning, uncertainty analysis and sequencing. And there are examples of all of these things at the various scopes. Here you make resource allocation decisions having to do with plants. You still have to do risk analysis on partnerships and so on. These are sort of, of course, on a bigger time scale than the uncertainty analysis that you have to make at the plant level. And the cell level, of course, you have to do things like dealing with uncertainty associated with employees not showing up, machines failing, running out of spare parts, and so on. And all the way down to the process.

And finally, you have to do sequencing decisions at all those scope levels. And the interesting characteristic here is the correlation of scope and functionality with time scale. So, essentially the time scale decreases, or the frequency increases, as you move down and to the right. So you make decisions much more frequency as you move down and as you move to the right.

And, as a matter of fact, this idea of connecting functionality and time scale and so on is a nested one. Even within some of these functional decision groups you may have further segmentation and further decomposition, and so on.

So, the idea is to understand how time scale varies with scope and functionality, and, for example, in the sequencing problem at the plant level, which is the problem of deciding how much you produce at what cell during each week, let's say, or during each characteristic time unit over your planning horizon, and the performance evaluation contingency planning at the cell level. This time scale is probably smaller than this time scale.

So, obviously as you move down, your time scale becomes smaller, and as you move to the right. But you may have a bigger time scale here than you have there.

So this is the example that I'm going to take and show you how a better structure and an improved exchange of information can result in substantial savings, relative to the current approach.

So if I were to show this now specializing into the factory level decision making or production planning, which is the sequencing planning at the factory level, and contingency planning and scheduling at the cell level, one can think of some decision node at the plant level deciding how much the cell should produce of each part type during each week in the future, and then each cell determining whether they can actually meet those requirements. If not, they have to return information as to how they were able to meet those production targets that were imposed upon them and provide additional information and sensitivity and second order information that would make the decision-maker here be able to make a better decision. And I'll explain what this information is.

So, the key here is to understand what sort of information has to go up and how this information should be time-averaged. So this information, which is calculated on a much finer time scale here, has to be time-averaged to the time scale, the characteristic time scale of decisions at the higher level.

So if I'm making production scheduling decisions that are weekly, then I have to time-average, over the week, my hour-by-hour or shift-by-shift decisions and analysis at the cell level. At the same time, cells have to coordinate themselves, because now they have to interact on a shorter time scale, which is the time scale characteristic of the cells, and they have to exchange information which is of the type of a schedule or probability distribution type of information, because they belong to the same time scale. And likewise with the process: under a cell, you have processes, each process has to tell the cell what its acreage performance is and what the sensitivity of its performance is with respect to resources that are allocated, such as spare parts, maintenance, and so on and so forth.

So let's look at this specific problem a little bit more carefully. So I'm looking here. I'm kind of specializing the problem. And I'm looking at the interaction of two modes, two decision-making nodes, or two levels of decision-making: one at the plant level, where let's say, let's assume that your time scale is that of a week, so you have decide each week how much each cell—and the cells are down here, so cell i-1, cell i+1, these are kind of interacting cells—how much each cell should produce of each of the part types that they're responsible to produce.

So these are kind of vectors. And these vectors are the production targets or the set points that are determined at the plant level, even down to the cell. And there is one vector for each one of those time scales, for each week. Week one, you need to produce so much of part type 1, so much of part type 2 and part type 3. Same for week two, week three, and so on.

So these production targets are passed down. And then the cells have to work with this information in the context of their own time scale, which is a shorter one, because at the cell level you have to determine how much to produce each week, how to load machines, how to establish your shifts and so on.

So, within each week now you have a finer time scale in which you make decisions. And you can think of this as perhaps a small network of machines. If you are familiar with queuing networks you can think of this as a queuing network, a controlled queuing network, perhaps, where you make decisions on how much to produce during each shift, on the basis of what your queues are, on the basis of whether machines are up, down, and so on. And, of course, you can come up with what we call contingency analysis and determine the production capabilities of this cell over the week.

And once this is done, in addition to what people have been doing in the past in existing practice, you tie up this architecture in a more efficient fashion and deal with the complexity in a more productive fashion—in other words model the complexity more effectively. The additional requirement is now you need to determine not only capability, but sensitivity of capability with respect to the set points that came up from the upper level.

And those set points, the most important determinant of those set points is the lag in production at each cell, and that's a very important problem.

That is known as the lead time problem that enterprise integration software has been trying to take into consideration through material requirements planning and scheduling at this level, at the level that captures the delay.

So, for example, cell i is supplied by cell i-1. However, in order to meet the final demand, because there are delays at each of those cells, the production of upstream cells has to be made in a timely fashion, so that the delays are taken into consideration and so that demand is satisfied in as much of a just-in-time fashion as possible, and with the smallest possible inventory.

So this is the efficiency gain opportunity that we have here, because in most of the existing techniques this delay information is not being modeled in an interactive fashion, and is not being modeled as it should be. Namely, the fact that these delays are a function of the activity level here, or the loading and the production mix, and so on, is something that is simply not done.

So then if additional information is returned, we may be able to capture these sorts of second order effects that can only be recognized and analyzed at this finer time scale, whereas, at the coarse time scale, if we average, they're completely lost. And, as a matter of fact, there are very important analogues between this type of an approach in manufacturing systems and in computational fluid mechanics, where you have effects that take place close to the boundary, where the mean distance between collisions is very, very short, and you can't use the average flow equations, and where sort of viscosity effects and so on are completely lost if you simply tried to take averages.

So there is second-order variability, here, which imposes nonlinear expectations and responses that have to be communicated, or that may be communicated. And if they are communicated, you can improve in your performance.

So let me show you on a simple system, and of course we've tried this on much bigger systems, what the possible gains are if this second-order information or these delays are communicated from the cell levels to the plant decision-making level.

So this is an example where you have three cells, two sub-assemblies produced here, and there are two components, and they are assembled at cell 3. And the production scheduling problem is, we have multiple machines or workstations in each cell, and we are producing multiple part types, and therefore we need to produce multiple components or sub-assemblies, three in this example.

So the production-scheduling problem is to ask each cell how much to produce of each part type, what the production mix should be over time, and that's at the coarse time scale level, let's say the week level. Then these cells have to interact, on an hour-by-hour or shift-by-shift basis. And you want to minimize inventory, working process, in each cell. And also you want to minimize backlogs in terms of meeting your demand.

So the problem here is to be able to capture the delays at each cell as a function of what the production rate requirements are during each week. If I have a high production requirement, then I have a higher delay. If I have a higher delay that means that I need more work in process. And so on. So, can we capture this relationship, and if we do, what do we gain?

The problem with the master level would be one of essentially minimizing over time, over different part types, and over various cells, work in process. These are the queues, and positive and negative finished goods inventories. Negative would be backlogs: you are not able to meet your demand.

Subject to the coarse time scale dynamics that tell us how inventories change, inventory at week t is the inventory at week t-1 plus the release into the cell minus what was produced. These are regular material balance equations subject to capacity constraints at each machine, and obviously we've been doing that all along.

However, the new thing is to be able to capture also the delays, or the lead time requirements, which is a very complex issue. You can't possibly calculate these relationships at the higher level.

So what is happening is that this relationship can be determined at the lower level, at the cell level, using Little's Law. Little's Law allows us to deal with lead time by essentially restricting the production that is possible to be a function of the inventory. That's not possible. So, since lead time increases with, or rather, as I get closer and closer to my capacity, then in order to produce at the higher level, I need more inventory.

And this relationship can be shown in a simple one part type production system, by this typical decreasing returns to scale, a relationship where in order to produce more you need more and more inventory. So this is inventory. This is what you produce. And essentially this decreasing return to scale tells us that at each point the ratio between inventory and production is the lead time. That's the inverse of this slope, and you can see that the inverse increases. So as the slope decreases, its inverse increases.

So how can we represent this information? It's not possible to recalculate it, especially since if you move to a more real system where you have more than one part type that you are producing, then you need to predict whole surfaces, whole response surfaces. So that's very difficult to predict.

However, it is possible to predict a point, here, and a sensitivity about that point. That's the additional requirement that we impose on the cells. Each cell does that and conveys it to the production scheduler.

So if the production scheduler then has this capability it's possible to learn and, in a few iterations, it is possible to essentially determine the shape of this surface adequately, in the small region of interest, which is the operation region of interest of the production system.

So let me show here three situations that we are comparing, and these are the following. Of course, in reality, this is a surface, but I'm showing it for a one part type example. So the production rate requirements at each cell show a decreasing return to scale with respect to inventory, and this is our true system. What most systems, in practice, have been doing is to say, look, I can't predict this surface, and I can't have frequent interaction between cells and the production scheduler, and, what's more, the decision makers at the cell level just don't have the capability of producing this information.

So therefore I'm going to do some sort of a static calculation, which is, what I'm going to calculate at some point of maximum production that I'm willing to allow. I'm going to calculate the lead time in the worst case situation, and impose a fixed lead time, or a fixed delay, at each cell.

Now, this is equivalent to restricting the production capabilities to be below this line. So now I'm doing production scheduling with a restricted cell. So this is the real one, and this would be an optimistic calculation which says that, well, in practice, I do not have much of a delay beyond the processing time. So that would be sort of an ideal system that assumes that essentially processing time is most of my delay.

And, by the way, if you are operating at very low capacity utilizations, it is possible to have ?? so your real system behaves more or less like the ideal system. And in some ways you can argue that this is what was happening in Japan in the last decade and so on, where because of cost of capital, there was much more investment than was necessary, and therefore factories were operating—and this is fact—they were operating at a lower utilization rate than factories in the U.S. So, you're down here.

So let's compare these three systems. So we did a study, and let me show you the ?? cell and the value of information. So, these are the costs of inventory and of backlogs, and so on, over different types of cases and over different horizons. This is a short horizon and a long horizon. Let's focus on the long horizon because the transient effects are not very important. Different cases. And this is the cost, up here, if you use the ideal system. And obviously if you use the ideal system then you plan on very short lead times and very short delays. You actually have big delays, and you end up with big backlogs, extremely high cost.

This is the situation that reflects more the current state-of-the-art, and would reflect a good system that takes capacity into consideration, but does not capture the full viability or the full nonlinear nature of the delays and imposes a fixed delay. Associated with the worst case, much higher costs. And this is how much you can decrease your costs if you actually recognize the true production capability.

Even at a very high capacity utilization where you are very close to that worst case position, just merely because of production mix changes—although production is level loaded—you can still get 30, 40% improvement, and if your capacity utilization varies, let's say between 80 and 90% by even a small sort of 10% margin, you can have improvements that are of the order of 100% in delays, and so on.

So that shows you that there's an important opportunity in using information, and that the value of information is very, very high. So if it is possible to generate this additional information, if it is possible to spend time and resources and effort in implementing a better architecture, a better allocation of decision-making responsibilities and in improved information exchange, rather than sort of a multitude of useless information—if I'm at the plant level I don't care whether a machine is down or whether a worker was late, I just care about how the high level second order, nonlinear response of delays, with respect to production targets that I'm responsible for determining, behaves. So this is the additional information that can come for each cell, and the potential improvement is just too high, I think, to ignore.

M: This is relevant to all the speakers. I'm thinking of the general case of whether I can get any insight about the problems I confront from the systems, as you designed them. It seems that we all know that the problem of physics and chemistry and biology is trivial, because they just have to interpret what the engineer is doing. But you, as an engineer, you are God, so you design the system, and you can make things do things. It's engineering, I think, from the outside.

Now, if you're in management, you're a member of that system, you're doing engineering from the inside. You're a participant, but you're not in control of everything.

Are there any rules, or any insights we can get as to how a participant in this system should behave, in order that the overall system will behave in a well-designed system, as you have just described?

And would the people—because it's people systems—would they be happy in that situation, or is there a conflict inherent in an optimal system that works well? Can we give any advice to managers to interact more, take into account more interactions?

Michael Caramanis: Well, that's precisely the qualitative lesson that comes out of this: improved interaction, which is something that's well-understoo. And you have those teams, you have the war rooms where everybody gets together every Friday, and so on and so forth. So obviously these things work.

What we can say over and above that is that it is very important to be able to bring into those meetings and into these interactions the right type of information. And the right type of information can be realized, can be determined, can be thought of if one looks at these, the classification and the time scale relationships between various decisions.

M: Something is between organizations, not within. They don't actually meet.

Michael Caramanis: Well, that's the information that's exchanged between teams, between decision-making teams. And there, it's very important to understand what you do is important, how you can summarize information that represents your performance and what you do, in a way that this can be used by others, by other decision-making teams.

So essentially, the idea is that the tighter this coupling is and the more you are able to translate local improvements into global improvements, then the enterprise can gain. Because otherwise, you may be able to have an incredible improvement at the local level, as long as that is not realized. For example, if you decrease your lead time, if you become more responsive at the particular cell. If whoever does the production scheduling doesn't realize that, then nothing is going to be gained, because raw materials and others are going to behave thinking that you're still in the old modus operandi.

And rather than waiting for a year for this to be realized, if the feedback and the cycle can be improved, then....

Dan Braha: Let's take one more.

M: Yeah. I thought you did a very nice job in decomposing complexity into its various levels and parts and issues. The place where I think I want to see you take the next step is to put it back together again, in an integrated way. Because it seems to me, so much of the firm decision-making process needs to come together into a vision of all of the interconnections both within and outside the firm, with the rest of the system.

And at that point, I guess where I started to go with that thought was into models of thought and framing issues in the Gareth Morgan images of organization kind of sense. And I wanted to hear you talk a little bit about how you put this together into an integrated vision, and what you would suggest to managers. I mean, I'm great believer in multiple models of thought, because it gives you flexibility. But I wanted to sort of push you into a more integrated dynamic systems vision of reassembling this into an integrated process.

Michael Caramanis: Well, this particular model and the results of this sort of integrated cell- and plant-level decision-making is an integration.

The integration at the enterprise level works in the same fashion, through the idea of state augmentation, but the state augmentation has to be very carefully designed, with the right type of information being communicated.

And what I try to argue is that if you can classify the time scale differences and the relationship between scope and functionality and so on, then you can implement such a state augmentation in a practical and doable and very productive fashion. So that's the integration.

Dan Frey
Complex Systems Integration

Dan Braha: Our next speaker is Dan Frey. He has also served as Assistant Director of the Systems Design and Management Program at MIT. He has won the Baker Award for Outstanding Teaching at MIT, and rumor has it that he's also an excellent pilot. He will be speaking on complex systems integration.

Dan Frey: Thanks. I started that rumor.

So, an engineer, a scientist and a mathematician were driving up to the NECSI conference the other day, and when they crossed over into the New Hampshire border they went into this valley, and they saw these cows. And the engineer looked at them and he quipped, "You know, I'm used to seeing the cows with the white and the black spots, but these cows are all black. It seems like all the cows in New Hampshire are black. Do you have any theory about why that is?" And the scientist said, "Well, you know, I could formulate a theory but, really, you've extrapolated far too much from the data. I mean, all we can really say is that all the cows in this valley are black." And the mathematician is just shaking his head. He says, "Come on, now, I'm appalled at your lack of rigor. All we can really say is that the cows in this valley are black on at least one side."

So, I mean, when an engineer comes and addresses an audience like this, with engineers and scientists and mathematicians, we ought to be cognizant of our differences in approach. And when I thought about giving this presentation, I thought about just coming and presenting about my research in robust design, but this is a burgeoning field in engineering systems. A lot of organizations are now making a lot of investment in this area. For example, MIT has just started a new division within the School of Engineering, called the Engineering Systems Division. And we only have one other division, as far as I know. I think it's Biotechnology. So that's a very big commitment.

And then, we're actually a little behind, because CMU and Georgia Tech and a lot of other institutions already have very substantial investments in this area.

So we have this burgeoning field, and I think it's worthwhile to step back and talk a little bit about method, how you evaluate the bases for an emerging science and engineering system. And I'm just going to try and make some comments about that. And a lot of my thinking in this area is really strongly influenced by Popper here. Popper was an influential philosopher of science, and he posited that the criterion for demarcation between what we would call a true empirical science and what he called—he used the derogatory term of a pseudo-science to describe the other ones—is that one is falsifiable in principle and one is not.

So he characterized things like Freudian psychology and Marxist sociology as unfalsifiable and therefore not very helpful, because you could look at some data and you could explain it, on the basis of these theories. But if the data had come out the exact opposite way, you could also explain it with these theories. So what good is that?

And, in fact, the better example is something like the first law of thermodynamics, where one and only one result from that experiment is consistent with the theory. So there's a strong theory, and the other ones are not so strong.

And I think that's a useful mindset for us to all have when we come into this area, and I want to take that and some other thoughts and just talk about one particular theory of engineering systems. So I think I'm paying Nam [Suh] a tribute, by taking his theory as the major case study for looking at a theory, poking around at it, and seeing what useful things emerge out of it. And I think a great many useful things do come out of it, and I'm just going to take it one by one. I'm going to talk about the independence and the information axioms, the two sides of this theory.

Now, first of all, I have to at least quickly refresh our memory about the basis of the theory of ontology, axiomatic design. The first concerns itself with the idea of mapping between domains. I'll just focus on the functional requirements, the DTs. It's represented mathematically by this model.

So the design matrix is mathematically a mapping between the two design domains and then is characterized by these partial derivatives. So we can clearly say what the design matrix is in which we define our DPs and FRs. And then we categorize designs according to uncoupled, decoupled, and coupled designs.

So these two categories are considered to be the acceptable ones where this is to be avoided. And we also define information content as log 1 over probability of success. And, in the special case of a uniform distribution, we can make it the log of this ratio of two different ranges.

And then, finally, Nam defines these two axioms: maintain the independence of functional requirements—that is, try to maintain that diagonal structure—and then, minimize information content.

So first of all, let me make a brief clarification so that the mathematicians in the audience don't get too upset with us. In the dictionary of philosophy we find that axioms are considered to be primitive propositions whose truth is known immediately. You don't require any proof; you just know that they're true. As soon as you understand the basic concepts of, let's say, what a line is, and what 2nd geometry is, and what straightness is, and what it means for two things to be parallel, it's just obvious that they don't meet. You don't need to prove it.

But mathematicians tend to use it as a statement that's just stipulated to be true, and then you construct the theory. And Nam is using it in a slightly different way: a truth which is observed, which is induced from observation, and there are no counterexamples. So just so the mathematicians and scientists won't be too upset, I'll clarify that these axioms here are to be seen in that third vein.

Now, one problem that comes up is that, if you define these as maintaining the independence of functional ?? and minimize information ??, those are clearly imperatives, right? I'm saying, do this, do this. So stated as an imperative, we can't exactly falsify it. We have to change it a little bit. We have to at least say, well, if we want this to be a testable statement, we at least have to say, "What would happen if you violated this imperative?" so that we can actually test it.

So let me start to do that. Now, through theorem, you start to define what the consequences are of breaking the axiom. It says here that if the a and the b matrix, somehow multiple, are not either diagonal or triangular, you're not going to be able to manufacture. So there's the consequence, you're not going to be able to manufacture the design. It just won't be

266

implementable. You won't be able to manage.

So let me think now like a chemist, and let me take some elements and bring them together and just observe what I see, in the light of this theory. So one thing I did this term was I brought together a group of sophomores and I asked them to do some design. I said, here, take this Dragonfly here— this commercially available, electric powered, radio-controlled airplane. And I'm going to give you a kit of components and materials so you can take this set of electric motors, this set of reduction gears, you can take foam, and you can craft it into any shape you want. You can take balsa. You can construct it in any way that you want. And make me an airplane that will perform better than this, according to criteria that I define.

So, I said, adapt it to improve, to meet certain parameters in terms of rate of climb, the maximum level of speed, the endurance, and the minimal level of speed. So I defined for them objective functions which they were to maximize. We set it up in the gym, and they went out and flew patterns in this gym, and I actually measured their performance along these criteria.

And here's the situation. I developed the design matrix for the original design for the Dragonfly. And let me tell you how I did that, just so you understand what I did. I took this commercially available simulation here which we believe to be about within 10% accuracy, in absolute terms, and perhaps a little better, in terms of relative performance, being able to evaluate one design against another. And here it makes a prediction, for a particular design, that my max speed is 32 miles per hour.

Now, if I go ahead and change something, if I decide that the diameter is 5.5, and then I reconsider, oh, I can go a little faster. So I can get that partial derivative, the dependence of maximum level speed on prop diameter, for that point design.

So I went through the simulation, and I extracted all those partial derivatives, and I constructed the design matrix, and here it is. So this is not the structure that we were looking for. I upper-triangularized it as best I could, but I still have at least one very strong -- Well, here let me say that big Xs represent clearly a very strong first-order interaction between the two, whereas the small Xs represent, well, a less significant interaction, but it's there nonetheless. And zero is really just negligible.

So, this is my first rough cut at it, and it seems to me that you can't just set these parameters in a particular order. This is a coupled system.

But, what are the consequences of that for the design? Well, I sent the students off; I didn't tell them anything about axiomatic design, necessarily. I just said, try to meet these objectives. And they started with that original design with these parameters, and it performed in this way. And this rate of climb was really excessive. They didn't need all that much rate of climb; they needed to trade it off against other things. For example, they needed to improve endurance, by trading off against things like rate of climb. And they wanted to be able to slow it down a little, because they were rewarded for that.

So they were able to move around within this parameter space, and improve matters, and get planes that scored higher than the original design, despite the coupling.

So, now we ask, okay, first of all, is it possible that in some scenarios we just can't come up with some uncoupled or decoupled alternatives? In some scenarios, are there subsystems that we just can't quite decouple? Is coupling something that we just have to accept some of the time? And I think that for this particular example it is. You just can't find a fully uncoupled design. Or if you did, it would be sort of clunky; it wouldn't perform well; it wouldn't be efficient.

And I think coupled designs can be executed successfully, but it's more difficult to do. You have to iterate. And simulations could help with that a great deal. Those are my top-level conclusions coming out of this, to pull something useful out of this.

So let me now switch my perspective from that of a chemist, where I'm just putting things together and making observations, to that of, let's say, more like a physicist, wherein what I'm going to do is I'm going to strip away the authenticity of this, in order to get more of the elements, the fundamental elements of this issue of coupling, to understand the atomic nature of it, to get to the indivisible elements of it.

So we developed this emulator of the parameter design process, and what I'd like to do is get an experimental subject, some volunteer. Is there someone willing to participate in this? Anyone at all? An intrepid soul somewhere? Please, will you? Okay.

I have this emulator. I'm going to bring up the figures. I'm going to bring up a new system. I want to bring up the system of a very particular type: Level 2, coupled, and there we have it, and start. Now let me tell you what the rules of the game are. Please come up and bring yourself in front of the computer screen.

Now, what I'd like you to do, your task is to get these little blue dashes into the red bins. So you're trying to change the parameters of the design to get them within these targets, and you're going to use these bars here to do that. You move these bars and they change the blue thing, and the targets stay fixed. So get the blue things in the targets.

M: And I don't know how this is all interacting.

Dan Frey: Well, all you have to do, for example, is you just take the bar and move it, and then you say refresh, and then you find out the result.

M: Okay.

Dan Frey: Just do your best to try to solve the little parameters on the problem. Click it, slide it up. You can click also in the white area, that will move it up by discrete amounts. Now just press refresh plot. You can see the effect. That's where you want it.

So, what kind of system is this?

M: Uncoupled.

Dan Frey: This seems to be an uncoupled system. Now, I can modify the kind of system that lies behind this. I can make it a fully coupled system, right, and that would make his task more difficult. So rather than belabor the point, let me say that what we did is we took a group of experimental subjects, and we gave them this task. So we gave them this task, and we timed how they did, and we looked at the number of moves that they made, etc., over time. And we came up with the following result: if we looked at two alternatives, full matrices and uncoupled matrices—full matrices being these triangles, and uncoupled matrices being this diamond—as you might expect, if we change the number of variables in this system, a 5x5 just takes a little bit longer than a 4x4. It's just a linear growth in the time to complete the problem.

So time to complete the problem, in this sense, is akin to computational complexity, akin to that. Now, for the fully coupled system, you see a very different sort of growth. If you analyze this growth it appears not to be polynomial but, rather, to be geometric.

So, here again, we see that people could solve 3x3 and 4x4 coupled systems without too much trainer coaching, but the scaling laws look really different. We know that for computers solving a linear system, the time to compute it goes like the number of variables cubed. But for humans, it looks like it's a much worse scaling law. I mean, vastly worse. So people start to saturate. If you ask people to do 10x10 systems, you can't find anybody with that kind of patience. It takes a week. Unless they just extract the derivatives and use a computer to do it. But we didn't allow them anything. We didn't allow them pencil or paper or anything.

So I think that this design matrix suggests that there are important inferences we can make about human capabilities and the way that we should structure designs.

And now let me think. I thought like a chemist; I thought like maybe a physicist; now I'm trying to think like an anthropologist. I'm going to sit in the bushes, outside Ford or Intel, and I'm going to watch them do their work. Then I'm going to make some observations in my notebook. And this is the way I would interpret Steve Eppinger's work. He's out there, he's watching. Is that fair?

Steven Eppinger: Okay.

Dan Frey: You're in the bushes. You don't actually use shields and all that. Do they know you're watching? Well, this is stuff that was put together by Dan Whitney. But it was thanks to the fact that you put the tool in the field, and since people are using the tool, you get data.

So the data seems to suggest that as the total number of tasks in the design project goes up—that's the number of rows in a design structure matrix—the average number of non-zero elements per row seems to follow a very linear scaling law.

Now, I can't explain this, but that r^2 value is something to take note of. So now that we're thinking about coupling as an issue in design, thanks to

Nam, and now that we have these design structure matrices, thanks to Steve, we have this tool. We can start to see what the patterns are. What are the strategies people in fact employing to deal with these human limitations in design? It seems that they are following a strategy, whether they realize it or not, of limiting the number of non-zero off-diagonal rows as the problems grow.

I don't have a theory for why that is, but here's one. Crawley, just yesterday, rolled this out in our session on complexity in aerospace. He said, well, here's my proposed 2nd law. So now we get a reformulation of an axiom. He says there's some essential complexity to get this product out the door, a minimum level of complexity that will allow you to accomplish this function at this level of performance. But we always seem to have more complexity than that, and the reason that we have more complexity, in terms of—I'm talking about descriptive complexity now, number of CAD files or number parts, some measure of that—is that we want to chunk it, break it up into these on-diagonal chunks that were evident in Steve's slides.

And in order to do that, we make sacrifices, in performance or in overall complexity or cost. We have to make the chunks transparent to some person that's executing that design or some small team, and we make sacrifices. We use descriptive complexity as a means, as a currency to buy us lower computational complexity. That's the way I would frame it.

Now we can talk a little bit about the other side of this, if we have time. There's this other type of complexity that Nam has defined, which has --

M: Can you go back and explain that, how you concluded that they were limiting the number of off-diagonal numbers?

Dan Frey: All right. Now, this is something compiled by a colleague of mine, Dan Whitney, and what you see on this graph is the total number of rows in the design structure matrix, that is, the number of tasks that has to be completed. And this is the average number of non-zero elements per row of a design structure matrix for a number of different projects that he's looked at, from Intel, from G.M., from wherever.

M: Number per row. Shouldn't it ?? with the number of rows, number of columns?

Dan Frey: Well, that's what this is saying, but you know, it might be that, for example, it might be that if people can only handle five inputs to them, for any given task, then in fact this would decline, right? The width of the band of the matrix would stay constant, but the matrix will grow, and this would curve over. This would not be a line; it would be something less than a line. But in fact, it seems like as the project grows, actually people take more inputs.

So, for a larger project you have to take more inputs for an individual task, but it doesn't grow exponentially; it just grows linearly. I mean, that's empirically what we're seeing. I don't have a theory that explains it, but here, these are data.

Now let me talk about a different kind of information content, something more akin to Shannon's entropy, but Nam defines it as log one over probability of success. So, as a simple example you could take just a uniform distribution. So let's say that your design parameter is distributed in a uniform distribution this way, but only the designs with DP values between here and here result in acceptable performance. Now your probability of success is the ratio of those two ranges.

So what axiomatic design proposes that you do is minimize information content, maximize probability of success. Sounds good. And all I'm saying on this slide is that Nam Suh's axiom is telling you to maximize probability of success, which I think we would all agree is a good thing to do.

I've approached the problem as an anthropologist, as a chemist, as a physicist. Now let me think about it as a mathematician. If all the FRs are probabilistically independent, we can sum information associated with the different functions, just sum them. But, it happens to be that in many systems the FRs are not probablistically independent. For example, we saw, I think, a lot of heavily lower triangular matrices in those design structure matrices that Steve Eppinger was showing. And I proposed the proposition that if the design matrix of ASD coupled, that is, its lower triangle, you can just prove that the FRs can't be probabilistically independent if the DPs, if the design parameters, by which you control the system are probabilistically independent. Those off-diagonal elements induce correlation among the FRs, and that just comes directly from this formula, which relates correlation matrix in one system to the correlation matrix of another set of variables, upon linear mapping through a matrix that looks like A.

So they can't be. So you have to take into account that correlation. And in order to do that, in general you just need to integrate, over the design range. You have to integrate probability density over that range, where you can define the design range—if you take this linear mapping literally—as this set right here, defined as a set, which will base this system of linear inequality constraints.

Now, in order to proceed with that calculation, in effect what you're doing is you're setting all of those DPs or setting limits on your acceptable performance range, and in order to integrate all the probability density within that range, the tricky part is defining the limits of integration. So what I did is I found a way to describe those limits of integration in closed form and nest them in such a way that I can use something like Simpson's rule or Gaussian quadrature to actually compute numerically this probability of success.

Now, why did I do that? Because I'm interested in knowing the relationship between information and coupling. So here's another way to do it. If you assume that you have a uniform distribution, you can think like a geometer and extend this formula. If i is the log of system range over the common range in one dimension, in n dimensions it's the volume of the system range over the common range, where the volume of, in this case, a convex polygon has to be computed, and, in n dimensions, it's a convex polytope. So if you can just compute the volume of that polytope, you can assess information. And I adapt this recursive algorithm due to Laserre, in operations research, in order to do that.

Now I have the mathematical tools I need to make an evaluation. So how are they related? First of all, let me think now like a complexity scientist and make a conjecture. Now we know that Shannon's joint entropy of two random variables is defined as this, and, it always obeys this inequality here. If you just add the two entropies, the joint entropy is always less, and it's equal only in the case that x and y are probablistically independent. So if your goal is to minimize the total information of the system, this would suggest that, in fact, coupling helps you lower total system information. You'd actually like to have some correlation between those functional requirements, if that's your goal.

So it seems like complexity theory is starting to give us a little bit of an insight here. Let me just try to make that more physical for you by showing you a quick demonstration.

Here's a system with uniformly distributed design parameters. They all fall within this box. And here's a Monte Carlo simulation. Now, if I map it across, to the FR domain, and say, only those designs that fall into this blue box are acceptable, and here's my design matrix, I get a probability of success of 85%. Now, if I change the coupling structure of this matrix, through a transformation matrix—so, I haven't changed the sensitivity of this system, I haven't stretched it or compressed it or anything; I just rotated it—then, in fact, I get much higher probability of success, by inducing coupling, oddly enough.

So it seems like that there may be this relationship between coupling and information that we begin to see, as we explore, in this theory. I think I jumped to my conclusions. I had an engineering case study in which we consider a passive filter. And we come to the conclusion that, indeed, in some cases, this more coupled solution has lower probability of success. I'm happy to discuss that with anyone who cares to, but let me just come to my conclusions.

First of all, the engineering systems community should really think about philosophy of science, in this framework of Popper about falsifiability. And by doing that, it's not that we're trying to go around falsifying each other's theories; it's that that framework allows us to come up with better things. And that's what I try to encourage.

Now, I would also say that this theory has really brought to light two very important issues, coupling and information content. And what I find is that understanding the consequences of coupling, especially to humans, in terms of our ability to deal with it in our minds, it's going to be a key to better architecture, and, I would say, information content. First of all, we have to do the math right, and then we have to explore its relationship with coupling, and see if there isn't some tension between the two. If they aren't two things that need to be traded off, one against the others, as engineers are wont to do.

So those are all the conclusions I have, and I'm very happy to take any questions you may have.

M: I think you are doing exactly the opposite that Professor Suh was trying to do, by decoupling the systems in order to understanding them. Do you believe that emergent properties might eventually be able to be designed by understanding coupling and where emergence is actually a desired property rather than something we want to avoid?

Dan Frey: I think that's a really interesting question. Dan Hastings, at our conference at last night, began to talk about this a little bit. It's true that engineers are not in the habit of trying to design emergent properties. In fact, we think of them as something that pops up unexpectedly and is to be hammered back down.

But, as you begin to think about designing, for example, clusters of satellites, you'd better start to think in the framework of getting the desired properties of the system to emerge from functions distributed across those satellites.

So I have not done any research in this area, I'm not aware of any, but I know that it's an important question and I think certainly it should be studied carefully.

Nam Suh: Just so you know what axiomatic design is really saying and what you heard, it's very interesting. Dan and I have been talking about these issues for some time.

The consequence of this independence axiom, information axiom, is that if somebody can come up with a design that's coupled that behaves a certain way, the consequence of these three axioms is saying that, if you can come up with an uncoupled design, that will be a faster ??. So, in other words, one of the theorems—there are a lot of theorems—one of the theorems says that if you come up with a coupled design that performs in a certain way, and then if you can come up with an uncoupled design, it will be far superior.

So if you go to that throttle body design that one of our students looked at it, and after she worked on it for about ten weeks or so, and I pointed out that she should come up with the uncoupled design, having shown that the current design is highly coupled, she came out with a design that seems to make a lot of sense. So she called Ford, and, in fact, the Ford people are now working on that kind of next-generation throttle body design.

So, the point is, if somebody has a coupled design, one should always try to come up with an uncoupled design that satisfies functional requirements in the design range that will be then superior. That's what they're saying.

And then, as far as this thing about the information content, there are two things about what Dan said that you need to understand. The consequence of having a fully coupled matrix—think about that now, okay?—initially you can start out, say, with 5 FRs. Then you came up with a design that's fully coupled. That means the matrix is fully loaded with x's everywhere, right? What's that equal to? Think about that. That means he reduced the problem with 5 FRs into one FR, because he's willing to accept five FRs having a definite relationship amongst them, right?

So, you have to understand this axiomatic design from the very beginning. So, FRs are defined as being a set of independent requirements. So that means, if I change x—suppose you start out with x and y, and then if you say, I want to be able to vary x and vary y, but along the way somewhere you say, oh, I came out with this design whereby each time I move x, y moves by 2. So now I make y to be equal to $2x$. And if you are willing to accept that, what you have done is reduced the number. You are saying to yourself, oh, I only need to satisfy one FR.

So when you have a fully coupled system, and if you are willing to accept that, now what you have is you really are dealing with a one-FR problem, and that means you do not have the independence of FRs anymore.

And then, the circuit diagram that he was going to show, some of my colleagues looked at that very carefully. I don't know which one he was going to show, but the tolerance was different. Two different tolerances will apply, two different problems.

So, the point is, it takes a while to understand all these implications of axiomatic design, okay? So the first point, if I may repeat myself, is that if you come out with a coupled design, and then, if you are happy with it and you can satisfy that within the engineering range, then one of the conclusions of these two axioms says that if you can come out with an uncoupled design that satisfies the independence axiom, that will be far superior. That's what they're saying.

Number two is that, if you are willing to accept fully coupled design, that means FRs are interrelated to such an extent that you really end up with one independent FR, right? So in that case, like this ??-moving thing, once you figure out what the relationship is, all you need to do is move one bar, right? Then you get a solution.

So one has to be very careful in interpreting some of these things, and it does take a fair amount of effort to do that.

M: I wanted to comment on your slider experiment. The first time I encountered a digital speedometer in a car, I thought it was a terrible idea, and I still think it is. And the reason is because an ordinary speedometer is also an acceleratometer. So what you saw with the exponential performance of the human subjects was, essentially, the parameter space in the experiment has a maze-like structure imposed upon it. The human beings are doing a back-tracking search, which you would expect to be exponential in time and complexity.

If you were to provide a more elaborate interface, where they could move the sliders simultaneously and get differential information, particularly if they could move the sliders simultaneously, I would conjecture that they

274

would be able to intuit from that, even at an unconscious level, the structure of a search space, and converge on a solution to the slide problem, probably in a time that might even beat the computer in a certain sense. Because even with a computer you can do slightly better than cube time, by amortizing your matrix operations.

I suspect that when human beings integrate the differential information that they would be getting, that they would do something approaching that amortized matrix operation.

Dan Frey: I feel a little like the engineer looking at the cows, right? There's always that danger I'll extrapolate too much from my data. So I tried to be a little careful, and when I put it on my slide I said, for this particular task, it looked like it was exponential. And clearly, is this task much like engineering design? Maybe not. And could you devise different tasks that would in some respects be more like information design, by providing slope information immediately? Perhaps. I think it's well worth considering.

Chapter 7

Banquet Session

Yaneer Bar-Yam
Session Chair
Kenneth Arrow
Economics

Introduction

Yaneer Bar-Yam: Two and a half years ago, the first International Conference on Complex Systems took place, in this facility, and there were really two objectives for that conference, and the first was to start to develop a real foundation in understanding of a unified study of complex systems in all fields, all disciplines; that there was a systematic and effective way of thinking about all of these systems together. The second objective was as a launching event of the New England Complex Systems Institute.

Now, at the first conference, and also at the second conference, there was one thing that disturbed me somewhat, if not a lot, and that was that there were a number of speakers that independently—they had not heard each other—but independently stood up at a session and said, "And now for something completely different."

I don't know who was here and remembers this, but

it showed me that the speakers clearly did not understand the connection between what they were talking about and what the previous speaker had said. And there were many people, also, who came to me during the conference, the last conference and this conference, and said, you know, you should really spend some time explaining why those people were talking together. There weren't a lot of them; there were a few of them, but even the few of them that came to me and said, why don't you explain this? Again, there was some concern, for me, about this.

Now, I'm very happy to say that this conference has been different. It has really been different because it has been clear, as far as the discussion of the speakers and the discussion of all of us, that we understand that there is this understanding—that we are approaching the same problems. No one has stood up and said, "And now for something completely different," because it wasn't the case, in their eyes.

I knew they understood. I mean, the last two conferences, it was clear to me, and it was clear to a sizeable number of people who were participating, but it really has propagated outward and become part of our general understanding and foundations for complex systems.

Now, the other aspect of the first conference, the launching of the New England Complex Systems Institute, had a very unintended effect, and the unintended effect was that NECSI became a society which is an international society. We formed NECSI originally to coordinate and develop research activities in the greater Boston/New England area. But after the first conference, it became clear that that was not bounding our role in the community that we had formed. And so we developed additional activities, and I am very pleased you are all here, and I can welcome you to continue to develop the activities that are being created not only here in New England but really all over the world.

Now, in this context, there was recently a meeting that we organized in Seattle, and this meeting in Seattle was sponsored by Boeing, by Microsoft Research, and by the Group Health Cooperative—three very different industries. It was oriented towards the public. There were a number of lecturers that came there, some from here, some from other places, under the auspices of the New England Complex Systems Institute.

I have to tell you that the feeling of our participation in that meeting was quite remarkable. The feeling was that we were really changing perspectives on the world, and that the nature of the ideas of complex systems had a tremendous amount to contribute to society at large. And we really felt it, it was a remarkable feeling.

Now, the feeling here, at this conference, is different. This is a scientific conference. But it is really a pleasure, to me, that the atmosphere here has been one of the excitement of discovery. When I recognized in myself that I was engaged in the study of complex systems, some 10 or 12 years ago, there was an ongoing and continuous and, until today, tremendous joy in the discovery of new ideas that I was borrowing from different disciplines and applying all over the place.

And it's very clear that that has become part of what we are engaged in doing. The way I sometimes think about it is that, at some point, I and a number of other people discovered that the scientific endeavor was getting into some maze of concerns, but there was some path that led you to a path that led you to a place where there were many gems just lying on the ground that one could pick up and say, hey, here's another one.... And that somehow has become possible for us to propagate outward. So that is by way of introduction of the meeting and how we've come to here.

Now it is my great pleasure to introduce Kenneth Arrow, who has been patiently waiting for us to complete at least the main course of our meal and hear my remarks.

One of the things that I am greatly impressed by the work that Ken Arrow did many years ago, is a willingness to take theoretical descriptions of complex social systems seriously, the proof theorems, that are paradoxical. And recognizing the paradoxes is an important and essential part of understanding the interactions between many systems of many parts. There are many paradoxes that we are discovering in our understanding of these systems.

But in order to discover a paradox, you really have to be serious about what you are describing, both the real system and the mathematical description of that system. And, of course, the particular result, which is so exciting, is what's called Arrow's Theorem, or The Dictator's Theorem, which I can tell you—if Ken will allow me, I'll describe it briefly, or are you going to describe it in your remarks?

Ken Arrow: No, I'm not going to describe it. Please, go ahead.

Yaneer Bar-Yam: You can correct me, please. I don't know the original formulation. I will tell it to you as I heard it. David Meyer who is on the executive committee has done work in this area. Unfortunately he had to go Washington today, and I don't know why even, but he's unable to be with us, so I will take the responsibility of describing the theorem.

So, as I understand it, the way it works is as follows. If you take a group of people that are voting, and in particular one might imagine the House of Representatives or the Senate, voting to enact some legislation. And now there is a possibility that there will be votes to change the legislation by amendment. Now, the process of changing it by amendment means that some people prefer some option, B, to the current option, A, which is the bill which is in front of the House of Representatives, say.

And then, there's a possibility of a second amendment, and a third amendment. Now, we know that in Congress, indeed, there are many amendments that are attached to bills, but what Ken Arrow demonstrated is that, in principle, if you write the process down mathematically, there is generically no stable state, so that you will end up going in the following way. You will go from state A to state B to state C, but then actually it is very likely that you will go back to state A. And the reason that is is that the people who preferred state B to state A maybe are outruled by the people who prefer state C to state B. But then once you're in state B there may be a majority of people who would really prefer state A to state C.

So you will end up looping around and around, and therefore the conclusion is that you can only have dictators that presumably have monolithic decision power in this process. And this is quite an interesting paradox.

Of course, as far as I know, no one talks about what's going on in the dictator's head, and of course we know that even individually, we might go around in loops: oh, I would rather go to this job or to that job or to the other job, and we can go around in rings. Presumably there is some way to resolve this, because otherwise in our heads and otherwise in Congress we wouldn't be able to do very much. And I don't know whether there's been a resolution of the paradox, but if I've described it correctly, maybe Ken will remark about what has been done since then, perhaps in the question section.

Now, in 1972 Ken Arrow received the Nobel Prize for pioneering contributions to general economic equilibrium theory and welfare. Of course, a little bit prior to that he received his Ph.D. at Columbia University. And his interests are in general in the economics of information and communication and the economy, in equilibrium under monopolistic competition, shifts in income distribution, and I've already told you about his willingness to discover paradoxes.

Thanks, Ken, for your patience, and I'll turn it over to you.

Kenneth Arrow
Economics

Kenneth Arrow: I want to say a few words about this sort of a meditation on the concepts of simplicity and complexity. At this stage, I'm really a consumer of all the work that people have been doing, not a creator, and I in some ways have followed the field, and I'll give some background as to why I was kind of predisposed to take a favorable latitude and to be very interested in the implications of the idea of complexity.

What's interesting is that complexity has become such a favorable word today, an OK word. For both of the last few centuries, at least, if not before, simplicity was the desired thing. I mean, there's an old proverb, "Simplicity is the seal of truth." We all know about Newton. I'm not sure, I think I have the quote right: "Nature affects not the pomp of superfluous causes."

So there was a great tendency to regard a simple theory as being an ideal theory, and that somehow, if you really understood nature, and by extension the cause of human and social interaction, you would be looking for simple systems.

Now we seem to have moved, perhaps in the last fifteen or twenty years, into a new realm, in which we think of complex systems as encompassing, at the very least, an extremely important part of both natural and social explanations, and also as a basis for decision-making, for policy, and for prediction.

Murray Gell-Mann, who, along with his many other attributes, has a fantastic command of languages, points out that the words "simplicity" and "complexity" have very different derivations. They're not parallel words—one is derived from the Greek, the other from Latin—but they do ultimately have a common Indo-European root that Murray would know about: *plek-, as I understand it, which means something like "to plait." Kind of the ultimate simplicity and complexity connection, even though they're opposed to each other.

But let me say a few words where I encountered these concepts, in a very early stage. For some reason or another, even as a college student, I was much taken with the uncertainty of the world—remember, this was Great Depression, and it's not surprising to see very remarkable changes in all sorts of things—personal lives, the economy as a whole, sudden changes occurring with no apparent reason. Now today, we'd be looking for explanations in complex systems, talk about chaos, or whatever you want.

But no one understodo those concepts, at least I didn't, nor did the next person. And I was much taken by another approach to this world of the unpredictable, namely, probability theory and statistics. Somehow, I encountered them as an undergraduate, more or less reading on my own, and very impressed with the idea that you can take uncertainty and make statements about them, and say how often things happen and make predictions. And with statistics, you might comb out the uncertainties and try to get to at least the parameters that characterized the uncertainty. Even if you couldn't eliminate the uncertainty and unpredictability itself, at least you'd have some control of the situation.

Now, through some early stages, and probably through an accident-- and of course one of the things that we've learned that has been emphasized in complex systems analysis is the role of path-dependence. Well, my career was certainly path-dependent. I went to Columbia originally to study statistics. I found that it was located in the economics department, and there was no degree in statistics. The only way I could carry on my studies was to enroll in economics, and that's how I got into the field of economics.

I was also interested in economics as well as in statistics, and I remember saying then, that we shouldn't be disturbed so much about the fact that economics was not doing so well in getting ?? the regularities as, say, astronomy or the physics that derives from astronomy. After all, if the solar system had revolved about a double star, the orbits would have been so complicated, that Kepler could never have inferred any simple laws, and therefore Newton would never have arrived at the law of gravitation.

It was very comforting idea, although it wasn't exactly helpful. In economics there were no single source of overpowering gravitation strength corresponding to the sun. It was going to be, according to this, rather hard to make any derivations. And while it was discouraging, from one point of view, at least it meant that, really, that it wasn't our fault.

I was studying economics at the time, trying to bring some intellectual order to what I was reading. I thought it was a problem, really, that I thought this is what it is, it isn't really a correct statement for what physicists or chemists do, but it was my impression.

And when you look at the individual parts, the households, the firms, and so forth, try to get the laws of their action through a mixture of theoretical work and "econometric" investigation, and then put them together, aggregate them, say through the idea of equilibrium on markets, or whatever concept should replace that. So we just needed better theory and more observations.

But I was well aware, of course, what a poor job was being done now, how little we really understood, but I thought this was somehow an objective for the future.

Well, I did something which should startled me more than it probably did—it did startle me, but it didn't lead to any particular action on my part. I left graduate school to enlist in the army, and I was sent for training. From economics to weather. Meteorology was rather interesting. Here, I moved from a subject of social science, with all its imperfections, with the lack of understanding at the micro-level, and I moved to a field in which, in some sense, the micro-phenomena were very well-understood. Not completely, there were obscurities, it was recognized, even then.

But, on the whole, the laws of thermodynamics and hydrodynamics told us what was going to happen to individual particles as they rose and fell in the atmosphere, as they moved under the influence of wind, as the energy was transformed, as the rain fell and water vapor turned into rain. All these phenomena could be quantified. And so we were given quite a course in the basic physics of this.

Well, it turned out that all this basic physics, as I say, was well understood, not at all like the micro-phenomena I got. Really had relatively little to do with the actual practice of weather forecasting. At the theoretical level, we observed that the phenomena, at least in what I call the middle latitudes, were essentially derived by the formation of cyclones, let's say, areas where there were was a sharp temperature gradient. And it was observed that around these lines, which bounded hot air and cold air, would sort of fold up, and as they kept on folding up the storms associated with that, what was called an extratropical cyclone. And these were the major weather events of the middle latitudes. The equatorial regions were something else, the Arctic regions were something else. But these were the areas where most of our military flying was taking place, where we were supposed to do our forecasting.

In the first place, at a theoretical level, you really didn't understand why these storms formed. There was a huge volume by some Norwegian meteorologist, written in German, which I didn't understand too well, which I don't understand too well, but it was pretty clear that they had a lot of trouble going from the simple principles of thermodynamics and hydrodynamics to the derivation, even to a theoretical derivation, of why storms should occur.

From the point of view of practical forecasting, some of the ideas from the physics did find their way, in a purely qualitative sense, into practical work, and they informed it. There was no sense in which the formulas of physics were being used in a quantitative way. Instead, we observed that there were storms; we observed that they tended to move from west to east; we observed that they tended to, once they formed, continue for a while and then eventually go through various stages in which they attenuated; and then new ones would form.

So these were the practical implications of forecasting, the practical implications of theory for forecasts. There was a very remote—I wouldn't say zero connection—but a very remote one. In fact, the ability ?? engaged in a project to verify how good forecasts worked, ?? statistical training, and my colleagues and I found that there was essential zero ability to forecast after 72 hours. By that I don't mean that a meteorologist couldn't do better than a layman. What I mean is, if you knew the norm for that time of the year, for that month of the year and in that place, then at 72 hours you could not do any better than just predict ??. So there was no skill. And the forecasters, of course, differed very much in their ability. Some forecasters ?? after 36 hours. I can imagine they weren't all that much better than chance, for shorter periods.

It was a huge project in which I participated, trying to forecast for next month. Well, I can only tell you, after several years of effort, gave essentially no value to that, and even today just fragments, so ?? associated with El Niño. El Niño, by the way, was ?? that we were taught about, but we had no understanding of why it occurred, and so we have a lot of theory, but none it could really explain El Niño. That was partly, I would say, lack of observations, in this case ?? temperature.

We do know that part of the reason, of course, why formulas were not used more sensibly is the computational problems involved. So one aspect of complexity is certainly computational complexity, and of course this has changed. We do have model-based forecasting. I'm not really up on how good it is, but my impression is that it's not remarkably better than forecasting by old-fashioned methods, but at least it's not any worse.

Well, that's our point. But I'll tell you about a famous study. In 1911 or 1912, a physicist named Lewis Richardson, mainly famous early on for his work on viscosity, and later for work on international arms races, decided to actually use all the physics possible to make a 24-hour forecast. The forecast itself took about three years to make, and that was a question of principles, not a question of practicality.

Well, after these three years, he found that the forecast was not only wrong, but ridiculous. It was far outside any normal variation. It had pressure about twice the normal atmosphere pressure. Of course, what he showed—remarkable—he actually showed that the known errors of observation in some of the ?? was ?? this discrepancy.

And you can see, this is a remarkable forecast, a remarkable precursor, I should say, of work such as Professor Lorenzo's. What it really showed is that minor errors—essentially, wind directions could be off by five degrees— were sufficient to get a very big error in the forecast. And that hint was not taken up, I don't know for what reasons.

So we see here that one aspect of the complexity was, let's say, the computational complexity, just the sheer question of doing the job. And, in a way, that's been, to a significant extent, not by any means entirely, resolved by simply lowering the cost of this complexity, as a proven computational method.

It also shows something else: that, in some sense, complexity, like simplicity, is a somewhat relative matter. But it's not clear to what extent it's a statement about the world, or a statement about ourselves, in our capacity as scientists, scholars, or just ordinary human beings.

Let me take simplicity for a moment. I've always been struck by the fact that in spite of the emphasis that philosophers of science have placed on it, it's really more a matter—and this says as much about the scientist as it does about the science—but in something as simple.... Some phenomena that do not appear to be simple become simple with a considerable amount of iteration. You keep on learning, you learn how to handle it. It's a question, in other words, of your state of knowledge. I think you'd find a lot of parallels in work in complex systems. The question of what is considered to be a simple thing depends on what you're accustomed to. After a while, what seemed to be a complicated idea becomes a simple one, because you've learned to handle it. You've become familiar with it. You've drawn through implications, and now it's become a simple concept.

Let me give you one example from my own work, the notion of general equilibrium in economics. Now, this is not my idea, the concept generally goes back more than a century now to the 1870s. But it was still not an ordinarily used concept, even to most economists. To a mathematically trained economist, it seemed like second nature, and to people like myself, or my collaborator, Gerard Debreu. In a way, general equilibrium theory was the simple way to formulate ??. Everything was consistent. We could handle all the cases in a uniform way. There was not a lot of ad hoc statements, and to us it was the natural, normal way of thinking. This is the result. Yet to the average economist, it was considered very esoteric. A typical thing was to think in terms of a couple of markets and then wave your hands as to what happened to the other markets and, say, there are interactions. And to us, Gerard and myself, this was very hard to understand. To the average economist it was simple. What we were doing was considered complex.

It was only complex in the sense that it involved mathematical formulas. It was pretty hard to discuss it without some higher notation. The theorems required sometimes advanced methods, like fixed point theorems, which were, then, considered very advanced.

But it was not, in a real sense, complicated. It really was simple, the general equilibrium theory of economics is a simple theory, and yet it would not be perceived as such by most economists at the time. Now I think it's part of every student's kit-bag and is now what we thought of as, perhaps, the oversimplified way of doing things.

So, in the same way, I wonder whether complex systems aren't systems that we're not accustomed to. Let me turn to one particular aspect of that. If you seek a definition of complexity in terms of something called emergence—and of course the concept of emergent properties is not a new one, it goes back 60 or 70 years, and, of course, it especially has been applied to biological evolution.

The idea is that, somehow, something new has occurred, something that was not expected. Somehow, there's something in the conclusions which was not in the assumptions. Well, that's very funny. In some sense, if something is a logical conclusion, it must have been there.

It reminds me of a story I read in my logic textbook in college. After all, logic never gives you more than the implications of the assumptions, and therefore, what you already know.

Well, a story was told about a priest who had been in Paris and was sent to a French provincial town where he took on his role as a parish priest. And the Marquis gave a party, for a month or so, welcoming him, and at the party the priest was approached by someone who said, "You must find it very dull here compared to Paris where everything is so exciting." He said, "No, you'd be surprised, in this pleasant town, what ?? deaths there are. Well, you know, the first person who confessed to me confessed to a murder."

At this moment, the Marquis comes over and says, "Oh, I'm glad to see that you're talking to a Father. You know, the Father and I have been old friends, now. Why, do you know, when he came here he was the first person I confessed to."

Now, is that an example of a complex system, where something unexpected emerges? It's clear that if, after a while, you kind of remember these stories, somehow ?? unexpected—in other words it's not one that you can easily repeat. And then it's a question of what we mean by complex systems, and the definition, no doubt, I'm sure many of you have thought of these things and have good answers to what I'm saying, but this is just a suggestion. I've been reading work by some of the creators in this field, and I find that there's always a slightly unexpected element here, which, it seems to me, has more to do with knowledge and psychology than it does with statements about the real world.

Of course, that doesn't mean they aren't important. In some sense, the whole process, whether it's science, or management, is a process of interaction between human beings, as they are now constituted with their current levels, and an external environment. It may be an external social environment, it may be an external physical environment, or a biological environment. So maybe the added complexity is the interaction of the actor, the human actor, and the world about him. It's not necessarily inappropriate.

Let me turn to another aspect of complexity, which is that one of the chief implications, not the only one, but one of the implications that's most striking, is the fact that it shows that small changes may frequently have large effects. Prof. Lorenz made a famous comparison about the effect of butterfly wings on weather. Stuart Kauffman, my friend, whom you heard, has certainly shown how simulations can bring about quite startling surprises.

But let me say, supposing that's true, supposing that we accept that a considerable part of the world—of the implications of complex systems—is that the surprises we're talking about, the emergence, is in the form of relatively large changes from what seem to be small causes.

But, if that's true, I'd say it creates a slight gloomy aspect for decision-makers, whether public or private, or prediction forecasters, and, as far as that goes, ?? turn around, for discovery—generalizations, useful generalizations about the natural or social worlds.

Let's take the first way. Assume that we have a typical situation in a complex world—the weather, meteorology, where essentially each particle in the atmosphere is a different object subject to local forces. Now, the individual relations governing these forces may be quite simple. You understand what happens in pressure if a particle rises so it's facing lower pressure, the effect on temperature, and so forth.

All these things are quite simple. But there are a lot of them, and they interact in a very complicated way, which is also uncontrollable.

Well, we're interested in the consequences. Just as a predictable point of view, we're interested in what the predictions that flow from these relations are, as in weather. We may be also interested in controlling the situation, as, say, taking a policy step, which will change things. But therefore, we have to predict what the consequences of our change are.

Well, if, in fact, it may depend, in a major way, on relatively small effects—say, small errors in observing the initial conditions, or random fluctuations that we can't observe individually, or, in fact, errors in the modeling process—even though our model may be, so to speak, basically correct, it's very unlikely that we got all the factors right. We've got a few somewhat wrong. This shows up all the time, in the most practical way, in, say, ecological forecasting, weather forecasting, and certainly in business forecasting.

Then, how ?? how the world can make useful forecasts. Occasionally, one can observe patterns. Even if you're doing a simulation, so you are controlling completely the inputs, you find that the output is some big mess, and even to sort of understand it, you would search around for patterns. Now, occasionally you find these patterns: Gell-Mann, for example, in his simulations ?? the existence of frozen accidents. They're accidents, the result of the sensitivity to initial conditions. But also they're frozen. I'm talking about patterns that can last for considerable periods of time.

The storms that we observe on weather maps are these kinds of patterns, and, as I say, the constancy—that's too strong—but the fact that they don't vary hugely, over a few days, gives us most of our hope for good forecasting.

So, it is true that among the many implications of complex systems will be, so to speak, by good luck, accidental pattern formation, and sometimes we can explain the pattern formation in a deeper way, sometimes just sort of the way we can. But on the other hand, we can't count on that. Many times, we do not see these patterns, and yet we're making forecasts. And forecasts from complex systems must be inherently difficult to make, to the extent that conditions are path-dependent, as the expression is sometimes used, or sensitive to initial conditions.

And then we come to the other point I mentioned before, namely, ?? observe the outcomes of a complex system, and we want to find out what the system generating it is. As my professor Hotelling put it, when Kepler was observing laws, difficult though it was, difficult though the observations were, he was able to infer, particularly from observing the motion of Mars, regularities. They were subtle, but not simple as Claude Ptolemy thought, but they were observable and, once he found them, they could be verified with a high degree of accuracy. That's an extraordinarily simple system, mainly because of the dominance of the sun, as I mentioned.

In a world in which many things are happening simultaneously, such as the weather, or such as the economy, it becomes harder to pick out these key relations. And essentially knowledge, at all levels, that is, direct knowledge, as is possible in social affairs, direct observation of how people behave, is a supplement. And of course, it's subject to biases, and micro-observations are a substitute for data, but they too can complement each other.

And I've seen this. Now, one of the examples, the best study in economics, is in the securities markets. The securities markets have the advantage of very good data, fairly homogeneous conditions, not really, of course, because the companies involved change, and even the companies that stay in existence change their character, but at least there's some kind of continuity in this.

And one of the implications of general equilibrium theory that were drawn was the so-called rational expectations model. Now, a rational expectations mode does really pretty well. There certainly are exceptions. Unfortunately, some of them are dramatic. But on the whole, a model based on that has not done badly. You can't make any money following them, because everybody else, by definition, that's what the market is following. So the only way to make money is by beating that, and people have beaten it.

Nevertheless we have extraordinary events. You see that, in fact, it's been fitting. One of the consequences tends to be fairly smooth movements in the market. However, and, if you looked at, let's say, the late Eighties, you'll find things were creeping up. Individual stocks might vary violently, but the averages were fairly stable. And suddenly, in October 1987, a 20% drop in one day. No explanation consistent with the model is possible. The model would say that prices should only change when there's new information, and there certainly was no new information of any significance on that day. Nothing happened, in the world or any particular industry. The rise the market in the last five years has similarly been inexplicable.

And this is a disturbing thing. That is, supposing you predict a model based on some kind of equilibrium. You expect, therefore, things to be changing smoothly. What you observe is, indeed, for long periods of time, things sort of fluctuate mildly, but then suddenly and inexplicably there's a big deviation. We know, of course, in chaotic systems this is just the sort of thing that happens all the time.

For example, a group of the Sante Fe Institute developed a system to explain the stock market, not based on rational expectations but based on the assumption that the agents in the markets were engaged in adaptive learning processes. But of course they were interacting with each other and that changes, modifies, the wording.

The kind of anomalous behavior that we discussed can be the observed model. You have the model going along nicely, just wobbling around slightly. Suddenly, without any change from the outside, suddenly plummeting, or rising sharply, big deviations occur. Qualitatively , most of the phenomena that can be found in the real stock market are observed in this artificial simulation of the market.

Of course, it doesn't reproduce the actual market. It's another market. it's not based on the factual information in the real market. Nobody's tried to assimilate, nobody has ?? which will simulate the real market, and, indeed, if small causes have large effects, you would not expect to duplicate the actual market. You'll only be able to duplicate its "kind" of behavior. Nor, of course, do we really know that this particular model is the true model. It is a model which will give you the qualitative results, I don't doubt. People find distinctly different models which would produce similar results.

The world of complex systems is not the one that I've done most of my work in. I've worked in closed, logical universes where deductive methods of econometric methods based on stationary processes are dominant.

Here in the complex system world, I think we have, rather, a much more open, I might call opportunistic, attitude toward data and to the systems which generate it. Solutions which may be valid over relatively short periods of time, though, claim to be universal. There are things which fit, and like the storms, which come and go, they'll be valid for some period of time. They'll disappear after a while, and we have to adapt to new regimes.

It seems that the idea of "understanding"—I put that term in quotes because I don't know what it means, exactly, but I think we all intuitively know what it would mean—is very difficult to come by, and, similarly, the idea of permanence. In that respect, of course, the prototype of complex systems, biological evolution, poses the inspiration for the ?? thinking, and shows exactly the same properties. You have stasis. You have something that changes in series. We have new forms evolving which didn't exist before, so we have novelty, at least in some level, even though there's continuity also.

All of these factors may also be prototypes of complex systems, of the management world, or of the world of social or even physical sciences.

Simple systems, on the other hand, yield some comfort in being big, giving you the sensation that you understand the world. Of course, they suffer on the difficulty that they don't fit very well.

So I hope you have reactions to these reflections on complexity and simplicity, but I think it's a tribute to the whole complex systems environment that it has created these much deeper issues in what have previously seemed to be certainties. Thank you.

M: My name is Jaime Lagunez. I come from Mexico. Right now, the problems in economics in Mexico seem to be related to new liberal policies and globalization-type attitudes, among the people that make the decisions in the government and people outside of Mexico. I wonder if you would be able to make some comments about what's going on now.

Kenneth Arrow: Well, a little remote from my theme, I think. Maybe not. Everything's a complex system. I don't really feel I know in detail what's happening in the Mexican economy. The Mexican economy before the liberalization was extremely stagnant, and there is no point in comparing the present problems with some mythical past which didn't exist. Let me put this way. Every economic regime creates problems; every social regime creates problems. There's no such thing as a world without problems. If you have the liberal globalizing dynamic economy, you're going have problems of adjustment. If you have a closed, internal economy, things may be more predictable. This may be a way of handling the chaos due to complex systems by suppressing them. That can be done. You can force stability, generally at the expense of great stagnation.

The Mexican economy, until the 1990s, or earlier on it had a period growth, but it had become quite stagnant and was clearly suffering from a kind of sclerosis. I think you find, if you look at the figures, that since the crisis of a few years, was it, I forget now, '95 was it, that the Mexican economy has done quite well, and that there has been an incredible increase in industrial output and in employment.

There are problems of adjustment, and I'm not really an expert on the Mexican economy, but my offhand view is that, in a way, the price of joining the international economy is one of instability. The payoff is that the averages tend to be much higher. Thank you.

M: Prof. Arrow, could you comment on the current state of the increasing returns issue and assumption in economics, à la both equilibrium theory and complex systems approaches?

Kenneth Arrow: That's a big topic. The statement that there are increasing returns in economy, and that this is inconsistent with competitive equilibrium, is a very old one. I can date it, at least, to a book 162 years ago, to give you an idea, Cournot's *Mathematical Principles of the Theory of Wealth*. So there's no questioning systems can affect modern information technology. Software's an extreme case. In software, there's nothing but increasing returns. The cost of producing software once you've designed it is, of course, essentially zero. So the cost is entirely the fixed cost of development.

There are many other industries in which high fixed costs and low unit costs are a typical example of increasing returns. Similarly, not so extreme but a somewhat similar thing occurs in the pharmaceutical industry, where, well, their cost production's not insignificant, but nevertheless the bulk of the expenditures still are research and development costs.

The implication for the standard questions economics, like, what are the prices that are going to prevail? what is the distribution of property? I'm embarrassed to tell you—and not even from a theoretical point that would have been worked out, let alone from a complex systems point of view—even if you were assuming as much equilibrium as you'd like and as much rationality, and complete adaptation, it's not very easy to figure out what the optimal responses are. You have not just in a single innovation but ?? innovations; you scarcely expect to have optimal adaptation to any of them.

So the really interesting question, in my mind—and there's been quite a bit of discussion on this—is what's called path-dependence. One of the arguments is, if you start with increasing returns you're liable to develop, for example, suppose you had two products which are not the same but sort of compete with each other. Then the cost of increasing returns, one is likely to get produced, and not the other. If I have to share the market ?? cover fixed costs. So one will—that's an oversimplification, but ??.

But once one has developed, the future course of the innovation may be affected. It may be cheaper to, say, improve that product than to improve the other. At some level, this has been argued for videocassette recorders. We had the Betamax versus Sony formats. It was argued. The typewriter keyboards. It's been argued at an economically more significantly level for automobiles. It has been claimed, by some, that if you go back to 1900, well, a fact in 1900 is that there were probably as many electric cars and steam cars as there were gasoline-powered cars. They were about equal. The three were about equal in number.

Within a decade, gasoline-driven cars essentially drove the others out of the market. Now, one argument is that this is an increasing returns. Because of the system of servicing, there were increasing returns on ?? drive out the other. And that ?? which one was something of an accident. We could have had steam-driven cars; we could have had electric-operated cars.

And, at that point, of course, the difficulty is these big cases are by definition speculative. We only have one history. The theory tells us there will only be one history. The others are only potential histories. And we have no real good way of testing it. And of course, others will say, yes, you have path-dependence within limits. It'll be like two different kinds of videocassette recorders. Sure, one of them drove the other out, but they weren't that different to begin with.

Similarly, with the typewriter keyboards, if we had had a different typewriter keyboard, maybe it would have been 10% ?? than they are now, maybe not.

That is actually a subject of debate. So, I've talked to some ecologists, and there's a similar argument in ecology. For a long while there was the following view. At a given location there was an equilibrium vegetation: forest, bush, brush, grass, whatever—it was at equilibrium level. This is called a climax forest. Now, if you burn it down with a forest fire, it would grow back, and it would grow in a certain sequence. But eventually you'd have the proper distribution; this is, because the different species interfered with each other, they would have influenced each other, interfered or helped each other, and it was therefore an equilibrium. So the forest had fir trees. The next time it grew it would have fir trees, but then there would be underbrush of a different kind, and so forth.

Now, as I understand it, ecologists say that accident in the growing process may produce pertinent results. But, you won't get radically different kinds of trees. If you have fir trees, well, you may get pine trees instead, but you'll get ??.

So the question is, will increasing returns lead an economy or world economy to a very different path, by small variations, or will lead it to a path that's only a limited degree away from the other ones? And that's, I'd consider, an open question.

M: I was very interested to hear of your extensive background in the area of weather forecasting, and it inspires a question about economics. As you pointed out, weather forecasting has developed in two directions. There's work of Lorenz, which you referred to, in which the focus is on remarkable behavior from very simple causes, like the butterfly effect, and this has led to a body of study on the stability properties of the atmosphere, and so on.

The other direction, which you also referred to, has to do with increased computation power and, basically, straightforward number-crunching approaches, which we see now, where as the computers get bigger and bigger they process more and more data. They are different approaches, one focusing on issues of stability and qualitative behavior, the other using greater computational power to crunch more and more data.

This is analogous to what goes on in many other fields, but I'll just ask about economics, because some of the phenomena that you referred to, such as stock market crash and then its inexplicable recovery, and so on, sound a great deal like the sort of effects that Lorenz spoke of, arising from instabilities in the marketing and rapid changes, perhaps catastrophic, as opposed to the procedures, which I understand many people use, and just putting in vast amounts of data about different stocks and other economic indicators, to just crunch their way through to make predictions about how economics is developing.

The question is, I'd like your thoughts on the general dichotomy, in many fields, between the one approach looking at qualitative behavior, looking at stability, looking at interesting behaviors, for us, and the other approach of using greater computational power, whether it's in weather forecasting or ecology or in finance, simply to crunch more and more data to see what comes out.

I'd like to know your thoughts, for one thing, on your ideas of stability and complexity as applied to these, but also your view on the general picture of how in different areas of science these two approaches might balance.

Kenneth Arrow: That's a very interesting question. Well, the two approaches clearly are complementary, and they obviously both need to be pursued. But, in a way, the qualitative results, to the extent that they're validated, of course, must in some sense take priority. They imply limitations on the data-crunching, on what data-crunching can do. And I think in the case of weather I think it's fairly clear we're talking about what the horizon is for forecasting. I mean, how good is the forecast going to be?

Unfortunately Stanford doesn't have a meteorology department, so I have to depend on contacts I make on occasional trips. As I understand it, at least as of a few years ago, the model-building for weather, really, it doesn't pretend to make forecasts of more than four to five days in advance. These instabilities, in other words, are going to take over, and it's recognized they will, so we're talking about essentially two different time periods, that is, the instability about what really should be done over weeks, which was, as I say, the goal of all our work during the war, and a goal never obtained, to get any genuinely good forecasts.

And it's ??, of course, how much you want to invest in higher and higher accuracy for a two- or three-day forecast. Presumably, I assume there'll be forecasts out to 96 hours today, instead of the 72 hours that was true to World War II. It also means you're getting better forecasts of 24 or 48 hours. So presumably there's a certain amount of additional complexity, and, of course, as the cost of computation ??, the costs in operations become relatively low. And I'm sure the research done in the course of it is illuminating. So I don't think there's really a contradiction between these two points, they just point at different things.

In economics, interestingly enough, the heyday of the large-scale number-crunching model has passed. I don't say it won't come again. It will come again, I'm sure, in a different form, but it has passed. There was much more going on even in the 1960s, and certainly in 1970s and early 1980s, and one of those thousand equations or more were being routinely fitted. They still exist for commercial forecasting services, but I don't think anybody feels they're much better than relatively small models, which accomplish basically the same thing.

Sometimes, you want the bigger model sooner, because you want the detail, not because of greater accuracy. You're not interested in just forecasting total consumption or price level, you want to know something about a sector.

But in terms of the idea that you get accuracy by disaggregation, by building up the totals from small pieces, as I say, it really failed because for empirical reasons, it just turned out not to be very successful. And therefore that's not the issue in economics, as opposed to the situation in ??.

Now what you get is extremely detailed modeling of the securities models, not of the economy as a whole. There, you have extremely sophisticated models as a basis of measurement for price and derivative securities. And that is a remarkable accomplishment. It has enabled these markets to exist, because, in a sense, if you couldn't market the model, nobody had any idea how to price these things. And it was a creation, from the Black-Scholes model on, of specific models, which actually enabled ?? to exist, enabled reasonable forecasts to be made, so people had some idea what to do.

Here's a case where predicting turned out to be necessary for the phenomenon ?? that exists. That's a rather different nature from the large scale. It's not really parallel to the weather model, which I think is more like ??.

M: I have one question and one comment. The comment is, you made a distinction between simplicity and complexity, and you also mentioned the emergence of patterns and complex systems. And I think if you can recognize patterns in some complex systems, then you have actually a simplified description of the system, so I think there is no contradiction between a pursuit of simplicity in the description and the study of patterns in complex systems.

My question is, I mean, you are in California, and you were talking about the weather, which I really appreciated, atmospheric phenomena, and in the news we have now this forecast of this 30-year drought that they expect for California. So there is a direct interaction between economic factors and atmospheric factors, and there is a lot of discussion going on, between global warming and what to do about it in terms of the economy, right now, for individuals.

So my question is, do you have any ideas about how, say, recently ?? especially global connectivity through the Internet and recent results in complex systems theory could maybe contribute to solving, or at least ameliorating, the problems that we are facing for the next 50 or 100 years?

Kenneth Arrow: Well, let me just say, it was a good question, and I don't have a real answer to it. I've seen a fair amount of the climate model, in both modeling of the climates, of the impact of ?? gas formation on the climate. I've seen a number of those models, and then, the implications of that warming, for the economy.

Now, these are complicated systems. I don't whether the word "complex" for these is correct, in the sense that the logic is pretty straightforward. There are a lot of details, but that's not the same as a complex system ?? cereals, what are the effects on non-agricultural production, and so forth. And there's been quite a bit of work of that kind. Different researchers disagree sharply about the implications. But in terms of ?? interactive system, which is generally complex, I've not seen really any very good usage, so far, in this field, and I don't have any particular ideas myself. Thank you for the question.

M: I have two interrelated questions. Do you know Prof. Jack Cohen, from England, who's the famous sex biologist?

Kenneth Arrow: I'm afraid I don't.

M: Well, my question relates.... He's been constructing limericks during your talk, and he's come up with one that's like this, and I wanted to know whether you think this is appropriate behavior:

At a popular meeting on complexity
We found we were wooing simplicity
When diversity grew, from principles few
We wandered about, lost in ecstasy.

Kenneth Arrow: Lost in what?

M: Lost in ecstasy.

Kenneth Arrow: Well, I'm not sure of the ecstasy, but everything else I'm fine with.

M: My second question is, can I have my photograph taken with you, sir? Have you heard of the Kevin Bacon index?

Kenneth Arrow: You use the word—pardon me?

M: Have you heard of the Kevin Bacon index?

Kenneth Arrow: Yes, I've heard of the Kevin Bacon index. Six degrees of separation.

M: I'd like to start the Kenneth Arrow index, and if I could have my photograph taken with you, I would have the Kenneth Arrow index of 1.

Kenneth Arrow: Now, you use the word "can," which, among others, implies a technological problem.

M: I can do it, with this ?? in front of me. You don't know it, but I do.

Kenneth Arrow: Go ahead. [applause, mixed with laughter, as limerick reader has photograph taken with Arrow, twice]

M: I was wondering if you could comment on the loss of lag time in communication between business due to the Internet. I'm thinking specifically of the beer game, for instance, where lag time between consumer/retailer, etc., down to producer, in many situations creates chaotic behavior, and how that certainly would be significantly affected by real-time information flow on the Internet.

Kenneth Arrow: That's an excellent question. There used to be a belief that these lag structures could produce instability. It was never very clear whether it was true or not. And reducing the lag time could work in either direction. I have little doubt that if it increased instability people would find ways to handle it. I mean, typically these lags, these models in which lags led to instability were ones in which people, let's say, didn't grasp the problem, and they could easily have put in dampers, lag adjustment mechanisms, which would have obviated the issue.

Whether these were really ?? in any significant way, I don't know. There's no question that, given that there was a lag, people were using those lags for smoothing purposes. They were filling in the gaps, in one way or the other. In fact, when you remove the lag—assuming you really do, by the way, because I believe there are plenty of lags left in the process, of which the delivery of the order is really one element of the lag. There are response lags of all kinds.

I'm sure it will turn out that there's a value in that lag, but people find some other ways of capturing it, that's all. I think this is the kind of adaptation I expect firms to be able to respond to reasonably well.

It's been a great pleasure. [Applause.]

Yaneer Bar-Yam: Maybe before we close I will finish the statement that I made earlier, if you all care to hear the end of the story [scattered applause], and this is really a complex systems story, as you will see.

There was a story that I told in the first banquet of the International Conference on Complex Systems. I told a story about the difference between Heaven and Hell, and I think I would like to repeat it here, and then tell you something that I've learned since then. So the story is about a rabbi who, for whatever reason, went to Hell, then went to Heaven.

The rabbi goes first to Hell, and there he sees many people seated around tables set with delicious food on fine china. However, no one is eating, everyone is crying and screaming. And he looks closer and he sees that their arms are bound so that they cannot bend their arms and therefore they cannot reach to bring their food to their mouths.

So, after seeing this he says, okay. So now they go to Heaven, and there he sees a room full of people seated around tables with fine china, with great food on it, and everyone's laughing and enjoying themselves and talking. And then he looks more closely and sees that everyone has their arms bound and they cannot reach to bring the food to their mouths. And of course, I'm sure you can all know what the answer to the puzzle is, because everyone is of course feeding their neighbors. [puzzled amusement, applause]

Now, we have learned that the problem of altruism was very difficult for people to understand in biology, in economics, but we learned that the process of creating altruism involves people choosing each other, in such a way as to create communities in which they're all mutually supportive. And I'm very pleased that our community is such a community, and I look forward to the continuing growth and development of this community in the future.

Thank you and good night.

Chapter 8

Pedagogical Sessions

Temple Smith
Biocomplexity
Leroy Hood
Complex Biological Systems
Jeff Stock
The Bacterial Nanobrain
Stuart Pimm
Biodiversity and the Construction of Ecosystems
Jay Lemke
Multiple Timescales in Complex Ecosocial Systems

Temple Smith: Welcome to Biocomplexity this morning. My name is Temple Smith from Boston University and I will be chairing the sessions this morning.

And I'd like to take just a couple of minutes in the beginning to perhaps play the devil's advocate here a little bit before the speakers explain to us how complex biological systems are.

Nobody argues over the last probably century that biological systems are complex, independent of your definition of complexity. And people have traditionally talked about the neural networks in the brain and their millions and billions of interconnections and cells and so on. We've talked about the complexity of ecological systems and so on.

And being in the area of molecular biology, the one thing that I think is probably most evident to molecular biologists or genomicists or whatever the current word is, over the last 10 years, is the more data we've accumulated. And we've accumulated now over 30 complete genomes. We have easily 3,000 protein structures that are basically distinct. Things just get worse. Instead of being able to look at all of this data and say, oh my gosh, it begins to make sense, I see how it all fits together, in fact what we begin to see is that a lot of our understanding of the regulatory networks, the feedback systems, the developmental systems, in fact just get more complicated every day.

I know that the graduate students and postdocs in my lab sort of fear the next issue of *Nature* or *Science* because instead of being illuminating, it's going to point out that the problems are getting worse.

However, all of biology at some level is driven, designed, and created by an extremely non-complex algorithm. So if your definition of complexity had to do with the algorithm that creates the system, and you want to talk about the number of bits required to generate the system in any classic sort of information theory measure, why it's not very complicated.

However, it's quite clear that the particular form of the creatures in front of me in this strange collection here are a result of an absolutely very complex random number, about three billion years long. So by classic measures of complexity, there's nothing more complex than an absolute random number because the only way you can specify it is by giving the entire sequence.

So it's not clear how our speakers will define what's complex and what's not and how they will describe these things, but part of the problem in biology is at one level in some sense biology's simple, but the results are extremely complex.

And what I would like to do is just draw your attention as an example a few months back with a *tour de force* in the crystallography business, two groups worked on basically solving one of the major machines in living organisms. And there were quotes like this in *Nature* and *Science* that this may be the most complicated machine ever made.

I sort of like a more recent article by Kornberg, which took on another one of the major machines in living organisms, and the reason is because it has sort of complexity for dummies. You've all seen those yellow books, right? Well complexity for dummies I thought was very nice. Here is the structure, and nobody's going to argue, at least at one level, the description's complex, but the machine's not. This is the RNA polymerase. The machine takes the DNA information and does the first step in translating it into the machinery for living things, but hidden here is complexity for dummies, and you can see that it's greatly simplified.

If you haven't read this article, even if you're not in biology, this is a bit of a *tour de force* in terms of the crystallography, but it's also an extremely well written article in *Science* just a month ago, and I would greatly encourage one to look at that.

The last sort of thing I want to point out, the sort of standard thing that people talk about is that, as most of you know, the fruit fly sequence was completed, that is, in a Pittsburgh meeting not so long ago. *Drosophila*, the thing that flies around your kitchen whenever you leave apples or oranges around more than about a day, its entire genetic sequence is known. And there's almost a hundred years of genetics, a hundred years of biology also available for the fly.

And taking these two data sets, putting them together, is sort of the challenge of this community. And I would say that probably 60 percent of everything we know about developmental biology, including for mammals, in some sense was all done originally in the fly. And that's one of the major challenges, is to understand that complex system, the development from the single cell up through the organism as complicated as the fruit fly.

And the anecdote that I want to give you before we start is that over two years ago effectively, another small organism, complex organism, higher organism, had its complete sequence done. That's the worm, *C. elegans*. And its organization, its biology, appears to be encoded in 16,000+ pieces of gene code for protein.

And so people expected that the fly, which is more complicated—worms don't fly, they don't drink alcohol and have all the dehydrogenations on board and so on that the fly does. And they also don't go through as many complicated developmental stages as the fly. And yet, when the fly sequence was done it was only about 14,000 coding blocks, genes. So on the surface it looked like it had less information to code for a more complicated organism.

Well people who'd worked on fruit fly genes sort of were suspicious to start with because it turns out in the fruit fly on the average the genes we know about, each of those genes codes for about 2.5 proteins. It's the same region. By a very complicated method of alternate splicing, the same region can produce multiple machines, multiple proteins, where that's much less true in the worm.

Well that stimulated the debate that was on the web and in *Science* this week about how many genes there are in you and me. Well at least you. I'm from Alpha Centauri. But at least in you guys.

And various counts have ranged from 70,000 to 150,000 genes, coding blocks. And the high numbers come from looking at actual raw data, what's called ESTs. People actually measure pieces of genes and extrapolate. But two human chromosomes are done. Actually three. The Y-21 and 22.

Now the Y is unusual. As most women know, there's no information on the Y gene chromosome. And in fact there's not. There's, what, a dozen genes? Very few genes on the Y. But on chromosome 22 and 21 there's only about 600 genes apparently or on that order. If you extrapolate that to the

298

total human gene it would predict only about 40,000.

And so yesterday I was just interviewed for one of these articles, and there were two articles this week saying, wait a minute, it's 150,000, it's 40,000, what's the number? Well we have no idea in fact. And so if you think you're complicated it may turn out that we're not as complicated as we think, at least in terms of the number of genes.

It is my extreme pleasure to introduce what I consider in the area of modern molecular biology and the understanding of genomics one of the world leaders, Dr. Leroy Hood, whom I was about to introduce from his old university and fortunately this morning he corrected me. He is now at the Institute for Systems Biology in Seattle.

And I will just take one second to point out that a number of years ago at one of the major conferences in this area at Hilton Head, I watched him get up and explain to us the entire immune system in terms of its DNA sequence. And at that time the audience was dead silent for the whole talk, just sort of sitting there. How could anybody have done this work and then be able to tell us what it meant?

So I'm looking forward to Lee's talk, and I will turn it over to him now.

Leroy Hood
Complex Biological Systems

Leroy Hood: Well thank you very much, Temple.

As Temple told you, I've recently started an Institute for Systems Biology, and a lot of what I'll talk about in the lecture today is really talking about the philosophy and the approach that we're using with systems biology.

But if you had to capture the essence of what's it all about, it's about deciphering biological information. And this is an example I like to use in this regard.

So this is a wild-type fly, a natural fly, that has a single pair of wings, and this is a mutant fly that Ed Lewis discovered probably 20 years ago that actually has an extra pair of wings. And that mutation curiously enough arose in a single gene, and it arose in such a manner that it converted these small balancing organs called halteres into a second pair of wings.

And what that mutation says about biological information really is two things. One, it's hierarchical. That is, there are units of information that have a profound impact on the phenotype of the organism. And curiously enough what it also says is information can be historical because the mutant fly is the analog of the evolutionary antecedent to the contemporary fly. We went from four wings to two wings. So accidents, mutations, can reveal interesting past histories in this regard.

Now what has transformed contemporary biology has been our view of what biological information is actually all about. And I would argue that there are three major levels of biological information. So the first of course is the four-letter language of DNA. That's basically a linear code. Variation in that code generates different types of information. We only know some of the types of information that are present in the chromosomes, and certainly one has to do with the coding for genes, and another has to do with the coding for the regulatory machinery that turns genes on and off.

You express genes in a quantized manner. That is, the hundred thousand or so human genes can be expressed in different patterns and different cell types, and that gives them the distinct phenotypes through messenger RNA, again, a four-letter language, into protein.

And I would argue that protein is really the second major type of biological information. And its language with 20 letters of course confers this incredible capacity for folding what is initially synthesized as a strain into a complex, three-dimensional molecular machine.

There are two problems about proteins that I think are intriguing. One is the so-called protein folding problem. Can we decipher from first principles and/or from experiment the rules whereby a particular sequence actually dictates a particular three-dimensional fold? And that is, as I've said, part computational and part experimental, and it's a very, very difficult problem obviously.

And the second problem I think is even deeper and more profound and that is, given the three-dimensional structure of the protein, how can we figure out what it does from first principles?

And one can do binning and general classifications by saying the structure of this protein looks like something else I know, but in fact there are five or six hundred protein kinases in humans and the real question isn't, is this a protein kinase, but it's, what does it do, where does it specify its activity, what's the cell type, what's the nature of that activity, and that's a very deep and very challenging problem.

Now as we know, biologists have spent most of the last 30 years looking at proteins and genes one at a time. And in fact the reason that Temple is so confused about this complexity issue is for that reason. Looking at genes and proteins in isolation, even though you may learn a lot of interesting things, doesn't tell you about systems. And that's the essence of what we have to start sorting out.

And of course systems are what biology is going to be about in this 21st century. So the human brain with its 10^{12} neurons and 10^{15} connections has these fascinating systems properties of memory and consciousness and the ability to learn.

The fact is if we take one neuron and study it so we know every input-output response it makes, study it for 20 years, that won't tell you one iota more than we already know about these systems properties because the only way you can study systems is by looking at the system as a whole.

So what we need in doing systems biology are global technologies that will let us first define the elements of a system and then interrogate the elements of a system as you perturb them biologically to carry out their functions. And that's what systems biology is all about. And we'll talk about that in considerable detail.

Now the real revolution that has transformed biology over the last 10 years—and it's really remarkable to think that the genome project has only been around that long—is of course this effort to decipher the genomes of the human and a whole series of different model organisms.

And as we all know, it's reaching in a sense its culmination probably in the middle of this year when the federal program and Celera as well will announce that at least they have a draft of the human genome done that will give us in all probability at least 90 percent of the human genes.

And I think very few biologists realize what a profoundly transforming event this is going to be. At least most people I talk to don't realize how much this is going to change the kind of human biology that they do.

I was at the very first meeting in the spring of 1985 on the human genome project. Bob Sinsheimer, who is Chancellor of UC Santa Cruz, invited five or six of us. I don't know, maybe Temple went to that meeting, too. But Wally Gilbert was there, David Botstein was there, Charles Cantor was there. And we went and debated this proposition of whether he should spend $35 million on an institute to sequence the human genome.

And I came away impressed, really impressed, with two things about this potential new project. One was the enormous technical challenges that it presented. I'm a technology guy, so I enjoy those. And the second was that it introduced to biology, at least contemporary biology, a completely new kind of science, and it's a science that I call discovery science.

And the idea about discovery science is you take an object, and you define its elements irrespective of any questions. You create an infrastructure on which then the more classic, hypothesis-driven science can be built.

And what I would say is it turns out that the critical aspect of being able to do systems biology is to be able to effectively integrate discovery-driven approaches to science with hypothesis-driven approaches to science, and that is something that very few biologists really understand today. And I'll make that much more explicitly clear in the future.

Now the genome project, as we all know, is the world's most wonderful software program. So what it allows us to do is to take a single-cell fertilized egg and transform it into an adult organism with 1014 cells of incredible complexity. It carries out this chromosomal choreography where different subsets of genes are expressed in different cell types. And of course the genome project is transforming our ability to carry out developmental biology.

I think a critical question that most people don't appreciate—in fact, it was really driven by a comment that Jim Watson made maybe 10 years ago now—he said, man used to think that his fate was determined by the stars, and he knows now that it's determined by his genes. And of course the really interesting question is to what extent is that really true?

So here are the left index fingerprints of two nine-year old identical twin girls, and you can see the fingerprints are absolutely different from one another. So that says that this program, this genetic program for creating fingerprints, creates different patterns either as a consequence of different environmental stimuli during the developmental process, or as a consequence of stochastic events that are a natural part of this developmental process. We can't really distinguish between those two, and it may well be combinations of both.

So the critical question you have to ask of any interesting trait is to what extent is it nature and to what extent is it nurture? And I'll just say we have very poor tools for making that ascertainment today, but it is a critically important question to keep in mind.

Now what I'd like to ask is, if we look back from 10 or 15 years in the future at the genome project, what were in fact its fundamental contributions? And I would argue they are these three.

So one is the this idea of discovery science and the imperative that you need to integrate it together with hypothesis-driven science to do this systems biology.

The second is it's created this marvelous parts list of human genetic elements, this so-called periodic table of life. It will let us define the hundred thousand or so genes that exist.

And I would second Temple's point, just to underscore it, what perhaps is less important than the absolute number of genes is the multiplicity of ways in which each gene can be used through this process called alternative splicing, because the same gene spliced different ways can create quite different kinds of functions. And that's a very hard question to get at. There is one new technique that may let us get at it in really interesting ways, and we'll talk about that later.

A second benefit that will be derived from the human genome project is it will define the associated sequences with genes that are the substrate or genetic regulation for controlling the temporal, spatial, and amplitude of gene expression. We'll talk about that in just a moment.

The third idea is it gives us this wonderful lexicon of building block components of genes and proteins that allows us to take genes and proteins and de-convolute them into motifs that constitute some of the fundamental building blocks. And the motifs can be small, or they can be large, and people are arguing about the nature of the motifs, but their existence is unquestioned. What represents an enormous computational challenge in the motif area is the degeneracy of the motifs. How we decipher that is an enormous challenge.

And the final opportunity that the genome project has given us is it's given us access to human variability, the differences between you and me. And on average one in 500 letters in the DNA code differ. So that's six million differences between the two of us. Most of those variations make no difference physiologically, but a few make some people tall and some people short or fat or thin, and a few others predispose to late-onset diseases. So being able to access human variability is I think the second generation of genomics and what it in fact is going to be all about.

So the third of the benefits arising from the genome project is this idea of catalyzing paradigm changes that lead to this view of systems biology, and that's what I'd like to spend a few minutes talking about now.

So this idea that biology is an informational science and you have one-dimensional, three-dimensional, and time-variant, and four-dimensional information, and DNA proteins, and systems is something that we already have discussed.

What I'd like to stress is this idea that the information is hierarchical in nature, and that as you move to successively hierarchical levels the units of information require additional complexity that at least at this point in time can be explained by knowing what precedes it.

So if you go in a simple way from a gene to a protein, proteins fold in three dimensions. They interact with other proteins. They have turnover times. They can be modified by chemical substitutions to alter their biological behavior. Most of those things cannot be predicted, at least today, by knowing the gene sequence.

So additional information accrues to the problem, and it's information that in large part comes from other genetic elements in this complex phenotype.

And likewise, when you go from a protein to an informational pathway, signal transduction, apoptosis development, knowing the features of the individual proteins and what they do doesn't necessarily let you predict the systems property of a given informational pathway. And then the same is true when you hook together the informational pathways into informational networks.

And a really important point that I'd like to make is the fundamental unit I think of looking at biological complexity is the cell because that circumscribes the networks of information that allow it to carry out its particular phenotype.

So a driving compelling force in contemporary biology is, can we obtain homogeneous cells to characterize them or can we characterize single cells? So those are very, very dominant themes in the biology of the future.

Now what is also very interesting is that the second type of biological information that we talked about, namely regulation, is an enormous and fascinating complexity in biological organisms. And gene regulation, how you turn the genes on and off and so forth, operates by virtue of sites on the DNA adjacent to the genes that combine proteins, which are called transcription factors, and arrays of those transcription factors then operate on a huge molecular machine called the basal transcription apparatus that is responsible for the synthesis of the messenger RNA from the gene.

The systems properties of regulation then are three. They are when in time does information get expressed, that is, the developmental stage? Where in space does it get expressed, the particular cell types or tissue types? And what is the amplitude or magnitude of expression? So one can take beautiful systems approaches to looking at this whole process.

And the really critical point that I'd like to make is, it is this type of information that really is the major substrate of biological evolution. Changing structural genes is really a minor perturbation.

So if we go back to these two- and four-wing flies, we see the gene that was mutated in the example I gave you initially is in fact the transcription factor. It's a Hox transcription factor. And it's a transcription factor that has evolved to suppress the operation of six different genes on the haltere that make it a wing. So when you knock that gene out those six genes can manifest their potential activity, and you create a second pair of wings.

So you can see then that evolution has transpired primarily by a fundamental change in regulatory patterns. And of course nowhere is that seen more clearly than in human evolution.

If you look at the chimp, our closest relative, the divergence was about six million years ago. If you look at the DNA, less than one base in a hundred is different between the chimp and the human. And you know that's really remarkable if you think that we differ one in 500 from one another.

So the really critical thing in evolution is not changing the structural genes, it's changing the regulatory patterns that control a morphologic apposition of cells. And the human brain has many more cells. They're much more densely packed. And all of that, those are regulatory phenomena.

So with that type of introduction then, let me turn to the second major paradigm change, and that is this idea of the genome that has spawned these high-throughput analytic tools for being able to decipher biological information at these various levels at a very high-throughput level.

So we all know about the DNA sequencer. This was an instrument that we developed in prototype form in 1986. And in fact the first instrument was a capillary machine similar to contemporary capillary machines except it had a single column rather than 96 columns or 384 columns.

And I can say that due to chemical instrumentation and software improvements, the throughput of sequencing from '86 to today has increased 2,000-fold. And there's a striking decrease in the cost per unit of information that you generate.

So high-throughput instrumentation is all throughput. It's all about quality of information. It's about cost of information. And there are really striking changes that are occurring in that regard.

There's a company in San Francisco that we've started collaborating with that has the ability to take a million sequences from a library, array each of them on a different bead, fix those million beads in a flow cell, and then interrogate with a CCD camera those million sequences simultaneously as they're analyzed 16 to 20 residues. So this is called multiple signature sequencing. The ability to do a million sequences in four to six hours transforms how we think about many problems.

For example, it in one operation can give us enormously detailed analysis of the entire information content of a single homogenous cell type. And that's actually what we're using it for.

But it can do many other things too. If you make these genomic fragments from an unknown organism, in a single run you get enough information to actually create a physical map that's better than the physical map humans spent the last 10 years trying to create before the sequence came in place. So high-throughput biology can transform strategies for doing biology as well.

The other classic global tool is of course DNA arrays, which we're all familiar with. They allow us to interrogate variation in humans. You can make a chip that has in principle a hundred thousand DNA fragments, each of which represents a different one of the hundred thousand human genes. And in principle you could look at a normal cell and a cancer cell, and it asks how all of the gene patterns varied in that transformation event. And that is a classic global kind of technology that's going to be possible with inkjet synthesis and so forth.

But there are also transforming technologies that are coming along now. So there's a company called Illumina in San Diego that is using optical fibers to create DNA arrays in the following sense.

You take a one millimeter optical fiber and you put 10,000 wells in the end of it, and you can put into those 10,000 wells 10,000 beads, each of which has affixed a different 20-mer. And by any one of several different assays, then you can either interrogate single nucleotide variations, these genetic polymorphisms, or you can interrogate patterns of gene expression.

If you put together an array of these coils, if you put together 100 or 1,000, you can then develop instruments that will have the ability, for example, to do two to four million SNPs a day. So that's an increase in throughput. Again, that's about 2,000-fold over at least the commercially available instruments that exist today. I will say there are other technologies that are very, very high-throughput employing mass spec techniques and so forth. But this gives you an idea of some of the kinds of global technologies.

Now again, what people are just beginning to realize is genes and messenger RNA are really terrific, but the players who really do all the work are proteins, and our ability to do protein analysis, called proteomics, the analysis of many proteins, is circumscribed by the enormous complexity of this problem. The dynamic range of expression of proteins is enormous, 106 or greater. The fact that they can be modified, they fact that they can be compartmentalized, the fact that they have turnover times, all of these introduce absolutely fascinating complications.

So let me take you through some of the classic approaches to proteomics. And the first was developed by John Yates in our department, and other people have done similar things. But the basic idea was you take a separation technology like two-dimensional gel electrophoresis for an organism whose entire genome is done, and you can then cut out individual proteins, digest them with a protease, and then measure precisely the size of those fragments in a spectrometer any one of a number of spectrometers.

And what you can do then is match the experimental profile against the theoretical profiles that you would predict, knowing the specificity of the problem, and for yeast you can in an incredibly short period of time correlate a gene with each of the protein spots.

So in the beginning, proteomics meant not very much more than being able to generate a road map where you correlated proteins with genes, but Ruedi Aebersold, who is actually one of the co-founders of the Institute for Systems Biology, has made a dramatic advance in that now we can look in a quantitative manner at the proteins that are present in two different cell types.

And the approach he's taken is to create a reagent that has a biotin tag on one end, a linker to which can be affixed either eight deuteriums or eight hydrogens, so that makes a light and heavy form of this compound, and then a thiol reactive group that can couple this compound to ascertain residues.

And the basic idea here then is to take—as we are now doing this experiment as we talk—cell surface proteins from a normal prostate cell, label them with the light reagent from a cancer prostate cell, label them with the heavy reagent, mix together equal aliquots of the cells, purify or not the proteins, it really doesn't matter, and then you proteolyze the mixture of these proteins. You use the affinity tag to pull out all of those peptides that have cysteine residues to which has been affixed this compound. So that reduces your mass by more than 90 percent. Important technical point.

And then you can put them in the mass spec in two modes. You can interrogate the ratio of the light to heavy compounds to one another or all the different proteins whose peptides you have in the mixture. And that ratio to within five to 10 percent tells you the quantitative expression ratios of the proteins. It's much better than DNA arrays where if you're lucky you get a factor of two to three that you really believe.

Then you can take those same peptides; you can fragment them still further; and you can determine their sequence. And hence, you can identify the gene if you know the genome, or if you have extensive EST expressed sequence tag data, you can identify the gene that corresponds to those things. So this gives us the capacity to do precisely what we can do with DNA arrays, that is, quantitative expression.

And the important point then is—and these are experiments we've done in yeast, Ruedi's lab and my lab—we can interrogate the relative expressions of protein and messenger RNA, and when they're in accord with one another, they fall on the diagonal line.

And you can see that in many cases, and in fact this is somewhat of an exception in that there are many more on the line than we usually see. But when they're off the line in every case that represents really interesting biology.

So the really important point again is you have to be able to integrate information at these different levels, information from the RNA with information from the protein, before you move to the ultimate systems biology. And we'll again talk about that.

So it's not only proteomics that have these global technologies, but it's cell-sorting if the cell is the unit compartment of information. Cell-sorting is going to be critical. So we again have at the Institute, Ger van den Engh, who is probably the world's premiere technologist-biophysicist, who's developing new approaches to multi-parameter high-speed cell-sorting and so forth. But visualization of information, high throughput—there are a whole series of technologies that in the future are really going to be important.

We've intimated that if you have the ability to define a system you want to look at and carry out discovery research and define all its elements, then what you need to be able to do is in model organisms perturb that system so you can follow the flow of information through the system. So these model organisms turn out to be the real Rosetta Stones for deciphering biological complexity.

And as you know, yeast, a bacteria, the nematode, Drosophila—all of those model organisms have been sequenced. And again, for Drosophila, with its 13,500 genes it turns out about 50 percent of them show homologues that are present in the human.

So in many cases we can take human genes and go to these genetically biologically manipulable organisms for interrogation, both of how the individual genes work and the context within they which they operate in informational pathways.

And likewise the mouse our own genomic complexity. Its genome will be done in the next couple of years. And then we'll have model systems for more complicated phenotypic patterns, immunity aspects of development, nervous system, and the like.

One of my favorite Max Delbrück quotes is this. Any living cell carries with it the experience of a billion years of experimentation by its ancestors. It was reflected again in this fly analogy that we learned earlier.

What is interesting is we have the means now to make explicit a lot of that history. And even more important, the evolutionary history of how information evolves gives us fundamental insights into how it must operate. So the integration of evolution with development is absolutely a key feature in contemporary biology.

The requirement for computational tools I think goes without saying. We are accumulating enormous amounts of information, and how we acquire and store and analyze and model and display and distribute that information of course requires the tools of applied math and computer science and statistics and the like.

A point that is interesting is that living organisms have had about four billion years now to learn how to manipulate their digital streams, their DNA, and they've actually evolved some very clever strategies that I think are going to have a real impact on how computer scientists think about the organization and deployment of their simpler digital language. So it's a two-way street, and there are opportunities in both directions.

We can think about the convergence of information technology and biotechnology in two different ways. So one way is to say, look, all of the major problems that the computer scientists have we as biologists have, whether it's data-capture of warehousing and data-integration, the dealing with the integration of heterogeneous databases, one of the major problems of contemporary biology, data mining, the modeling. Problems of scale are real interesting and fascinating kinds of points.

But what is again a really fascinating argument is, because biological information deals with an objective and measurable reality, we can test very complicated new kinds of things and see if they're consistent. Now that's a hard thing to do with a lot of other things that computer scientists are interested in because there isn't necessarily that objective reality for the check.

Another way of slicing it is at every level of biological information there are enormous computational opportunities, whether it's looking at genomics and trying to extract from DNA, RNA, biological information, whether it's this whole problem of proteomics. And the way I look at proteomics is it's both one-dimensional and three-dimensional, but the problems are very different in the two cases. Whether it's thinking about systems. Or whether it's thinking about how we put the data together from all these different levels. And indeed, ultimately, how we can take this whole variety of different kinds of technologies and integrate together their kind of information. So there are enormous challenges that exist there.

So let me talk about our approach to systems biology. And in doing so let me just give you the algorithm by which we think of systems biology.

So for most systems they are so complicated you can't take them on in full. So the immune system, with 1012 lymphocytes in humans is unapproachable in a frontal attack. So you have to use biology to define a subsystem that is tractable in an experimental sense.

Once you do, you need to use discovery science to define the elements in that system. You need again to use discovery science to interrogate the flow of information through that system as you perturb it in biological organisms. And those perturbations can be biological, genetic, or environmental. Everything falls into one of those three classes.

And then ultimately you have to be able to integrate together the information you've ascertained from each of these different hierarchical levels of biological information to begin modeling this biological system with two objectives in mind. One, the model should give us insights into the structure of this informational pathway. And two, it should give us the ability to predict systems properties given particular kinds of perturbations. And that is an enormously tall order as you will obviously see.

So let me make really explicit how we're approaching this. A very, very simple model system. And our whole objective is to develop tools that will allow us to do this.

So this is in yeast where the genome is known. It's the system that metabolizes galactose, a sugar, to glucose-6-phosphate, which is usable in the general metabolic machinery of the organism.

It's a pathway that employs at least three kinds of problems. So it has, as I've indicated here, four different enzymes that operate in this pathway. It requires a transport molecule that can bring galactose from the outside environment into the yeast. And it requires a series of transcription factors that operate on the major structural genes, those enzymes, to allow it to express itself, manifest itself, in two contexts.

So if there's no galactose in the system, two of these transcription factors sit down in the promoter region of those structural enzymes, and they shut them down completely. When galactose comes into the system it interacts with a third transcription factor, which in turn modifies the behavior of those first two transcription factors, and for example the first gene in that pathway is up-regulated a thousand-fold and the other genes are up-regulated 300- to a thousand-fold. So you have the system off, or you have the system on.

So the approach that we've done then that Trey Ideker, who was an MIT undergraduate in computer science and is a graduate student with me now, and Vesteinn Thorsson, who's a theoretical physicist—what they've done together is taken all the players in this pathway, and in yeast they've either knocked out or obtained knockouts for each of those players. So that defines 10 different genetic perturbations of the system.

And then we interrogate those perturbations again biologically in the context of the system running or the system not running. And the approach that we take, we've taken initially, is to use this DNA array technology where you can compare quantitatively the expression patterns of all 6,200 genes in yeast as you test these genetic perturbations against the system running or the system not running. So we've gathered all of that data together.

And I'll say but won't talk about it, we're doing the same with these proteomics technologies, because we need ultimately to do this integration process that I've talked about.

So the idea is you put the genetic perturbations through the biological system in those two contexts, and you collect data. And at the same time we're modeling the system to be able to begin predicting theoretically the nature of the system in the context of the various genetic and biological perturbations. And what we want to be able to do ultimately then is to compare the theoretical with the experimental, and then we use the disparity there to develop two kinds of software.

So what initially happens with the perturbations is you get information that's consistent with many different models. So you like to optimize the process by which you eliminate those models. So we're developing software to pick automatically the next set of perturbations you need to cut down those ambiguities.

And at the same time we're developing software to much more automatically begin picking model parameters and the nature of the models so we can bring into juxtaposition the theoretical and the experimental.

So these are approaches through which we hope to develop tools then that can be applied to far more complex biological organisms.

But let me give you a really quick once-over example of the results. If you look at the 20-some perturbation conditions, what you find is about 500 of the genes present in the yeast with 6,200 genes that are differentially expressed in at least one perturbation. And we set the threshold very high. It's a five-fold difference. So there's no question about this statistically.

So then the next question we ask is, can we take those 500 genes and classify them into categories according to the sets of genes that respond similarly to all of these different types of perturbations? And what was really interesting when we did that is that we found, in addition to the genes of the system itself, that all of the rest of the genes fell into somewhere around 10 to 12 categories, and each of those categories represented one or two biological systems, amino acid transport, amino acid metabolism, and other aspects of sugar metabolism.

And what we provisionally think is those different classifications, those 12 classifications, actually represent interconnected informational pathways that are themselves perturbed by the perturbation of the initial pathway. So that makes us optimistic that we can begin not only to understand the essence of the system itself but a lot of its interconnected pathways.

Now how you go from that general statement to modeling how each of the genes affects the other, or affects each of these major categories of genes, is a very, very challenging question. And I am really coming to the conclusion that we may really need new kinds of mathematics both for this integration process and then its assimilation into models. So we've done kind of simple, straightforward Boolean things and differential equation kinds of things. I think it's going to be much more complicated than that. And so we're really interested in getting good mathematicians involved in this whole process. And again, what's nice about this is there is good quantitative data that one can use to play with these kinds of questions.

So let me run through very quickly just a couple of the kinds of systems we're beginning to think about using these tools on to give you an idea of the power of these applications.

So we recently have sequenced an organism called halobacterium. It's an organism that lives in 4 molar salt. That's 10 times as much salt as we live in. And it is a fascinating organism from the point of view of biotechnology.

And one of the most interesting aspects of this organism is it has a big bacterial chromosome of two megabases, and then it had two smaller things that were called plasmids. Now usually plasmids were thought to be dispensable, but when we sequenced the first of these plasmids we found two things. One, it had a whole bunch of essential genes on it. And two, if you looked carefully at the sequence you could reconstruct how it evolved. And it evolved with two kinds of events, namely a series of three different smaller plasmids were all joined together, and it actually recombined with the major bacterial chromosomes to generate a whole segment of genes in contiguous apposition that were absolutely essential. So we think this is really a mini kind of chromosome that's going on.

So let me just say then that we're thinking about the immune system and how it actually works. We can get homogeneous cell types that carry out these basic systems properties. And we can do all of the following types of analysis on those in exactly the same fashion. And so we can do it on hematopoietic development. Stem cells give rise to all the different blood elements. We can now separate homogeneous populations at each of those critical stages of development, and we can interrogate the kinds of systems changes each of these cells go through to get to these various states. And we've sequenced rice, and we don't have time to talk about that now.

But let me leave you with a couple of thoughts in closing then.

Moore's Law—Gordon Moore, the founder of Intel, suggested in 1970 that the amount of information you could put on a computer chip would double every 18 months, and it has, and that's been the basis for the IT revolution. Even steeper exponentials are seen for DNA sequence information.

And of course one of the really interesting questions is how you convert sequence information into knowledge. And our argument is that you integrate it together with the information from RNA and the information of proteins and the information for pathways, and you begin doing this kind of modeling, this reiterative process where physical and computational scientists work together with biologists in close apposition. And that really is a critical point of this that allows you to attack really challenging problems like this wiring diagram of human development. And you do it by starting at the terminal branches of this developmental process and working inward and integrating together your analysis of these different kinds of systems.

So then the final point I would make is that at the Institute for Systems Biology, we have indeed tried to put together—are in the process of trying to put together—both the people and the technologies and the technology development we need to do this kind of systems biology that we've talked about here.

And because we are going to remain relatively small we're really interested in forging both academic and industrial partnerships. And so if any of you are interested in thinking about that in the future, we certainly would be willing.

And the final point is we are intrigued by the idea that as we develop these tools for successively having the capacity to integrate biological information at these higher levels, that the applications that we're developing here may well be applied in more general terms to some of the higher levels of organization, organs and individuals and populations and ecologies.

And I don't know how many of you have read E. O. Wilson's book on Consilience, but this idea of the grand unification of knowledge, in this particular case biological information, seems like a terribly compelling and attractive idea today.

Thanks very much.

Jeff Stock
The Bacterial Nanobrain

Temple Smith: Our next speaker is Jeff Stock from Princeton. And the title of the talk is "The Bacterial Nanobrain." The bad joke: I hope he's not talking about some of my students.

And what I would point out is that one of the fascinating systems that's been studied for a long time is the ability of bacteria to find food, move away from toxic things, with a very complicated feedback system and propeller on the back which they can actually swim with. So nature did invent the wheel in spite of the comments by some people.

Jeff Stock: So you just heard about the universe, and I'm going to tell you about a hydrogen. And the message that I'm going to try to leave you with is that we don't really understand even very, very simple biological systems, although we have all the sorts of information that Lee told you about.

And over the past year, although I think the major laboratories in most universities or many universities in this country at least, at Harvard, Princeton, Yale, Caltech, Berkeley, work on this very, very simple problem of how E. coli move around in their environment and go to places where they want to be for nutrients and to move away from environments that are deleterious, we still don't understand how that works. And over the past year we've profoundly altered, I think, or begun to alter, our view of how this system works. It's become much more complex.

So this problem of bacterial chemotaxis starts out as using the approach of classical biochemistry to try to investigate the problem of intelligence. And biochemistry has been triumphant over the last century in resolving the problem of metabolism, of how different nutrients and different cells interrelate to each other, and has worked out really the structure of DNA and how information is passed and how heredity works. And so we applied these methods to the problem of intelligence.

And the rules of biochemistry, the fundamental methods of biochemistry, are in large part encompassed in these two quotes here.

The first is that an E. coli is like an elephant, only more so, by Jacques Monod, who in many ways founded molecular biology. And the idea here is that all of biochemistry is unified, that we can pick the ideal model system to study a problem, and we'll get information by studying that model system on all of biochemistry in all systems, many of which are very intractable to biochemical methods.

And the second quote here is from Arthur Kornberg, who worked out the mechanisms of DNA synthesis in large part and received the Nobel Prize for that work, and he said, don't waste clean thoughts on dirty enzymes. So the idea is that you purify, you analyze the protein or the component that you want to work with, and you determine its structure, and you determine its nature, and then there's a reconstitution process to put these elements together to try to figure out how a system works.

And then the third rule of biochemistry, which I haven't put here, is never try to outthink an E. coli. That's something you should always remember.

This field really began in the '50s and was initiated by Julius Adler who was a student at Stanford in the Biochemistry Department of Arthur Kornberg. So there's been a tradition in this field beginning with Arthur

Kornberg, who also has a tradition of American biochemistry.

And Julius Adler really started in the field of E. coli neurobiology. He had conceptually the first idea of studying neurobiology using E. coli. So the idea is you'd go to this simple organism, and by looking at the fundamental biochemistry of sensory motor regulation in E. coli you could understand the fundamental biochemistry of the way the brain works.

So now I want to tell you a little bit about the way E. coli swim and the way they move in their environment.

And the first point I want to make is that they move using flagella, but their flagella are quite different from eukaryotic flagella. E. coli are very, very simple. Here's a sperm, and a sperm swims using a flagellum and the flagellum beats, all right. You wave the flagellum, and the flagellum moves.

E. coli flagella are very different. They're composed of just a single protein which makes a helix, and that structure is relatively fixed. There's no energy utilization in the E. coli flagellum. And the energy, the driving force for this, comes right at the base where there's a nanomotor. There's a little electric motor that rotates the flagellum, and the flagellum, because it's helical, the rotation pushes the body. If the rotation is in a counter-clockwise direction it turns out it pushes the bacterial body through its environment.

And this is a comparison of the complexity of the bacterial flagella to the eukaryotic flagella. The bacterial flagella is just composed of one single filament made up of one protein.

The mechanism by which the bacterium moves through its environment in a desired direction relates to the simple notion that this rotation of the motor can be either counter-clockwise or clockwise. So it's a simple two-state motor. And by alternating between clockwise and counter-clockwise the bacterial cell can control the direction in which it moves. It's an incredibly simple system.

So when the rotation is counter-clockwise the flagella push the body, and all the flagella come together into a bundle. They work together, and the body moves at a rate that in terms of body lengths per second would correspond to a velocity of about 50 miles an hour for us, a pretty good rate of motion.

And then after a second or two generally the rotation on the flagella at the flagella motor is reversed to clockwise and the flagella fly apart and generate a tumble, which is a reorienting motion that lasts generally about .2 seconds. And so there's a tumble.

And then the rotation reverses again, and the cell swims off in a random, new direction. This was all defined by tracking by Howard Berg in the '70s. And the cell swims or runs for a period, and then it tumbles again, and then it runs again, and then it tumbles again. And this generates a random walk in a uniform environment.

But the cell has a system which determines whether as it's running things are getting better or whether things are getting worse. And the cell determines that if things are getting better, it should suppress the probability of a tumble. It tends to keep going.

It's just like we should behave basically. If you're going in the right direction, as you determine, if things are getting better, you should probably keep going. And if things are getting worse you should probably go in some other direction. And that's exactly the way the bacterium works.

And by doing that you can watch in real time a population of bacteria moving toward a source of nutrients, for instance oxygen, or away from a repellent.

The system that determines whether things are getting worse or better, that makes that qualitative good/bad decision—and it's a decision interestingly. It's a prediction; it's a question. The bacteria has to determine, are things going to get better so that I should change direction or are they going to get worse? Are they going to get better so I should keep going or are they going to get worse so I should change direction?

Now the genetic approach was used to determine the components of the system that mediate this process, and this was begun by Julius Adler and continued by several groups, but most notably Sandy Parkinson who is at the University of Utah. And the method that he used to isolate mutants, to characterize mutants, was the swarm plate. And this is an interesting system, and I wanted to show it to you because it's fascinating. You can get into looking at swarm plates forever almost because it's sort of a paradigm for a lot of biology I think.

Here's a wild-type cell, and this is a petri dish with loose agar poured into the dish, nutrient agar. There are amino acids in the agar, and this is an inoculum of bacteria that starts in the center. And as the bacteria grow, they use up the amino acids and the nutrients in the agar. And that creates gradients of nutrients, and they chemotax—they move up the gradient and swarm. It's called a swarm plate. And they move out very, very quickly, relatively quickly. Over a few hours or several hours, wild-type bacteria will cover a whole plate.

And these rings turn out to be the utilization of different nutrients in the media. The outside ring, an E. coli, corresponds to the utilization of serine. So there's no serine inside here, and there's serine out. And this ring, these bacteria are following the gradient of concentration of serine. And this inside ring of cells is following the concentration of aspartate it turns out. So there's no aspartate inside, and there's a significant concentration of aspartate outside and in the medium.

Okay. So you make mutants of E. coli, and these fall into several categories of aberrant behavior on swarm plates. Some of them, a large fraction of them, are immotile, and these are cells that have a disrupted motility system. They lack flagella, or they have flagella that can't rotate.

They're paralyzed flagella. And there are about 50 different genes in this category.

The second class of mutants are generally defective in chemotaxis. They're able to move. They sort of diffuse through the agar. But they can't effectively measure concentration gradients. They're mentally ill individuals.

Some of them are always tumbling. They're the eternal pessimists, right? And others are always smooth. They're overly optimistic. And either phenotype has either always tumbling individuals or always smooth individuals, are unable to chemotax, and make this sort of small dense colony of cells.

And then there's another class or a group of classes that are blind to specific stimuli, to specific amino acids in this case. And this is a mutant. This example here is a mutant that can't detect aspartate. It's lacking the receptor for aspartate and that receptor is called Tar, the taxis to aspartate. And this also mediates responses to some repellent. And so it's Tar ("taxis to aspartate and repellents").

So this basic strategy was used to define the genetics and the components of the chemotaxis system in E. coli. And that's been done with several thousand different mutations all having different behaviors. And this system is totally defined and of course sequenced because we have the E. coli genome.

So this is E. coli chemotaxis genomics, and this is the system. It's a very simple system for mediating all of this really, I guess, simple behavior. And it turns out that there are five different receptors, and each of them is very similar. They represent a family of proteins. They're all homologous. This one for instance detects carbohydrates, ribose and galactose. And this one's the aspartate receptor. And this is the serine receptor. And air detects oxygen, or actually redox potential.

So these are the receptors. And they're scattered around the chromosome. It looks like three are located randomly around the chromosome. And two, the receptor for peptides and the receptor for aspartate, are located in a little cluster of genes which ?? chemotaxis system, the system that is required for sensing and making the decisions about all chemotaxis responses.

And there are six proteins in this system, and they have random names, A, CheA, Che for generally non-chemotaxis. CheA, CheW, CheR, CheB, CheY, CheZ.

So what we've done—and "we" being the field at large—we've cloned these genes. This was done about 15 years ago. We cloned all these genes, expressed these proteins in high levels, purified the proteins, determined the enzymatic reactions that they catalyze, determine their structures by x-ray diffraction, and put the system together to show how it works.

And this is the system that mediates this decision-making process.

The receptors at the cell surface. And those are those five proteins that I told you about. And these interact physically and make a complex with two other proteins, one called W, CheW protein, and the other called the CheA protein.

The red components here are absolutely required for making a connection between information in the environment and the control of flagella rotation. The blue components are modulating agents that put memory into the system, that regulate the flow of information.

So the W protein and the A protein interact with the receptors. The A protein is a protein kinase that phosphorylates itself at a histidine residue. This phospho-histidine is a high energy phosphoric bond like ATP. And the phosphoric group is then transferred to a small protein called the Y protein, the CheY protein. And the CheY protein binds when it's phosphorylated, only when it's phosphorylated, to a switch apparatus at the flagella motors and causes clockwise rotation, causes a tumble.

Attractants like serine and aspartate bind to the receptors and inhibit the kinase, which is associated with the receptors, and stops the flow of phosphate so that CheY is de-phosphorylated. And there's another protein Z, which catalyzes, facilitates the de-phosphorylation of the Y protein, lowers the level of phosphate, and suppresses the tendency to tumble. So it's a very, very simple system.

Now what I'm going to do is go backwards from CheY, what we call the response regulator protein, talk about CheY, and then I'll talk about CheA, the individual pure component, and then I want to talk about this system here, which it turns out is altogether one large complex, and I want to argue at the end of my talk that that constitutes what should be regarded as a brain.

This is the CheY protein. It's the simplest and in many ways the most elegant protein of the chemotaxis system. It's a perfect example of what's called an alpha-beta protein, a doubly wound alpha-beta protein. The sequence goes beta-sheet, alpha-helix, beta-sheet to alpha-helix, beta-sheet to alpha-helix, and you wind this up in this beautiful, elegant little structure.

It only has 127 amino acids. It has an active site, which is indicated here, which has a little cluster of aspartate residues called the acid pocket and that binds the magnesium ion. And this helps to catalyze the transfer of the phosphoryl group from the histidine phosphoryl group in the kinase, in the A protein, to one of these aspartyl residues.

And that makes an aspartyl phosphate group, which is also a high-energy phosphoryl protein, an intermediate like ATP. And it's interesting that that phospho-aspartate is similar to the phospho-aspartate that's in the sodium potassium, ATPase, and the calcium ATPase, that is the energy-coupling stack involved in human thought, the what I think is Maxwell's demon, that makes high gradients that we use to send neural messages.

And in fact that protein has a similar alpha-beta fold and a similar active site. And you can make arguments that the two are doing a similar job. They're coupling energy to the transfer of information to control of motor activity.

So this is this protein. And when it's phosphorylated, it's activated to cause a confirmation of change that causes a reversal of flagella rotation, that causes the cell to change its direction of motion.

And we can characterize this by rigging the system. This was a study done that we did in collaboration with a group of physicists at Princeton, Stan Leibler and his associates, and what we did here was we rigged the system, and we can do anything we want with the chemotaxis system. We can put genes in, we can mutate them any way we want. It's very, very easy to work with E. coli. We work with 1010 cells at a time or one cell. We can do anything we want. And it's easy. It's cheap. We have a little lab.

So this is in the absence of CheY in the cell. This is a two-second frame, and we're looking at individual tracks of swimming cells. And this is done in a computer, and then we analyze the data from all these swimming cells and make conclusions about the relationship between phospho-CheY in this case and swimming behavior. And we're just increasing the level of phospho-CheY; we've rigged the system so all the CheY is phosphorylated, and we change the level of CheY. Or we can make a mutant CheY that's activated without phosphorylation, that's always turned on, and that works as well.

And we get the same picture. As you increase the amount of CheY phosphate or activated CheY, the cells go from a smooth phenotype to a more and more tumbly phenotype. So they change direction more and more. They become more and more pessimistic about the direction in which they're moving.

And this shows the relationship between this tumble frequency and the level of CheY phosphate in terms of this curve.

And this is the wild-type cell which has this level of CheY, total level of CheY, and this level of CheY phosphate. And it turns out that about a third on average of the CheY in a wild-type cell is phosphorylated at a given time. And then that level of phosphorylation is fluctuating, and that fluctuation controls the probability that a cell will change direction and go off in some other new direction.

Now I want to look at the kinase. The kinase is the most complicated protein in the chemotaxis system. When you purify the kinase, it's a soluble, cytoplasmic protein. We purified it. And its structure has been determined. It's been determined in parts. It's made up of several domains. The domains have been cut apart. And different laboratories have determined the structures of different domains in multiple times. There are a lot of people who work on this system.

We start at the end terminus. And the first domain is an alpha-helical bundle. These are alpha-helices, and they make a bundle. And off to the side is the histidine. It just sticks out to the side of this bundle. And that histidine is the site of phosphorylation.

The histidine is phosphorylated by an ATP-binding kinase catalytic domain. It binds ATP. Here it's bound ATP and phosphorylates this histidine residue. So this is a very simple kinase phosphorylating of protein, but it happens internally within the structure of the CheA protein.

And then there are other domains. After the H domain, you go to a Y-binding domain, and this domain binds the Y protein, which then has to accept the phosphoryl group from the H domain.

And then there's a dimerization domain which is composed of two up-down alpha-helices and mediates the dimerization of this homodimer. And then you go to the kinase domain here.

The C terminus is an interesting domain. It's called an SH3 domain, and in fact it's two SH3 domains linked tightly to one another. SH3 stands for src homology domain. Src was I think the first oncogene discovered. It's a protein kinase, and it has a regulatory portion that's called an SH3 domain because other proteins were identified that had that same domain, homologous domain. And that domain is called an adaptor protein because it mediates protein-protein interactions.

And this domain or this set of domains is responsible for the interaction of the kinase with the receptor. And it turns out that the Y protein has exactly the same or almost exactly the same fold as these SH3 domains here in CheA. This little part here is in the CheW protein. It's the same structure.

The only other thing I wanted to say here is that the kinase domain turns out to be analogous to a very, very important family of ATPases. They're not kinases, but they're ATPases. They just hydrolyze ATP and cause big conformational changes in macromolecules, in either DNA. And in that case it's called a topo II, topoisomerase II, a very important molecule for unwinding DNA.

It looks a lot like this part of the structure and it hydrolyzes ATP. It's a dimer. It hydrolyzes ATP and causes a big conformational change that brings DNA in and out of itself, to unwind DNA.

And the other members of this family, they interact with proteins to pull proteins apart and control protein-protein interactions, and that's called HSP90. It has the same basis structure and binds the proteins and pulls proteins apart in some way. It's called a chaperone function. And it hydrolyzes ATP and uses that energy to cause that conformational change.

And there's a huge conformational change in this kinase because the kinase has to interact with the H domain and phosphorylate it, and then the two have to come apart, and this phosphoryl group then has to transfer to another protein called CheY.

So this whole thing is going through what we think is like a sewing machine operation. We think it hydrolyzes the ATP, flips up one side, phosphorylates the histidine on one, presents the opposite histidine, phosphorylated histidine, to the CheY protein, then goes like this. So it's going through a big motion and that could be important for the way it's regulated.

Now it turns out that this purified kinase is almost completely inactive. Although we spend a heck of a lot of time studying its kinetics in pure form, it's very, very, very slow. It has a rate of auto-phosphorylation when it's alone of only about one phosphorylation event per six seconds.

So you can just count, one thousand one, one thousand two, one thousand three, one thousand four, one thousand five, one thousand six. Boom. Phosphorylation. It's too slow for chemotaxis because the cells are changing direction every two seconds.

And it turns out that when this protein plus CheW is in combination with the kinase, its activity goes up about 500-fold. So it really only functions when it's with the kinase and it makes a tight complex, when it's with the receptors and it makes a tight complex with the receptors.

And this is what the receptors look like and, it's an amazingly simple structure. It's all alpha-helical. It's a long coiled coil, right, what we call a coiled coil. Coiled coils are just alpha-helices which have hydrophobic faces, and they just come together either in parallel fashion or anti-parallel fashion.

The outside, which is variable between the different receptors, is in four out of the five cases of this family of proteins a 4-helical bundle. So you can start with the end terminus. This is the membrane; this is outside of the cell; this is the inside of the cell. You go across the membrane, and then there's this 4-helix bundle. And then you come back across the membrane, and then there's this anti-parallel coiled coil structure, and its x-ray structure.

The outside and the inside were determined separately by Sung-Hou Kim at Berkeley, and this is its basic organization.

The inside has this anti-parallel helix. And then there are sites. There are two parts here, actually three parts. There's the linker region, where the structure isn't really known but is largely alpha-helical, and then there's this hairpin, and the hairpin has two parts. It's actually a continuous helix here except for at the bend, and you can't see it I guess because of this. Just focus on this internal part. And it's at the hairpin that the W and A protein, these SH3 domains, bind, right at the hinge region here.

And then the hinge region is flanked at the C terminus and at the end terminus by a region which contains sites that are subject to modification

320

by these modulatory enzymes. And there are at least four and sometimes five or even six of these sites in heptad repeat on the sequence, a sequence along the external face of this hairpin. And these sites can either be charged when there are what are called glutamate residues, or when they're modified by a methylation reaction, when they lose their charge. So you can control the charge density in here. And there are at least four sites that can be either charged or not charged.

So I want you to think of this as just plus or minus, charged or not charged, and it's just essentially four bits of information per fiber. And this is just essentially a fiber going in and out of the membrane.

If you want to think about alpha-helical coiled coils think about your hair, right. Your hair is made up of an alpha-helical coiled coil. So this is just a hairlike structure. And in fact in the initial papers talking about the bacterial brain, I called it a "hair brain" because of this.

We've looked at the composition. We can reconstitute this signaling complex in membranes. And I apologize for the complexity of this slide, and I won't go through it. All we did is we take membranes that are enriched for one receptor that we've genetically engineered so that we have a lot of in this case the serine receptor in these membranes, and then we add our purified W protein and our purified kinase. And they all bind together, and they make a structure. And we've characterized the composition of that structure as a function of the concentration of the purified proteins. And we measure the activity of the kinase in that structure.

And it turns out you have to have W for kinase activity. And there's this very simple relationship between the amount of W bound and the amount of kinase activity. And it turns out that the kinase-binding, this SH3-domain-binding, and the W-binding compete with one another at high concentrations. You can saturate these hairpin sites with W so every monomer of kinase, of receptor, can bind at W, but you can't saturate with A. A is a big protein, and it can't apparently fit on all the receptors within what turns out to be a cluster of receptors. And I'll talk about that in a second.

So anyway, we've characterized this binding. And here's another awfully complicated slide, but I just wanted to give you an idea of what we do. And there are two parts on this slide. This now takes that complex that we've isolated, this highly active kinase in the A-W-receptor complex.

And there are two points I want to make here. The first is up here. And I guess you can't quite see it. Here we're measuring the methylation. We can add all the components together and reconstitute the whole assembly in vitro under defined conditions.

And I just wanted to point out here, the bands here on the gel measure the methylation of the receptor, and that's converting minus charges to zero, to no charges. So we're introducing information into the receptor by methylation.

The amount of modification is dependent on serine. So with the serine receptor, when it binds serine it increases its level of modification. In other words, it puts in a memory the serine has been added to the media.

And then the other factor I wanted to show here was down here is the response of the kinase to serine, and what you want to do is, if you remember, you want to inhibit the kinase to suppress tumbling when you detect that things are getting better.

So what we've done here is we're measuring kinase activity here against serine concentration, and the red curve is the inhibition with serine of the kinase in the unmethylated receptor. And then we encase the level of methylation of the receptor, and it takes more serine to inhibit the kinase when the receptor has been modified.

So the introduction of those methyl groups modulates the response. So it has the characteristics of memory. You want memory to be introduced, and you want it to do something. And the level of methylation change is relatively slowly compared to the contraction of serine when a cell is swimming.

And then we fit these curves, and these curves have been fit to a two-state model. And probably most of you aren't aware of the two-state model, but to me it's one of the most fundamental concepts in biology and maybe all of human thought. When you look at biological regulatory systems closely, they're not really two-state systems, but they look like two-state systems. And at every level of the chemotaxis system, you see this idealization of a two-state system, run or tumble.

And this is a kinase on state and a kinase off state. So you have two states, and you operate between these states with dividing of different ligands which favor one state or the other state.

And we can model this system by a two-state system. And there are three parameters which control a two-state system classically. One is the equilibrium between the two states and the absence of any ligand, and that's called L in this equation here. I had to include an equation for this stuff.

And the other one is the affinity of the ligand for one state usually as opposed to the other, and that's this K which is the KD for the binding, the hypothetical binding of the serine to the serine receptor when it's in the off state, because the serine binding causes the receptor, according to this model, to go into the off state.

And then the third component is this N here, which is the number of units within this regulatory apparatus that flip between the two states, that are linked together in a cooperative fashion between the two states. And for instance, hemoglobin has four units linked together. So that when it binds oxygen it flips from what's called a T state where the oxygen is brought down into the R state, and that's how you breathe and transport oxygen.

Now to summarize, this is what the receptors seem to look like. You have a cluster of receptors, and the number we get is that we have a minimum of six receptor dimers per CheA dimmer, a unit cell essentially in the membrane.

And this cluster binds CheW and CheA, presumably the same site because they compete with one another. And so you have this kind of arrangement with these potentially charged groups in between.

And aspartate binds in the dimeric outside of the cell. And that's supposed to be an aspartate bound in this case, and that in some way causes a conformational change or some change in the structure that causes some change in this assembly—we don't really know what it looks like—and that turns off the kinase. Everything remains together. We do know that. You don't change the stoichiometry. You don't change the nature of the complex itself, aside from the activity of the kinase.

We've studied this then by cutting off to get at the structure of the complex. Now we want to know what the structure of this complex is. We've genetically engineered this part of the kinase, the signaling part that interacts with W and A, and we've made a complex of this W protein and the kinase and the receptor in a soluble form because it's impossible almost. Almost. It's very, very difficult to resolve a structure in a membrane. So we want to get this out of the membrane and characterize its activity.

But this works. And we were able to get a complex. And it turns out to be enormous. And we get a well-defined structure that looks like a phage, and we're studying that structure now.

And this has 28 receptor monomers and about six CheWs, and two dimers of CheA, and this is what it looks like. It's a bundle of receptor fibers, and in the middle of the bundle, this is an EM image enhancement from frozen grids done in collaboration with David DeRosier at Brandeis. And this is the structure. CheA, CheW are in the middle. This is this bundle of fibers, 28 fibers, and two monomers of CheA.

And this has the same activity, fairly within about five or 10 percent, as the CheA, the same specific activity as the CheA in the active receptors. And we can sort of begin to put this together maybe, how this might assemble in the membrane.

And these are the receptors, and we think that they probably bend, and they're brought together by A and W into a large complex. And what we've got is a ring that seems to come around and make a barrel of fibers in our isolated complex with A and W in between. So we're sort of beginning to piece together this structure.

And I wanted to point out the value of alpha-helices for nanofabrication of a large complex. And that is that if you have intramolecular interactions such as came out in the Sung-Hou Kim structure, so you have this hairpin, that implies the existence of intermolecular interactions like this. And this is what we think may be going on, that this hairpin is opening out, and we're getting these kind of lateral complexes.

And it turns out that the work that was done at Stanford in Lucy Shapiro's lab by Janine Maddock, that if you look to immuno-localization of receptors or CheA or CheW, in E. coli cells, all of the receptors—and there are about 10,000 receptors in a cell—cluster in a little patch, generally only one at one end of the cell. So all of these receptors—it's not just the six; there are about 10,000—come together in one complex.

And it turns out that all of the receptors, all of the different receptors, are together in that complex. And these are the five different receptors, and they all have similar signaling domains, and they detect different stimuli in a complicated way at the outside surface of the cell in the bacterial cell envelope.

So you have a system now, certainly a sensory motor cortex, with the receptors. And this is just small, about 10 percent. I mean it's a huge structure of the receptors organized in an array, and this has all five receptors in this array. There's a lot of evidence now that they interact with one another and in fact have to interact with one another to produce the responses that are produced. So these detect lots and lots of different kinds of sensory information.

And all of the chemotaxis components except Y are in this complex, and Y is a small soluble protein that then takes the information from the receptors and communicates to control the motor response. And there's a memory input into this system by the methylation, four bits per receptor at least. And there's a learning input into the complex at one level from alterations and gene expression within of the different receptors.

And so I wanted to argue that this has all the characteristics of a brain. So what are the characteristics? How do we define a brain?

A brain I would say is a central organ for sensory motor regulation. That's true. Our brains are anyway. And a brain has a capacity for memory and learning. And that's true of the bacterial sensory motor network. And a brain allows a distinction between good and bad, we hope.

And then there are two kinds of organisms. There are plants that don't have any brains. You can be a vegetable, or you can be an animal. And it turns out bacteria can be divided in the same way. With one exception so far, every prokayrote that's motile has this brain, the same brain with the A, the Y protein, and the fibers. And immotile bacteria never have it so far.

And usually motile bacteria are bacillus. They have a shape like a rod, and that's ideal for swimming. And immotile bacteria are usually spherical. They're cocci, right, which is ideal for growth. So there are two kinds of bacteria.

And from a bacterial perspective, I would like to just comment on the difference between chimpanzees and humans. I'm sure that they would seem to be totally identical. The difference between a rod and a sphere.

Now I tell this to my neurobiology colleagues, and they laughed when I told them about this. I didn't understand that. They just think it's the funniest thing in the world that bacteria would have a brain. And there are three objections.

One objection to this is that a brain has neurons. You know, you can't have a brain without a neuron. That's ridiculous. Right. It's just funny.

And the second point that they object to is that bacteria are too small to have a brain. That's ridiculous. You can't have a brain that small. I mean this thing is a femtogram, right. A bacterium is a picogram. It's only a thousandth away. You know, we have this huge thing on our head, but they've just got this little patch. It's too small to be a brain. Right? It couldn't process that much information. How could it possibly?

And the third argument is that we all know bacteria are primitive, right? They're just simple little organisms. Okay.

The other kind of brains that we sometimes think about—and nobody laughs at this—are intelligent machines, right? So I looked at the evolution of computers a little bit. And you know, intelligent machines start out to be really complicated. This is Babbage's difference engine. And made up of really expensive components.

It turns out that Babbage never even made this engine. He made a lot of the parts, and then they were melted down later because they were so valuable in and of themselves. Kind of like neurons, you know, they cost a lot to make, and they're made of very valuable material.

And then later as things got more advanced, you get these huge setups, and then it gets smaller, and then you end up with something which looks really simple and tends to be really small.

And the point I wanted to make here is that small is usually more advanced. Right? If you wanted genetic engineering where are we going to go? Why don't we make ourselves really small? There could be more of us. Take less food. Be more efficient. And the other thing is the material that goes into something changes, right?

Now the other argument. The other interesting thing is, well, bacteria don't have culture. I mean we talk about bacterial cultures, but we don't mean the same thing as human culture.

So this is to make a little argument for bacterial culture. And this is an observation by Budrene and ?? at Harvard. This is a swarm plate under a little bit different conditions than the other swarm plate I showed. And these are E. coli. It's a homogeneous population of E. coli, swarming out on a swarm plate, and they make these interesting patterns, and you get all kinds of patterns from bacteria.

And you should remember that inside us there might be 1014 cells. I think that's an overestimate. But there are at least 1014 bacteria. And that doesn't include the mitochondria. And all of these bacteria are talking to each other. We're starting to know that more and more. They send all kinds of chemical signals. And there's been a recent report that they actually do communicate with each other in some mysterious way across vast distances, but nobody believes that.

So I searched the Internet. I mean we know, physicists tell us all the time, that ecosystems, anything can make complicated patterns. It really just looks to the human mind to be interesting, but it's really pretty trivial. Crystals do it. Snowflakes do it.

So I looked on the Internet for examples of real bacterial culture, and I found some primitive beginning examples, so maybe it's happening.

This is some bacterial art that I found. This is luminescent bacteria that give off light and arrange themselves to identify themselves and beginning to make pictures and I think they're trying to communicate.

And then I found an example from nanofabrication of a nanoguitar. And so there are incredible possibilities.

Okay. That's it. And these are my collaborators at Princeton, Mikhail Levit, who's done all the work with soluble and now the composition of the membrane particle, and other members of my lab, and then at Brandeis, David DeRosier. Thank you very much.

Stuart Pimm
Biodiversity and the Construction of Ecosystems

Temple Smith: Our next talk is in one of the areas in which there has been lots of discussion, both in terms of the complexity and in terms of the importance of understanding these areas just so that the human race can continue to survive upon this planet, in terms of understanding the biodiversity of our environment, and various ways of trying to deal with the fact that in many places our interaction has reduced that considerably.

So I would like to introduce Stuart Pimm from Columbia University as our next speaker.

Stuart Pimm: Biology is the study of life. And so it seems to me that the salient question for the coming century is, what fraction of life will remain at the end of it?

There are approximately 50 million species on the planet and currently a third of them are threatened with imminent extinction, and that's as a consequence of the six billion others that inhabit the planet at the moment.

Optimistic scenarios from the U.N. suggest that there will be 10 maybe 12 billion people by the middle of the century. In addition to having more people, those people will have the same kind of aspirations that we do now.

So shall we look at the single most important statistic about the planet at the moment? In addition to the six billion people, we know that we take 50 percent or thereabouts of the annual plant productivity on land. A comparable proportion from the oceans goes to support our fisheries. We use half of the available fresh water. And as I've already said, about a third of all species are being driven to extinction as a consequence of human action.

So the great biological challenge for us in this next century is to slow or maybe even stop that massive erosion of the variety of life that is currently under way. As other people have put it, we want to avoid the sixth extinction, an extinction event which is already poised to become as big as the one that eliminated the dinosaurs of the K/T boundary.

The scope of our problem can be divided I think into two categories. One of them is we have to stop doing bad things, and then we have to correct the mistakes that we've already made.

As an example, we have already reduced tropical forests which contain the great variety of life on earth to about 50 percent of their original area. What we've got in their place is not new farmlands that will make the people of Brazil or the Congo or Indonesia particularly wealthy, but rather abandoned ?? which isn't really useful for very much at all.

The problem is not restricted to the tropics. We once used to have prairies. We don't anymore, and 99 percent of the prairies in the United States are already gone. So we're left with tiny remnants of areas that were once biologically rich.

So the second side of this problem is whether we can put nature back together again, and what are the issues of doing so? It is, I believe, one of the most immediate and practical challenges facing biological complexity. Do we know enough to go out there and assemble a complex system? Can we go out there and build something that will maintain the variety of different kinds of species?

Now there is a long history of studying ecological complexity. In the '50s and '60s, there was a strong sense that sort of complexity was good, complexity begat stability, that if you had a sufficiently complex system that it would work.

And that notion which came out of people who were studying ecological problems in the field was rudely interrupted when Professor May came up with what is sometimes called May's Theorem, which is terribly unfortunate because it's not a theorem, and it's not even mathematically correct, but it's broadly right.

You can probably tell from that that I'm not going to get my Royal Society membership this year.

His idea was broadly this, that we can define complexity in a very simple way. We can look at the number of components of a system, as the number of species in my case. We can look at how connected they are, which is just the fraction of trophic links that there are over what's possible. So the number of links over the number of possible links between those components. Connectance is a very simple measure of complexity. And then of course we can ask how powerful those interactions are, how strong they are, how dynamically important they are.

Well May's Theorem was basically that if you take the product of the number of species, the connectance, and the interaction strength, that if that product was sufficiently small, then in a dynamical set, a connective set of differential equations, that the system would probably be dynamically stable, and if it was bigger than that quantity, it probably would not be.

And what that simplifies to is the diagram which I've shown on the right, recognizes connectance as a function of assets, the number of links divided by approximately a square, the number of specific interactions.

And so it says that if you have a sufficiently small linkage density, a sufficiently small number of links per species, that the system will probably be stable if you pick the parameters in some random fashion.

The problem with that, other than the slight exaggeration of the mathematical rigor, is the fact that biological systems look as if they are hugely complicated compared to that criterion. I've just shown a couple of food webs. Food webs are the diagrams, the road maps, that show which species eat other species. And these are two examples taken from Stephen Carpenter's work of lakes in Minnesota so that there is an enormous amount of dynamic connectedness. The connectance is high. The number of species is high. The interaction strengths are high. And in fact if you try to build a mathematical model of that you'll have really a hard time of it.

And this notion that it's hard to build something that's complex is in fact matched by all kinds of experiences. People try to build a balanced aquarium for example. A familiar example, something that you probably did as children, or as you have children you're probably doing it for them now.

You try to put a bunch of species into a tank, and what you usually end up with is a lot of sort of green and brown sludge which isn't probably what you wanted.

Now if you were really ambitious, you build something like Biosphere II, and what they ended up with was a lot of brown sludge. And fortunately they got the people out of it in time. That happened before I went to Columbia University, and I take absolutely no responsibility for what happened.

If you do the whole thing in silico and you run your computer models, if you pick food web structures to be broadly sensible, and you pick parameter values that you think are ecologically sensible, you find that it's vanishingly improbable   that if you create a system that it will be dynamically stable.

The particular applied problem that I want to tackle is whether we can go and restore nature, whether we can go and assemble nature. And the particular problem within that subset is whether we can bring our prairies back.

We more or less have all the pieces. We for a variety of reasons have got most of the species that survived our plowing of the prairies in the last century. They survive in peculiar places, graveyards for example that nobody plowed. But given that we have those pieces, can we restore our prairies?

My colleague, Dr. Julie Lockwood, and I have analyzed the entire extent of ecological restorations attempts of people to restore nature. And if you want to get something living, restorations are usually quite successful, but if you want to get a particular set of species, if you want to restore complexity in that sense, restorations fail almost all of the time and usually quite spectacularly. It's extraordinarily difficult to get nature put back together again.

Well this is all really rather embarrassing because nature is clearly complicated. There's a lot of advantages to dynamical systems that are complicated in ecology. We know that they're resistant to change. They don't vary very much. They're quite hard to invade. And a whole variety of dynamical properties, all of which are explained in the requisite amount of detail in my book, The Balance of Nature?, available from the Chicago University Press at $25.95.

So nature is complex. The complexity has a lot of interesting properties. But that doesn't really tell us how we really get it.

While we're talking about this subject, I know you heard from my friend and colleague, Stu Kauffman, earlier in the week. You may recall that in Stu's book, At Home in the Universe, Chapter 2 deals with a recipe of how you create life. I think it's a wonderful recipe, and it's fascinating, but I think he is missing just one ingredient.

The argument that appears in the book is that if you get a sufficiently complex set of molecules, of chemical compounds, that they will generate more complexity, not less. They will breed a greater diversity of chemicals than when you started with.

If on the other hand you are below that critical threshold, the number of chemicals that you have will generate a smaller number of chemicals and yet a smaller number. In other words, there is some sort of phase transition.

If you're above that sufficient complexity then you will get more and more chemicals sort of ad nauseam. And then at that stage of the game, the chemical mixes that will persist will be the ones that persist. And you will get some selection upon those that persist because those that do will, and those that don't won't, and therefore you get life. I hope that summarized Stu Kauffman in far fewer sentences than he would.

So what we need to get is we need to get sufficient complexity. So how the hell are we going to do it?

The interesting question is then what happens when models fail? If you build Biosphere II and it starts to go wrong, instead of saying, it's gone wrong, and doing all the things that they had to do in Biosphere II, or the aquarium or anything else, wait until we see what happens. Let's just let the system fall apart.

And it's that notion of looking at what happens from a systems change as they lose complexity or build complexity that's going to be the theme of what I'm going to say because as you'll find out later it gives us an answer to how we get complexity in the first place.

I want to build a simple model. I'm very impressed by Stu's view on complexity of how we can build minimally simple models. So I want to build some Boolean models of how we might assemble nature and see what we can learn from them. Some of this will be terribly familiar to you all. I'm going to use some different language that's appropriate in the context of ecology.

The simplest Boolean model: you have a system that doesn't have a species in it, and species A invades. So I'm just going to look at models that are that simple, either no species or one species.

And of course we can make that model a little bit more complicated. We can put two species in the system. And in this particular case, you notice that we have what we in the ecology trade call alternative persistent states. We can either have a community with A in it or a community with B in it.

We can make things a bit more complicated and put three species in the system, and in this particular case there is a state, ABC, that is unreachable. It is an unreachable persistent state. It embodies the kind of wisdom that people have when, how shall I put it, damn Yankees come and visit people in my home in East Tennessee, and they ask how to get from one place to another, and the people in the local shops say, you can't get there from here. If you live in a sufficiently fractal landscape, you may not be able to get from A to B, and in this particular case there's an unreachable state. Okay? You can't get there from that particular set of configurations. So if you've got it you'll be okay, but you can't.

And that raises a final sort of piece of terminology that I want to bring up which I call the Humpty Dumpty effect. You may remember the nursery rhyme that all the king's horses and all the king's men couldn't put Humpty Dumpty back together again.

I've imagined now a four-dimensional model which is exactly the same as the three-dimensional model, but there is a cube that's exactly the same except it has species D present in it. And you can imagine that you get to the state, ABCD, and from there you get to ABC. So you would have a state which had D to get there, but you no longer have D—it might have become extinct at the K/T boundary—and so you therefore have a Humpty Dumpty situation, a system that you cannot reassemble from the pieces.

Now we can begin to do some very, very simple mathematics, minimally simple mathematics on this idea. If you imagined that you have a very, very large number of species in the system, S, you can calculate what the probability is that a particular state will be persistent in the sense that its species composition will not change. It will remain persistent. It will not become a tank of brown sludge.

– And the probability is a half to the power S. S species and each point is next to S. Other points, there are S arrows. If you pick the direction of change from one species to the next with equal probability, the chance that a particular state will be persistent is therefore a half raised to the power S, which is going to be very small, but the number of states, the number of binary combinations, is 2 to the S, and it follows that the expected number of persistent states is going to be 1.

Given that the number of persistent states is very small, and that every state that's not immediately adjacent to another one is statistically independent of that, you can show that the number of persistent states is essentially going to be Poisson-distributed. You're going to get a fraction of zeroes, ones, twos, threes, and fours. In other words, you get a fraction of alternative, persistent states. Some models have one; some have two; some have three; but the probability of a Poisson distribution with a parameter of ?? equal 1.

Now a very simple-minded view, a biology free view of assembling complexity, says that if you try to create a system, you have a vanishingly small probability of getting it to work. There is only an expected number of 1 out of any system of any size that will be persistent. It does say however that it's out there somewhere. There will be a persistent state, but you won't find it.

Number of persistent states if Poisson-distributed. And what it interestingly says is that if you create models at random, that they will be typically models that will have half the number of species in them that you started out with.

So if you do something like Biosphere II, you should end up with half the things you started out with, which was not exactly what they wanted because that means that when you as a Biospherian walked in it you have a 50 percent chance of dying, which is probably what they didn't tell those people wearing their nutty little orange uniforms.

All this makes sort of sense. It says, you know, that there is a system out there, but it's not going to have as many species in it as you put into the arena at the beginning, which isn't terribly encouraging in terms of how you build biological complexity.

Everything I've said so far has allowed one species to invade at a time. You cannot go across the diagonal. It's not like Snakes and Ladders where if you land on the wrong thing you go zipping down or zipping up. But there's no reason in ecology why you should have such a pedestrian rate of species invasion and extinction. Under different circumstances, you could have a circumstance where many species invaded at the same time, simultaneously, and you can have many species going extinct at the same time.

If you think about what that does, it means that at any particular state you have vastly more transitions coming in and out of that state. You have the same number of states, but a very, very much smaller likelihood that those states will be persistent. That says the expected number of persistent states goes effectively to zero.

I call this the 1066 effect because it reminds us how incredibly unsporting the French were in 1066. If you know any English history, you will know that the Brits had to fight a battle up in the north of England and the only reason that the French were able to invade was that they broke the rules, and they came in a few weeks later, and the poor Brits had to run like hell to the south of England, and of course they were defeated. It says it's hard to stop two invaders even if you're capable of one.

So if you have a 1066 effect, if you have a lot of species invading at the same time, it's going to be much more difficult to get any kind of persistence. So that makes it less likely, much less likely, that you'll have any kind of complex system persisting at all.

Now all that isn't terribly encouraging in a way. It says, you know, in some circumstances we won't get persistent states and in others we will, but we'll typically only have half the species in them that we started out with. So how on earth do we have all of the species which we need to get the complexity that's only half as big?

In a sense that's the same problem that Stu Kauffman is posing in terms of how we create life. You know, if you don't have enough diversity to start off with, you'll end up with less, not more.

What we've done is to convert these biology free models into some very simple dynamical models. The basic dynamics, for those of you who know the ecological literature, is Lotka-Volterra, a mass action dynamic between the abundances of the species. We start with a species, a community that only has one species in it. Species 2 invades. Species 3 invades. The implication is that this is a plant. That's a herbivore. And we get a predator coming in—a carnivore coming in. Many carnivores. It feeds on the herbivore. It also feeds on the plant. In doing so, it drives species 2 to extinction, leaving that system. Species 5 comes in. And so on. And there we have a recipe for an assembly, a dynamical assembly of an ecosystem.

The mathematics, this is very much sort of a standard equation for ecologists. It's a couple series of differential equations. We apply some sensible ecological constraints to that. They are that if a predator is feeding on the prey, the per capita effect of the predator on the prey is much larger than the reverse. When the fox is chasing a rabbit, the fox is running for its supper, and the rabbit is running for its life. We don't have any external limitation on the system except for the plants which are limited by access to space or nutrients or something like that.

For a species to be able to invade, as in example 2 and 3, it has to be able to grow when it's rare in the presence of the other species. So this is a very simple algorithm for generating, you know, complex systems within the laboratory. We can do that with in this case a suite of about 60 species. We pick 20 plants, 20 herbivores, 20 predators, and we let this thing go for a few hours on the computer.

I'm going to divide the results up into two classes. The first corresponds to those Boolean models where I did not allow diagonal moves. So I call that slow invasions. Only one species can invade at a time, and then only one species can go extinct at a time.

What I've done here is to show how long each species persists within the system. I've used this as a measure of success. So we measure how many years of the simulation does a species last. And these species are simply listed in descending order of success. So the first species, No. 1, goes better than the poorest plant, No. 20, simply because we've arranged them in that sequence.

If you look at the error bars on there, those in fact are ranges based on a small number of simulations. And I want you to notice how dramatically different the fate of a particular species can be. Species No. 1 is on average the best-performing species. There are times when it performs superbly well, and there are other times when it performs quite pitifully. Much the same is true of all the other species.

What this means is that there is a huge amount of historical contingency in the assembly, that history matters, that if you run the tape of life again—God how I had to have to quote Stephen Jay Gould—you get a different set of ecological communities developing.

In this sense it answers your question—and you did charge us with this—of asking whether ecosystems, biological complexity, whatever really is complex? The answer I think is emphatically yes, because it says that the initial conditions and the history are enormously important in determining what you get. These are identical model species. All we change is the order in which we allow them to invade the system. Whether you go down that pathway or that pathway makes a huge difference.

In fact, all of these species can persist some of the time. And in that, we begin to see a glimmer of how we can get complexity. If you run the experiment differently in enough different places, you will not go from 60 species to 30 species, you will allow all the species to persist somewhere. In some part of the system there will be a group of species. If you ran the system again in that place, you'd get a different set of species. But none of these species—all of which are picked at random, some of which on the basis of their parameters you would think would be unbelievably wimpy—all of these species manage to be somewhere some of the time.

What happens if you have a fast invasion? What happens if you have these diagonals so you can have species invasions happening quite rapidly relative to the time it takes to sort species in and out of the system? Those are the species ordered in exactly the same sequence as before.

What you find is that there's much less variability across runs, and you'd expect that because in fact what's happening is these species are invading many, many times, so you're taking an average. But you find that species that were quite poor under the slow invasions can be quite successful otherwise.

You notice Species 17 does really very, very well in terms of coming into the system very often, persisting for a relatively short period of time, going out again. Whereas under the slow invasion, it was a very poor performer indeed.

What this suggests is that if you have a very high turnover, species are able to invade your community very quickly. All of the species will be able to persist some of the time in the same place. So this is a view that says, you know, if you have fast invasions, everybody will be in the system. They'll differ in how much time they spend in it, but if you watch this spot long enough you will see every species present there for some of the time.

Just to give you a comparison of those two slides, I've just put them up simultaneously, and you can see there's essentially no correlation with the performance under one system and the performance under the other. So it says some of the species will do well in different places, some of the species will do well in a particular place, but all of the species survive in some places and at some time.

These are not species that were picked to be able to do this. These are just random selections. But it says that we have a model and it allows us to keep all our 60 pieces, providing we recognize that we have a diversity of space, that we run the experiment a sufficient number of times.

So if we view this in the metaphor of creating life in the test tube, if you create life in a test tube it isn't going to work. On the other hand, if you created life in a lot of test tubes and allowed them to mix occasionally, then indeed you might have a recipe that gets over this phase transition that Stu Kauffman talks about. In other words, it's the diversity of places and events that gives you the complexity that we have, which answers your question in a very emphatic way.

So here are my conclusions. With slow evasions all the species will survive somewhere. History, the initial conditions, what happens next is going to determine which survive and which don't, applied to large areas. If you have an arena, there will be different histories over here and over here. And so you will get different species over here and over there.

And for the ecologists in the audience, I know what your criticism is likely to be. Many ecologists look at the diversity of plant communities, for instance, that we see and assume that it must be that we have different environmental conditions here as opposed to there. I'm not denying that as a mechanism, but I think it's equally plausible that you get a different suite of plants and animals here than there simply because they have different histories.

If you have fast invasions all of the species will survive at some time. And this really applies to very small areas. A small area has a much smaller boundary relative to its area, and therefore the immigration and extinction across that boundary will be proportionately greater than the large area. So you get a constant churning, a constant mixing in small areas, and across the big areas you get alternative different communities.

What does this tell us in practice? It tells us a recipe, suggests a recipe, for how we can restore nature. And this is not a mere academic exercise. It has prompted a set of experiments that Stu and I set up at an institute in Kansas called the Land Institute with the very express mission of restoring prairie diversity.

If you get a prairie mix, if you grab a bunch of seeds, shake them up, homogenize them, and seed them over an area, we predict that you will lose 50 percent of your species. More or less, you do.

If on the other hand you deliberately have mixes of species that are very different and plant them in very different places, we predict that you will lose half of the species in each place, but overall you'll have all the species that you need.

We may not be able to restore nature to what it was, but we will be able to obey Aldo Leopold's famous dictum that the first law of intelligent tinkering should be to keep every cog and wheel. We want to keep all the pieces, a third of which globally we are poised to lose.

What we found from our six years of running this experiment in Kansas is indeed we can keep the pieces, and we do it by giving different areas different histories, recognizing that because they're in different places that they will have different histories by chance alone, different rainfall, different patterns of insect outbreaks, and so on, and the consequence of that experiment is that we catch all the species in the system. It suggests that we do have enough understanding of complex systems to assemble them.

Thank you very much.

AUDIENCE: So you told us a couple times that we're poised to lose a third of the species, but that nonetheless, in spite of all we've done, we haven't lost any species from the prairie. How can you feel confident that we're going to lose a third of the species if they're so robust that we haven't lost any yet?

Stuart Pimm: We probably have lost a few, and there are plenty of parts in the world where we have lost about 10 percent of the species. We've kept the prairies going because the destruction of the prairies is relatively recent. Much of it happened less than a century ago. They did survive in a few places where they're hanging on by their fingernails, and we can collect the seed. So the prairie example is interesting because we've just been really fortuitous in keeping the species.

There are plenty of parts of the world, the Pacific Islands would be one example, where we've lost 90 percent of the vertebrates. We're never ever going to get them back.

AUDIENCE: So I've been talking for the last year or so about analogy between let's call it the fault-tolerant computation in computational theory and resilience and persistence in ecosystems. Without trying to explain what I mean by that, let me just ask a question that that analogy suggests.

And that is that your 1066 principle will have less sway when you have spatial distribution, as you do in the examples you talked about at the end, and also that it will have less sway when you talk about introducing new species at the plant level, at the bottom level, than when you talk about it at higher levels in those graphs that you were drawing. Is that what you see or not?

Stuart Pimm: Very broadly, yes, although I might disagree with the details. Ecological systems occur across a huge variety of scales, from this big to the planet. What I'm saying is that small spatial scales—where plants and animals will define 'small' in different ways—small spatial scales will always be characterized by a constant churning of species composition. Larger scales will be driven very much by whether they are sufficiently large to have alternative histories.

AUDIENCE: But I have in mind specifically that it's more the perturbation of an ecosystem if you try to introduce a species at the carnivore level than at the plant level. Would you say that that roughly speaking is right or is that just completely wrong?

Stuart Pimm: That's a difficult question, and I don't think there's any simple answer. It's clear that those what we call top-down effects from carnivores can have very large spatial consequences. But on the other hand, introducing plants like kudzu, for example, can also have very large spatial consequences.

Jay Lemke
Multiple Timescales in Complex Ecosocial Systems

Temple Smith: I would like to go on to our next speaker whose title is Multiple Time Scales and Complex Ecosocial systems, which maybe at the end you can explain the stockmarket for us, too, which is clearly such a system. Jay Lemke from City University of New York.

Jay Lemke: Thank you very much, Temple.

And it's a great pleasure to be here. I was happy that Yaneer persuaded me to come to this meeting because I've been learning a great deal and it's been a great pleasure meeting and talking with so many of you over the last few days.

I am not a biologist, although I'm beginning now to think that maybe I should start to think of myself in a sense as a biologist. My interest is in the collective behavior of a certain biological species in its larger ecosystem environment. It's a species that is one of the most mysterious to us because of the perspective that we have in viewing it, our position relative to it. And of course it's one of those primate species. He walks around upright and makes a lot of strange noises and exhibits very peculiar kinds of collective behavior, such as this conference of ours, since the species I'm talking about is homo sapiens.

And I was thinking last night as I was trying to fall asleep about how one would analyze just the movements of us as individual agents or people around this building in the course of the conference. And I was thinking of taking a kind of bird's eye view of transparent walls and people moving around here from one room to another and through the corridors as a function of time, as if I had a videotape, maybe speeding that tape up so that something like that film, *Koyaanisqatsi*, where you see the patterns of traffic movements and cloud movements and so on, and thinking of the patterns that you would see let's say over the course of a day cycle in this conference, and in fact the repeating cycles that you would observe, and imagining how that would be modeled by someone doing the usual kind of dynamical systems theory, wondering for example how one might dynamically explain this room as an attractor such that all of these people seem to periodically head for this room, and it can't be a sort of density feedback effect in the sense that people are attracted to places where there are already more people because at certain moments this dynamic reverses and suddenly we all spread out from this center and go our separate ways.

I was also thinking just before the banquet last night that we -- there were a bunch of us standing out in front of the banquet hall waiting for Jack Cohn to give us the signal that it was in fact time to go in and get started. And as we were standing there there was a reception in our usual space where we go all the time in the courtyard area and yet there was an invisible dynamical force field of some kind which seemed to prevent us from defusing into the group of people who were already in that space.

And I began to think about how one would model what kinds of dynamical forces were at work that kept us as mobile mass units from entering into a space where previously we normally did enter, where there were all these -- already people, people enjoying appetizers that seemed far more attractive than what we have been able to afford here. It was the Nashua Chamber of Commerce that was occupying the space at that point. And I even thought about what would have happened if one of us had accidently wandered in to that group.

I think you can see that what I'm talking about here is the sense in which the modeling of even the spacio temporal mass dynamics, mass flows of human populations is dependent upon factors that have to be taken into consideration in an ecosystem that involves human beings and the kinds of things that we do which might not have to be taken into account or which play a lesser role when one is analyzing similar kinds of complexity on the part of other systems.

And I will come back to this issue towards the middle of my talk, but I think you can anticipate that the relevant factor that's been added to what I would call ecosocial systems or perhaps the original names for these were ecosocial semiotic systems, what's been added are the ways in which human systems construct meanings about things.

There was a sign there saying, Nashua Chamber of Commerce, which somehow seems to have an effect on behavior. There are signals that are given here that suddenly send us all running out of the room that we interpret. It's not a dynamical force that explodes us out of here, but there are certainly regular principles for our response.

And in fact if you wanted to understand the larger timescale movements of people around the building, it has something to do with a material document -- can you give me a program for a second. There's another material object in this system. It's a material object that is actually in wide distribution in the system. It is consulted on short timescales when people seem to begin their rapid changes of direction or switch from one attractor to another. And the role of this object is another clue in some sense to what is special about human ecosocial systems.

Let me put up a few transparencies and work through some bits of an argument here.

I've often found that in talks like this it's useful, instead of trying to do dramatic mystery and have my conclusions only unfold at the end, to give you advance organizers and some sense of where things are going. That was the first one, which is this question basically of how you analyze the collective behavior of human beings in terms of the fact that our communication systems involve responses to material stimuli.

After all, the sign and the marks on the pages of the program and the signals that we get here that a session is over and so forth, those are themselves material phenomena, but we respond to them not just for their material properties but also for their semiotic or symbolic properties.

So the issue arises, and this is perhaps the most consuming issue of interest to me, as to what are some of the general principles we would need to follow if we wanted to semiotic principles, that is, principles of the symbolic mediation of human behavior, human dynamics, with the kinds of organizational principles for complex dynamical systems that we have heard about and seen so many examples of for other sorts of biological and even non-biological systems.

I have not yet quite mastered the art of both engaging with my audience and putting on my glasses and reading from my notes because this is a relatively new handicap that I have acquired. I have to decide, having skipped ahead in my outline here, where I'm going to reenter it. Let's just let the overheads do some of this work.

This is a table only of approximate accuracy. It was designed to construct a range of timescales, effectively, starting around one to 10 seconds, the timescale of the utterance, and then working up and down from that. And the reason that I constructed this was because at some point in my efforts to understand the organization of human collective behavior I became interested in the role of timescales.

Very often when we analyze complex organization hierarchical levels of organization in other kinds of systems we start with the default assumption that there are predominant spatial scales of organization, which are the simplifying principle, so that roughly speaking the smaller systems are aggregated in their collective interaction in the larger systems and those larger systems in some sense physically contain the smaller ones in three dimensional space, and so on, to larger and larger aggregations.

And this is a reasonable approximation for many kinds of systems, but it's been pointed out that in some not so common cases in ecosystems, natural ecosystems, but in very many cases in human technologically mediated ecosystems that the topology, the compactness is not quite the same.

That is, for example, we saw last night with Kenneth Arrow on screen that there were direct interactions at a distance in effect between people here in Nashua and Ken Arrow at Stanford, such that someone might be interacting with Ken more intensely over a certain timescale than they were interacting with someone who in physical three dimensional space was much closer to them.

And there are many examples of this, from the Internet, to the telephone network, to the railway network, to river systems, and trade that is mediated by river systems, such that the connectedness of the topology of iteration does not have to be such that two points which are nearer in physical space are therefore that much more likely to have a close coupling with one another.

And this means in effect that you cannot always rely on the fact that shorter distances correspond to the lower timescales of organization. You -- the lower scales of organization. I tend to always say, timescales, because it turns out that a safer principle, a more general principle, that applies both to human ecosocial systems and to other kinds of systems seems to be to pay attention to timescales; that is, that the processes which take place on shorter timescales, faster rate processes, tend to be aggregated together in their collective behavior into systems which operate on somewhat slower timescales and often also on wider geographical or spatial scales, and so on, up and down the hierarchy.

But this leads to an interesting interpretation. In the timescales that you have here you can see that I was interested in one of my other areas of research. I was originally a theoretical physicist and not a biologist in that sense. I got interested in communication of science and in science education and science teaching and in the role of language, mathematics, symbolic diagrams, and so forth in the teaching and communication of science. And from that did a fair bit of research on what happens in science classrooms, in science lectures, in the exchange of ideas and information that was taking place there.

So one of my sort of laboratories for looking at the complex organization of ecosocial systems is the classroom, school, educational system. And you see here the timescales running up and down for that. I'm not going to go into that in detail now. I just want to sort of orient you to the sense in which you can look at these things in terms of timescales.

I'm not going to coordinate too closely between the overheads and what I'm saying. You can regard these as two parallel channels of information for you to be integrating as we go along. Some things will be closer and some things you'll have to do a bit more work on.

Looking at the top half of this diagram, this represents a fairly general model for looking at both the emergence of new levels of organization and at the relationships between levels. This model was developed sort of jointly in about five or 10 years of conversations between me and a developmental and evolutionary biologist and theoretical biologist, Stanley Solti, who was a colleague of mine in New York.

And the model essentially is a three level model, a three level paradigm. And I think that if the talk that I'm doing today has something to offer to various people, it's to set a fairly general but also fairly simple principles or things to remember, kind of checklist, in doing analysis of complex systems, of things to pay attention to.

One of those is paying attention to timescale, to relative time rates and how systems integrate or coordinate, create coherent behavior across relatively great differences in the timescales of their component levels of processes.

Another one closely related to that and in fact which preceded it in the development of these ideas is to always look at sandwiches, to always look at any level in focus, any focal level, in the middle, sandwiched between a level of organization above it and a level of organization below it.

These kinds of very general principles are particularly helpful when you are looking at human social systems where it's sometimes not at all obvious how to carry over your understanding of other kinds of complex systems into human systems.

The basic model here says that each of these different levels is characterized by processes which operate on radically different timescales, let's say one to more often two orders of magnitude difference in the typical timescale of the process.

The lowest level, the level below the level in which you're interested at the moment, is the constitutive level. These are the units and the processes those units are engaged in which give the affordances the possibilities for collective behavior to emerge at the focal level, at the middle level of the sandwich.

But what the model says in effect is there's always a level above, there's always some level of more slowly changing constraints which acts to select from all the possible self-consistent solutions of let's say auto -- collective autocatalytic sets of behaviors or processes at the bottom level, just those which are consistent with the slowly changing constraints at the top level of the system.

And this has at least one fairly strong implication. The implication is that in the evolution of complexity of systems as they go from having smaller numbers of levels of organization to having larger numbers of levels of organization, that new levels always appear intermediate between two preexisting levels, and the -- so that it's the intermediate level which is the new level.

So instead of a model that says, you know, we start out with A and B and then we add C to it and then we add D to it and so on up to Z, it says we start out in effect with A and Z, though you don't know they're A and Z at the time, you start out with A and Z and then you get maybe A, L, and Z, and then you get something, A, I, L, I appearing between A and L, and Q appearing between L and Z, and so forth, interpolating these levels.

Another way of thinking about this is sort of as cycles within cycles, longer time cycle processes and within them shorter time cycle processes, with the emergent behavior coming intermediate in the length of the time cycle.

This I think is not particularly unfamiliar. We've heard about this. We've heard many examples of it in the conference and even in this session. Notice that the -- another implication of this is that in some sense once the patterns at the middle level are established that variation or noise, if you like, at the bottom level is now being filtered through the patterns at the intermediate level such that the top level of the sandwich is paying attention only to those variations at the bottom level which amount to significant variation at the middle level. That is, in some sense each level looks one level down at what's happening and doesn't look two levels down at what's happening.

This is typical of many kinds of systems, but there are important exceptions to it in the case of human social systems, which is where I hope I will wind up at the end of the talk.

So in this three level paradigm, as a fairly general description of what's happening across levels in many kinds of systems including ecosocial systems, again the basic question comes, how do we integrate such a picture of complexity with the role of meaning, with the role of symbolically mediated behavior, with the role of signs, both in the simple sense of signs up on the wall telling you that it's somebody else's reception, and signs in the general senses of semiotics of (C.S. Purse) or (Suserian) European, Continental semiotics, which tells you that there is something a little bit different about sign mediated processes from simple material causality mediated processes even though sign processes are a special case of material processes.

That is, semiosis is done by material systems and a very fundamental question is, what's the simplest kind of material system that does semiosis, that is, that does the interpretation of a material phenomenon as something other than a response to its causal properties?

In the lower half of the overhead, what you see is an attempt to represent one of the conventional models of how semiosis works in the same terms that we've represented the organization across the levels for the dynamics of organization in the system. And I assume that these notions are less familiar to people here, although they're usually more familiar to most of the other groups that I talk to.

Essentially, the model of semiosis that is fairly standard today is one in which a system, called a system of interpretance, or it goes by various other names, interacts with a phenomenon and then -- and I'm going to state this in somewhat more biological sounding terms -- it then exhibits an adaptive response which is not just a causal response to the properties of the object it's interacted with, but is a response at, in effect, a higher level of organization.

So for example, if I -- if part of my olfactory bulb interacts with a stray molecule, an aromatic compound, and I then decide that this means that the coffee is ready and I go off and get it, I am not interacting -- I'm not interpreting the interaction at the level of the molecules, I'm interpreting it at the level of my whole organismic behavior in an adaptive way.

And of course there are many people in the field of biosemiotics who ask in just how simple a system biologically can this kind of phenomenon occur. I think it's an interesting question, for example, whether the nanobrain in E. coli is in effect doing a rudimentary form of semiosis. I certainly know people in biosemiotics who would say that it is.

And part of the issue is how is that different from the kind of semiosis that we do in human society? And I would say that in some sense one of the ways in which it's different is the sort of number of levels of reinterpretation, number of layers, in effect, of reinterpretation that one goes through with the accompanying possibilities of many to one mapping that you get.

To the extent that responses have a sort of one to one mapping to stimuli, it's harder and harder to distinguish semiosis from causal chains of behavior. But to the extent that there are one to many behaviors, that is, the same stimulus can be responded to in different ways depending on context, as we say in semiotic theory, and that is a familiar notion I think to most people.

If you interpret the stimulus differently in different contexts, this is an indication that there is semiotic mediation going on. And if you ask what it takes, in effect, or what it means to have a context dependent response to a stimulus, in some sense it means that you're integrating one stimulus with another stimulus, perhaps with a past stimulus, hence the importance of memory in many interpretations of how this goes on in biological systems.

And in means that in some sense you are integrating stimuli either across space or across time or both, and perhaps even, and I would say generally, integrating across timescales, that is, the integration takes place on a much longer order of magnitude longer timescale of integration than the actual process's interaction with the individual stimuli that are being integrated into the pattern which is the pattern to which the response actually is adaptive.

Mapping this notion onto the three level model of a dynamical organization of systems, the conclusion, as I think is fairly obvious from the way I've been interpreting the examples already, is that the system of interpretance, that is, the system which integrates stimuli over time so that the system responds to the entire pattern, is at a higher level of organization, that is, operating on the longer timescale.

The patterns which are the meaningful units in the sense, let's say, that a word or a sentence is a meaningful unit, they are the intermediate or emergent level here. And the bottom level is the level of the actual interactions at the much shorter timescales with each element that is being integrated into this pattern.

So if these two models are strictly mapped on to one another, then they say, for instance, that semiotic or intelligent behavior must be multi-scale behavior, that the signs represent the emergent patterns at an intermediate level of organization, and that may well be an intermediate level of organization in the neural cortex, but it can also be an intermediate level of organization in the social interactions of individuals in a community, establishing community norms or community values, then that the interpretant level, the level at which these things make sense, is the higher level of the system. So again, new sign levels emerge intermediately between the preexisting other levels.

Let me give you one -- that's not the one I was looking for, but you can look at that while I'm finding the other one. These are two hypotheses that have been of interest to people looking at these questions. One is the first one that I've just sort of summarized for you, the mapping of these two different sandwiches on to each other. The second lower one on here I won't have time to go into. It could be a whole talk in itself. Though I may be able to throw up -- here's a list of the examples that you can peruse.

It turns out that one of the additional difficulties in integrating dynamical perspectives with semiotic perspectives is that by and large our theories of semiosis rely only on categorical differences between signs, that is, the patterns are like two state systems, yes-no, on-off, and yet a lot of the variation in the environment that we respond to is gradient or continuous variation.

And so one has to basically add to the semiotic theory a model of how we make meaning by differences in degree, like differences in color or sound frequency or just in spatial continuous representations, as well as how we make meaning by differences in kind, by two state, on-off, systems of contrast.

It is in fact possible to do that. I have the overheads here, but again there isn't time to do this. But if you do this you get an interesting generalization of what in some ways is a very old conjecture and that is that in complex systems which do semiosis or which do interpretative behavior, and perhaps even in ones that don't, that there is a kind of alternation as we go across the levels of organization in which in the sandwich each new intermedial level is basically processing or reorganizing information or variety, let's say, at the level below as information for the level above.

And the principle of alternation says that the intermediate level reorganizes continuous variation, continuous variety, at the level below as discrete information for the level above and/or it reorganizes discrete information, that is, it averages over a lot of discrete information at the level below, restoring a quasi continuum of variation of degree as meaningful information for the level above, which in turn becomes an intermediate level for the level above, which now can reorganize this quasi continuous information as a new kind of discrete information for the level above that.

Which may lead in fact to a strict alternation as one goes from level to level in which one moves from reorganizing continuous information as discrete for the next level and then discrete averaged over as continuous or quasi continuous information for the level above that and so forth.

What you have on that overhead are a few steps in this process and somewhere here should be the overhead that continues that -- probably buried under here -- continues that on up all the way to the level of speech.

It starts with the organization of discrete quantum difference in chemical types up into the quasi continuous deformation shapes of proteins, on up through ligand cell membrane structures and so forth, on up to the nervous system, to the point at which discrete firings of neurons are coordinated at a higher level of organization to enable us to have smooth motor behavior, to enable me to be able to, in effect, put my tongue in any one of a continuous series of positions in my mouth.

And then in our collective social behavior, our languages decide that no matter where I put my tongue in my mouth I'm either pronouncing the vowel "a" or the vowel "a" and anything in between is either ambiguous or gets resolved as one or the other. It becomes like an on-off or two state system again.

Well if you try to continue this alternation model up into higher level human collective social behavior, it turns out that it doesn't work or at least it does not appear to work and this becomes a clue that there is yet one more kind of difference that has to be taken into account in the case of human ecosocial semiotic systems and that is in effect the fact that for us we are able and typically do organize information with shortcuts that go across multiple timescales in ways that other systems may also do, but it doesn't seem to play nearly as significant a role in their organization.

Normally you don't get this because there's a very basic principle in physics that roughly says that fast processes and slow processes don't efficiently exchange energy with each other and so don't efficiently exchange information with each other.

There's a basic exception to that which is matter; that is, if the information is coded in matter rather than coded in energy exchanges, it is possible for the information coded in the matter to participate both in fast processes and in slow processes.

So returning to my initial example of how our behavior is organized here in the conference in terms of our using and reading the program, the thesis would be -- and this is actually a single simple example of a very general principle -- that the program was produced in a long timescale process. All of the planning, all of the organization that went into it, for which we all thank Yaneer and his colleagues very much. And yet that same physical object when distributed around in our community we now can interact with that on very short timescales, altering our behavior as we move through the corridors of the building to the various rooms.

And the generalization of this principle is that semiotic artifacts, that is, human system created material artifacts, which encode information according to codes that are themselves the product of longer timescale cultural evolutions in the larger social group, become the principle technology of social coherence, coordination, and organization across multiple timescales in human social ecological systems.

So thank you very much.

Temple Smith: We can entertain a question or two before we go to lunch.

AUDIENCE: To me it seems like a very ingenious account of how you might account for evolutionary processes that have gone from the very earliest stages to the current human stages. And yet when you come to the human stages semiotic behavior, which sounds very much like what you might call consciousness, has some big gaps that I think are very important.

For instance, one of the gaps, as you study consciousness there is what is called the hard problem, which is that even though you may have a lot of information about the neurology and about the chemistry of how neurons work, how do you get from that into a real description, not just a semiotic description which is symbolic, but a real description of what goes on that converts all that basic underlying neurological information into the ability to draw in terms -- or the ability to walk or the ability to plan ahead or determine behavior in terms of a self-concept, for instance?

Because a self-concept is such a developmental idea that it begins in childhood and it grows and changes and moves and becomes something other than what it was continually.

Now the idea of recombination of course is a great part of that whole idea of developing a self-concept, but there's that gap between the sociology and the psychology of consciousness. And I'd like to know what kind of understanding you have of that.

Jay Lemke: Well I wish I had a good understanding of that, which I don't. But I'll say two things. One is I'm a bit of a heretic. I don't believe in the mind. I don't believe in psychology insofar as psychology assumes there's autonomous separate explanatory domain of the mind. I think these are all material physiological processes.

And that in some sense the gap -- the gap is not a gap that is as large as it's made out to be. I think a lot of these gaps are exaggerated in the interest of the honor of the human species and they are actually much smaller than people traditionally have believed. That these are still physiological, neurological processes. What is different about them is how they're organized, that they are organized in some sense across these levels of organization, that they do these transformations of information in various ways, and that the enormous resource that they provide us with comes from adding many more levels of these one to many possibilities that we have, many more forms of context dependence that become possible, leading to extremely elaborate networks of conventions about how to restrict that too large universe into something that enables us to coordinate with each other.

The other point specifically about the development of human identity, I did in a paper at Berkeley in March, which is up on my Web site, attempt to use this model to address the development of human identity to some extent. In fact, that Table I that I showed you with the levels of organization and classroom interaction was part of this discussion. And in effect what it said was that one had to look at the development of identity on multiple timescales, that identity is in some sense a kind of integration of your behavioral patterns from short timescales to long timescales, and that in the course of your identity development what is particularly important are the ways in which there is social support, social affordances in your interactions with other people and the environment for the maintenance and renewal and continuity of the elements of your identity over longer and longer timescales.

Chapter 9

Medical Complexity

Clay Easterly
The Virtual Human
Alan Perelson
Theory of the Immune System
Jim Collins
Dynamics in Multiscale Biology
Tim Buchman
Multiorgan Failure
Ary Goldberger
Fractal Mechanisms and Complex Dynamics in Health, Aging,
and Disease
Stephen Small
Medical Errors

Thomas Deisboeck: Good morning. It's a great pleasure to welcome you all to the last big plenary session. The medical plenary session—there may be a reason why it's the last one on the last day. I don't know exactly why. But I think it's not because it's the least interesting one. It's probably the last big frontier for complex systems and complex biological systems.

One of the reasons certainly is that medicine is used to look very, very straightforward and accurately on facts and data. There has been a great deal of reductionism approaches over decades. And lately, because of large-scale projects like the human genome project which nears its end, there is a lot of effort now to get into functional genomics, into data analysis, bioinformatics. And people realize that you only can conquer these big challenges with complex systems approaches. It's going to be very difficult, because most of us in the medical area are educated very conventionally. Conventionally focused, experimental approaches try to reduce the information parameters and get robust results.

So holistic approaches naturally take some time, take some time to write these grants, and take some time to convince the reviewers. So this is still some step down the road. But I think after the initial start of more awareness for complex systems and physics and computational science, mathematics over the last decades, it started certainly to invade economics and evolutionary biology; and now it's medicine.

And so I like to see this as the last plenary session, because it's the last big challenge and frontier. And we're not that far off. I mean, there has been research in the past. Obviously again, going back to the schedule of the conference, on Monday, Stu Kauffman received the award. And as I am sure all of you know, he is an M.D., not a Ph.D. So there are pioneers in that area. And that will help certainly to spearhead the medical effort.

The topics today that Yasha Kresh and I sort of tried to introduce range from theoretical modeling, applied modeling, all the way from immunology, HIV, science, and collaborations with experimental and clinical researchers to the search for fractal analyses in cardiac arrhythmia to very, very clinically focused and very interesting topics like multi-organ failure, all the way to things I thought we don't need statistics for, medical errors—because there are none, apparently not the case.

So it's a very, very broad spectrum. Telling you that there is a vast demand for complex systems science approaches in medicine. And we in the medical area have to learn to make these collaborations to form interdisciplinary teams. And this is one of the great advantages, hopefully outcomes, of this conference: to form these collaborations.

So Yasha, do you want to show some introductory slides?

Yasha Kresh: Thank you. I won't tax you too much. I just want to have my two minutes of presence. Well, I think you already convinced yourself, I'm sure, that the reason this is such a wonderful gathering is that we are all in the same room, and we probably come from completely different bias and orientation. And someone thought of how to actually organize technical knowledge and what embraces it. And if you think all of you probably will find yourself some place here, it's all here. (Laughter) It's in physics.

And then it looks like there is a void over there; there's something waiting to be filled. And you can imagine that I think what's waiting to be filled is medicine, is bounded and is well confined. Maybe for some people it's too confined by philosophers and psychologists. But nevertheless, I think this is a good way of seeing why medicine is such a fruitful discipline to hide all of your biases and put all of your equations into motion.

What makes medicine and biological complexity really difficult is that it's really not a hierarchy, a structural hierarchy. It's really not. It's really a functional order. That's what medicine is, or medical problems are. And as you get older, it's not that we lose components; the components fail. Really systems fail, but more so than that, it's the coupling that you lose. So you start being rather simple. Which is convenient if you're rocking in a chair, but not convenient if you want to drive on the highway system on the weekend back to Philadelphia as I will.

So I think that's what makes it exciting, that these networks are not hierarchical; they are really what I think are really heterarchical. And they are nested. So there is some of that and also some of straightforward hierarchy. And you might be surprised to hear that a cell is actually seeing the rest of the body, not just that the body is made up out of cells. And that's kind of a hard fact to observe, because it just signals to the next cell. But in fact, it really is in the world, some place there, deep inside.

Clay Easterly
The Virtual Human

Yasha Kresh: So with that in mind, our next speaker will tell us about the virtual human. And I guess what he will tell us is that one day we'll be on a chip; and if you're too busy to deal with yourself, you'll just tell your healthcare provider to look you up, find out what needs to be fixed and exactly what your phenotype is, and take care of it, and then call you in the morning, and drip, you know, drip at a ??, even if you have an IP address, maybe even plug yourself into it and take an injection. And then you'll be good for it, because you'll be completely virtually accounted for.

So without any further delay, Clay Easterly.

Clay Easterly: Thank you. It's great to be up in the New England area where you're still having springtime. We had spring, and then we had a really hot season. So the spring flowers I am able to see now are really nice.

An answer to the question, yes, I think so. Okay, so this morning what I'd like to do is just take a little time to share with you what our vision at Oak Ridge National Lab is about the virtual human, and maybe get some of you to embrace that. I don't want it to, for very long, to be just an Oak Ridge vision. Because for it to be successful really, something like this has to be embraced by a large number of us. We'll talk a little bit about what we're doing at the lab, and then about just some very brief thoughts as to how this community might fit together.

Well the idea is that we need to go past Hollywood. Hollywood has got great structure. It can have structure with all kinds of supermen and various kinds of heroes. But if you go into the back and you run the zipper down;

you look inside; there isn't anything. So we feel like it's the right time to begin to push the idea of structure and function, where we link chemistry and biology together.

So the idea in the long-term really is that we would like to be able to have our knowledge base that goes everywhere from on the left side, from data sources that we have now, beginning with the visible human project, and be able to transcribe relationships that take place when we interact with the environment, whether it's a physical force, or a chemical force, or something like that. And the reason is that even though we are developing a lot of tools in the functional genomics, that information tells us about the body's potential; and we've got to find some way to interact with the environment. And I don't see how to do that outside of a very complicated and complex model, which we're calling right now the virtual human.

So the idea is that we want to have really a complete system. And in some sense, we have to be able to work with the cognitive part of the body, because that influences how we respond to these external stimuli. I don't profess to know how to do that, but I think some of you have been dealing with cognitive responses. We think that it's going to be critical for this endeavor for people to be able to retain ownership of their work. And there may be as many as dozens to hundreds, and eventually, maybe thousands of people that collaborate on this endeavor.

So what do we want to do at Oak Ridge? Well, we're in the process of trying to catalyze this concept right now. Now how many catalysts do we need? I think we need lots of catalysts. And maybe I'll show you a little bit why.

Integration. We would like to be one of the integrators. How many integrators will there be? Many, I don't know how many. We certainly are going to be doing some specific modeling. We've housed the International Commission on Radiological Protection modeling system for radiation effects since the beginning of time. And then of course, being a Department of Energy facility, they like gadgets and things and ways to measure. And all of that fits together. But in the main, our role is a catalyst. And you know, we'll be a real tiny piece of any technical work, because it will be so broad and so vast.

What do we need to do? Well, let's say we got together and began the process. Well, in five to ten years, I think we really need to have sort of goals. And these goals would be that we are scalable by age and gender. Some of the tools to do physical parts of that are not too far away. We need to have all, or at least most, of our organs. And maybe we won't have the pineal body and a few things like that, but we'll have a vast majority of organs, and some limited pharmacokinetic capabilities. And that's useful, because some of the sponsors might want to be able to do some kinds of radiation and chemical risk evaluations.

Biophysical properties. Really, are only real sponsor right now is the Marines, their non-lethal program. Well, how do you evaluate new non-lethal techniques, even blunt trauma techniques? We've got to have physical properties of tissues, maybe as many as a dozen different physical properties which aren't being measured yet. And so there is a whole variety of things that will be needed. We'd like to have some specific disease information for simple, a very simple model to begin to describe and portray what this virtual human concept can do. If we can't duplicate physiological tests, you know, we're not going to be where we need to be. So that's an important characteristic.

I think one of the main goals is really to assist in diagnostics, as was heard at the very beginning. Because in the long-term, I see the virtual human as being something personalized, so that each one of us can have our medical history put together, not just in tables, sheets of paper, scraps that the doctor writes on and forgets, turns yellow, but actual functioning of our physiology with the nuances of susceptibility and the various times that we get broken bones, you know, playing soccer as youths and all of that.

I think the ability to fast-forward—if we can, predictive models will be something that's necessary—but being able to fast-forward has a lot of value, because that would help us in education. You know, it could show our children, well here's what happens to your lungs if you take up smoking. Here's how it affects your heart. As we progress in life, you know, we have different habits that we wonder well, you know, I like to eat. And how bad is this going to be for me in the future? How much wine should I drink today? I don't know to what degree our medical knowledge will support these things, but I think those are important visions.

Well, one of the things I've learned—and it's only partially—is that, see, our major sponsor at the National Lab is the Department of Energy. It's becoming less and less of a major sponsor as time goes on. Not because we want it, but they seem to be losing ground in the science world. But you know, the long-term perspective of the Department of Energy is improvement of health and environmental quality. And so that kind of trickles down to us, asking what sort of basis do we have for trying to catalyze this idea.

And so we have a fairly substantial science base, whether it's in systems biology or distribution systems. But really, that's only a small fraction of the kinds of capabilities that will be needed. There are a number of applications that we see are possible. And we are developing partnerships. And with some organizations, we are beginning to write proposals. But it's turning out that there seems to be almost no agency interested in integrative science. And that's causing us substantial trouble.

So in addition at the lab, one of the things I think that will be good to be one of the centers for integration—and by no means the only center, there may be dozens or hundreds—is the fact that we have long-term kind of commitment. That's the only job I have, working at Oak Ridge. And a lot of people are like that. So that you have this long-term memory there. We have a lot of user facilities, and we're quite accustomed to doing that.

The lab is making some up-front investment. And hopefully that will pan out. And for some sponsors, some perhaps military sponsors or private organizations, there's going to be a need to compartmentalize some aspects, classified aspects or business confidential. And that's going to be important we think in the future. So there are a variety of resource areas. So we're starting to begin to think about this big idea and break it down into smaller parts.

So whether we look at measurement systems, information mining, and computational hardware, these are some of the things that are going to be needed to build this resource before we can really make progress. Now there's probably a lot of other items in this visual that are needed and maybe a few that we don't need. But what we're thinking at this point is, not being able to find agencies that really want to tackle this yet, there are smaller opportunities for resource building. So that's one of the things that we're doing is looking for smaller opportunities.

Why can we do this now? Well I think one of the main reasons is that we've had huge developments in computational resources. We're promised Internet II and beyond. It depends maybe if Al Gore is reelected. We've got huge medical measurement in analytical devices. With the advances in computers now, things that I worked on in my dissertation with don't even exist anymore, because you can make almost every measurement you need to do—like I did in photophysics—a lot of those, you can very much automate it. The computer can be used as an oscilloscope. And there's a lot of new chip technologies that are making data analysis and data tools much more accessible. So they will be able to reach into the cell even and take data with our new nano-scale technology.

And even I've seen agencies are beginning to cooperate. And I find that really rewarding. Even our Department of Energy collaborated with the National Science Foundation, looking at computational needs for the future. So those things are very positive.

So we've got what it seems to me is just an enormous array of information. We've got billions of dollars of research going on each year now, a lot of information sources. But we don't really have the tools to deal with this. We don't have the infrastructure. And really that's the kind of thing where we think is a lot of need. Well, a long time ago, a few centuries ago, people were faced with the same problem. What did they do? We got all of our nice Latin names for all of the animal kingdom and the plant kingdom. And so there's an analogy here. We've got to find a way to provide this infrastructure.

So our view—and I hope to share that vision with others and maybe modify my vision—is really that we need to simulate biology, so that we can really get forward in advancing our knowledge about the complex biological systems, and then needs integration and infrastructure. I think those are elements that are very vastly underrepresented in science today.

Well, to bring this forward to some agencies, last year we had a meeting at the National Academy in October. And we had about forty-five attendees, a number of individuals. We have one here, Alan Perelson, who was present. A number of individuals spoke from academia or labs, and we had responses from different agency representation. And it seemed at the point that there was a consensus that we needed to move forward, move forward cautiously. And the Department of Energy sort of took the banner and said, we will begin to shepherd this idea. And so we're waiting on that eventuality.

We had a first road-mapping workshop in November, just after this academy. We really worked hard to get physics guys and biology and engineers and philosophers and medical guys together. And it was a tremendous opportunity. And one of the main things that we learned there was that while there were differences in how we thought about making something like this virtual human happen, it was a very significant consensus that we had to be very deliberate about the planning process.

We really don't want to go and have another spike event where there's a big flurry of activity and people get very over-encouraged. The Senate is lied to, and then you can't deliver. You know, like fusion. And there have been a lot of big projects that do that. So we've got to be very deliberate about what we do, in order to stimulate something that's going to have a good long-term life and be productive in the way that we want it to be.

So what we are doing in the near-term is coming up. We'll be having a number of focused workshops. The first one will be in July. And that will be in support of our sponsor the Marines. And we'll look at kinetic energy effects. So we're going to be looking at models and data that are being used, archived data, and the big issue of how do you connect the various levels of information from the physics of some kind of a soft kinetic energy round, to how does that fit in all of the various models—you have gelatin models; you have clay models. You then have some of what are called three-rib models. And you have some data on a pig and some data on cadavers. But none of those are connected together. So that's going to happen.

And then we want to plan a series of conferences where we get together for a week at a time and begin to build a process for laying out how shall we really as a scientific community attack this kind of big issue? So I'll give you a brief glimpse of what we're doing at Oak Ridge.

So we see that you can develop maybe some infrastructure that allows communication between models, that you can attempt to directly link models, or you can develop an infrastructure that will serve as something for the future, assuming that we make specifications for the future that everybody abides by. Well actually, I don't think any one of these is going to be successful, because there is a huge amount of legacy information. Individual models are in the range of thousands or maybe tens of thousands. And they are all very focused. And they don't need partners.

So I'll just talk very briefly about what we're doing. We have a laboratory directed R & D program where we are trying to link a model from one of the fellows at Vanderbilt, a model of lung edema with a very simple heart model. See, and neither one of these needs the other. Roselli doesn't even need oxygen. He doesn't need blood supply, because he's focused on response to some kind of stimuli and the alveoli level. Well, you know, that's the result of twenty years of research. So you don't think it's wrong to think well, we'll just make everybody redo their work or something. We've got to find ways to make use of people's entire careers.

And then, we're attempting a much more complicated system with a problem-solving environment. We have a proposal that's being reviewed right now with a couple of other organizations where we are looking to build a problem-solving environment. So the idea is, really, begin to get a problem-solving environment that connects some anatomy and physiology with models. So again, structure and function. There's our structure.

So what are the problems that we face? Many. We've got legacy codes. We've got C, C++. We've got Fortran. And Fortran doesn't fit in very well. We've got to retain ownership of these models, because the model owners are the ones that are improving, developing. And it's just the right thing to do, is to retain that. So we've got to have this means, this opportunity that allows new models to enter into the environment. And we've got to be able to take people where they are. You know, that's kind of the facts that we've been finding out.

So what we see are needed really are these standards, whether it's XML-developed, as Jim Bassingthwaighte has done at University of Washington, or these wrappers, COBRA, CORBA, component libraries of a variety of things in what we'll call our virtual human component library. And I'll show you a diagram of what we're doing right now. We're building on SCIRun, which is a problem-solving environment developed at University of Utah by Chris Johnson. And it's kind of like a really elegant MATLAB if you will. It's a visual language. It's got solvers. It can do partial differential equations, very elegant.

So we're developing what we call a portal right now at the lab. Really, it's going to be a window to many tools, whether it's just a training- or a simulation tool for very small things, or all the way up to something that will take supercomputing and network interfacing. So if you could visualize here on, sort of on your left side, we're talking about having sort of more simple sort of things. And on the right side, we have more high--end things, where as we're going to be opening up a website portal, where you can access through your web browser and do fairly simple kinds of model work. Or you can go all the way to needing high-end computation and visualization with what we're calling the problem-solver environment.

Well, a component library would take solvers for ordinary differential equations or partial differential equations, depending upon what the needs are. And for the sort of low end, we'll have a client-server relationship which uses Java RMI. We plan to have prototypes there using this early model that we'll show you here.

This will be like the opening window. And then we'll move, as an example, to a system of models where we have a very simple model of a left heart circuit connected with this fairly complex model of Roselli from Vanderbilt. What we have operating these two is an old legacy code that actually was a Fortran code that did react to accidents, sort of recycling code, because it has input/output capabilities.

So you can see now that what we're doing is we'll do a simulation that you can operate from your PC. And we can move around in terms of initial properties, states that you want to investigate. It could be a tool for learning, looking, having students go in and look at the role of blood pressure on the response of the lung tissue, whether it's interstitial pressure or whatever, and be able to have back on your own screens, information. This is just one example of what we think we'll have.

When we need to go to more computer-intensive modeling, we really need—and we've learned this—we really need to have what's called a problem-solving environment. Now problem-solving environments are like sort of a complete package, or almost a complete package. It incorporates a lot of features, whether it's an expert system, provides you assistance in working through and putting together your models. And it facilitates output, so that you don't have to do everything. Problem-solving environments are used in a lot of settings now, whether it's in science or industry. And it provides sort of a transparent means to deal with a lot of hardware and software sensors. So you may have a problem-solving environment for a particular experiment, maybe for a process, an industrial process. It could be for science.

And of course, that's what we want to do. We want to focus on our science. So a generalized problem-solving environment, which is what we're going to be putting up here shortly, builds off of the SCIRun, which is down at the University of Utah. And you can see it's kind of a visual object-programming language. It's got a lot of solvers for differential equations, both ordinary and partial. And it facilitates modeling to a great degree. And it's been used in a number of settings, a number of scientific settings.

We can start at your own PC and work through servers, where you may begin to take a couple of models, and you're trying to put them together. SCIRun is part of that. And then we go out to NetSolve. And NetSolve says, okay, well I understand you need this kind of software, this kind of hardware, and it exists at university X, Y and Z, and I'm going to go out and do that. But you don't have to talk to it. It's a process that's taking place, developed at the University of Tennessee.

Well, so we have explored a couple of options, whether it's modeling that you can solve right on your personal computer, or assisted by servers, all the way up to incorporating a problem-solving environment wherein one needs to be able to access supercomputers, parallel systems. And you may need compute power. And so, by connecting this with NetSolve, NetSolve will go out and find that compute power for us.

And the bottom line is that there is probably no single approach that's going to be needed or successful or serve all applications. We kind of ran out of steam with the Internet, the Java-related modeling, because that has limitations. It will get better as time goes on, because the capabilities will expand. But right now, it's difficult to work in a Java environment. It's too slow. It's good for certain purposes. But in order to do serious computing, we've got to go to a problem-solving environment, much more high-end.

So the problem-solving environment that we are putting together right now is really based on a couple of things that are really existing, whether it's virtual, a visual component composition library and intelligent resource, which is encompassed in SCIRun and utilizes a number of things like wrappers, COBRA, which is a wrapper that allows you to use legacy codes, if you will, and being able to be interoperable.

Again, why should we start this and why are we doing this? One of the big reasons is that there's really a confluence of technologies right now that's allowing this to happen. It's giving us the opportunity. And I think we probably will have to consider virtual human as Big Science, in a way that's analogous to high-energy physics programs, which are Big Science. It takes a lot of people to do things over a long period of time. What we don't have is we don't have a single big instrument, a big machine accelerator or something. But we have a distributed kind of instrument, which is really locked up in a lot of our heads and in the models that we are developing and have been developed over time. And this just illustrates a little bit of this problem-solving environment.

Okay, what about sort of the next generation? Okay. So the next generation is, maybe, let's think of developing a problem-solving environment, which is specific to a biological need or application, whereas SCIRun was developed to be multi-purpose and general. So we've gotten together with Chris Johnson, Jim Bassingthwaighte and a couple of other investigators to respond to an NSF information technology call. And this will look at vertical integration of the heart.

So by vertical integration, we're thinking, you know, going from molecular level up to the organ level. And this is going to involve a number of elements, some of which will be very much involved with complex systems, where you're talking about ion currents and ion motion and scaling processes and transport. And it goes all the way up to muscle motion. And some of the latter parts probably get out of the idea, or the need, the immediate need for complex systems. But there will be a lot of need in the early parts.

So we'll begin to look at modules of information. How do we design a modularity? Looking at a lot of time-varying operators. So there's you know, complexity coming in there. Some linear, some nonlinear response characteristics. Of course, the analytical goal really is to understand these integrating principles so that we can actually make better steps in the future. This is a step in the pathway to the virtual human.

So, we come to ask these questions about complex systems and the role in this virtual human idea right now. Do we need it for kinetics of motion? Breathing and coughing and walking -- probably not. I don't know that we have enough data to actually turn those things into a much more highly refined descriptions. Mass transport? Well, maybe not food ingestion. Probably not heat transport. But, okay, oxygen transport in motion, beginning to be. And then in information, it's very, very clear that complex systems are a vital part. This is not to say that a whole organism modeling won't be described in a complex way. It's just that I don't think we have the information yet to begin looking at that. But I think what we've seen is -- really just from our preliminary visions -- is that there will be a lot of different applications. And each application will have a certain degree and requirement of specificity. So that's going to guide, I think, development. For blood trauma and things like that, I don't know at this point that we can benefit. I mean, we have very crude data when you think about it. The old data from the Atomic Energy Commission where they had goats exposed to blast overpressure. And you know, did the goat die or not die? Well that's real simple data. We need a little bit more complex data to be able to model advanced processes. But when you get into biomedical data, EKGs, EEGs, time series analysis of our biochemistry, very, very clearly a need for chaotic analysis.

We're finding that with EEG readings now, we can predict epileptic seizures thirty minutes to an hour ahead of the event. And in fact, the same kind of prediction is done for seizures with EKG data, which is rather surprising. So there's information in the noise that we just don't fathom yet, because the tools that we've used, mostly our eyes on data, and we've gone to great lengths to smooth out the data. It's kind of like throwing out the baby with the bath water, but we didn't know that. We're learning that now.

How do you approach these problems of building the virtual human? Well, we know that most data is just linear accumulations. And we're organizing it better and we're providing relationships. But it's really old-fashioned still. Do we do it top-down or bottom-up? And there are proponents for both. You know, there is a lot of benefit I think to be gained from top-down systems approach, the blunt trauma issue. A lot of applications can probably go from there.

Well, how do we do it from the bottom up? Yes, that's very intellectually pleasing, but our database to do that is very small right now, and we can't make much headway. In fact, you know, I did my research in a research group that said -- I come from a physics background -- we're going to learn about the interactions of small molecules and radiation, and be able to eventually predict disease, cancer. Wow. That research has been going on since the early Fifties. And we're not even -- there's no prayer of doing it at this point. But I think that our functional genomics is going to move fast, much faster than that. Because atomic level interactions are just so vastly different than biological interactions.

So I think there is a role for complex systems approach in looking at the different scale levels, bridging across scale levels. My perspective is that there's about $20 billion worth of research going on in the United States per year. And pretty much, those are all stove-piped. That is that you can take that research from a lab up to the funding agent, and you can just transfer all of that information. And the funding agent might include your review committee. But there's really not much bleed-out. And certainly, it's not possible, even if you could have bleed-out in a mass way of all of that data to other people, we're just not able to comprehend that much information anymore.

And so we need to find ways to allow us to use information that everyone else is gaining, but that it's impossible to know personally. And one of the ways that I think that's possible is through the medium of some set of interconnected, interlinked models and data, where new information can influence our overall vision.

Okay, what can we do here? And I think this is my last slide. Well, you can join this catalytic effort. And it's not a fast catalyst. It's a slow catalyst, okay. You can identify sub-models in the areas where complex systems work. And you're already doing that. You know, the [conference poster] walls are full of those things. And join in this process to really help develop the principles for integrating in the future, for all biological kinds of research, so that at some point in the future, these stovepipes will kind of break open, and all of that data will have ways to be turned into knowledge, not just for a select few, but for larger numbers of us.

And that really concludes my prepared presentation. And I'm open for questions. [Applause]

GARY: Hi. My name is Gary and I'm from Cook County Hospital. I enjoyed your talk very much. I apologize if my question is already answered. It stems out of my ignorance of the analytical tools that you've talked about. But I'm wondering whether or not there is an agent-based component or track in your project. And the reason I ask that is it seems to be that from the flexibility modularity available in that technique, if you could establish a common descriptive grammar that can be used for individual domain specialists to translate their information into this common grammar, it might lead to a great deal of modularity and flexibly in collaborating it.

And one of the reasons I ask about agent-based modeling is one, that's what I can do. I can't do any of this other stuff. But the modeling workshops that I have been to have emphasized possible benefits of parallel tracks of model development. And if the parallel, I mean different approaches could actually do the same types of results, it could lead to improved assumptions regarding validity of your models. And so the modeling process artifacts might be cast out. Could you comment on that please?

Clay Easterly: Just a little bit. It's a little outside of my area of expertise. But the Physiome Project has begun to do a little of this, where people that are brought together to work on a particular organ, agree on how they will relate to one another. I think your question is really targeted at the need that we've identified. And we believe it's going to be a strong need to have these Gordon-like conferences where we can sit together with not just our modelers, but the people that are actually going to be taking information and developing information, where we can slowly develop sets of interaction rules, if you will. And I think it's critical before we're able to move forward in any kind of fast way.

And my belief is that without these kinds of agent-based relationships, getting down to the basic research level of all of the NIH grants and things, I don't think we'll be able to do what we want to do.

Nigel Gilbert: My name is Nigel Gilbert and I'm a sociologist, which perhaps accounts for my question, which is: isn't your virtual human going to be very lonely?

Clay Easterly: No, because my virtual human is going to have a spouse and three children and a cat. [Audience laughter] One child is thirteen and one is twenty-one and one is twenty-four.

Nigel Gilbert: The serious side of my question is that of course, is it in fact sensible any longer to consider health, indeed medical issues, without considering the kind of social situation of the person that you are thinking of? And it was conspicuous to me that really there weren't any what we might call social variables in the kind of things that you were talking about, nor really any space for such knowledge.

Clay Easterly: We actually have on that, on of my figures -- I don't know if we can back up to see it. But we have a group at the lab that deals with social issues. And we've identified that as an important area to begin to explore. Because in a sense, this idea is like the health card that sort of failed during the last couple of years in terms of gaining approval within society. We certainly will have to deal with those issues.

There are many complex social issues associated with having a personal model. My personal belief is that the value is so great that we need to struggle to develop resolution to those issues, rather than say, let's not embark on this initiative. And it's an area that is very critical. It can kill an initiative at the onset, if it's not addressed properly. And so I think you're a good candidate to be sitting together with us as we begin to map out how to address these issues. We don't want a big initiative to start before those things are thought through.

Yasha Kresh: You could investigate that. If you thought that the genome was a big thing; the physiome is beyond the big thing. So there's much work to be done. Thank you very much. [Applause]

Alan Perelson
Theory of the Immune System

Thomas Deisboeck: It's a great pleasure to introduce Dr. Alan Perelson. Alan is a leader of the Theoretical Biology and Biophysics Group in Los Alamos. And his research interest is in modeling the immune system. He has quite interesting collaborations. I think one is with David Ho in New York. And I hope to hear about their work on HIV.

Alan Perelson: Well, thank you for inviting me. I actually had planned to speak about agent-based modeling, which I will, and some aspects of the immune system. But Yaneer asked me last night if I could say a few words about the work with David Ho and HIV. So let me start more with just a story about that.

So for those of you who are not familiar with this story, you all know that AIDS, we believe, is caused by the virus HIV. And if one looks at the time from infection with this virus, until one typically comes down with the symptoms of AIDS, that's running on average of about ten years. During that period, if one tries to measure the amount of virus in a person, or the decline in CD4-positive T cells, which are cells of importance in the immune system, what one sees is it's a relatively slow process. And in fact, when we started this work around 1995, the technologies for measuring the virus were just coming online.

And typically, one didn't see very much change in the virus, if one looked during the intermediate years of infection. So typically what you would see is, if you caught somebody early after they were infected, you would see a lot of virus in them for the first few weeks. And then the virus level would go down, and in many assays, also undetectable with new PCR-based assays. After two or three months, the level of virus would stabilize in a person. And as far as we can tell, it stays relatively stable for years.

And so one question around this time is should you treat people? I mean, somebody in their first year, second year, fifth year. Their virus wasn't changing much. They are still relatively healthy in terms of not having opportunistic infections, and it looked like the virus wasn't doing very much. So the work that I got involved in with David Ho is we asked whether or not that's a true picture, whether or not this is really a very slow process. There are some slow viral diseases, things like BSE, scrapie. A number of true slow viruses are herpes infections.

And the mathematical work involved looking at experiments where people were given new drugs. Drugs were coming online for HIV. The protease inhibitors that you now hear about were being put into clinical trials. I had been developing models, simple, sort of predatory/prey dynamical systems models of HIV interacting with the immune system, and in fact, gotten in contact with David Ho's group to get data about that.

And he sent me some data from what's called a phase one and phase two clinical trial. So there were one of the new protease inhibitors given to patients, and they measured the amount of virus in their blood using these new PCR-based assays. And what was noticed is that, first, it was a placebo-controlled trial. So for people who had been given placebos and their blood was sampled every few days, the virus didn't change, an essentially constant level of virus. And the patients that were given the drug, what we saw was that the viral load fell very dramatically. And in fact, it fell about two logs, a hundred-fold over the first two weeks of therapy. These drugs weren't all that potent. And in the majority of patients, the virus would start coming back up after about two or three weeks.

And so from the point of view of just looking at the drug itself, this drug wasn't "a cure," because after a few weeks, the viral loads were starting to come back. But the real impact and the insight that I had is you could do a very simple analysis. One, you could say before you gave the drug, patients looked like they were in some sort of quasi-steady state. So a timescale of a trial which was about a month, the virus wasn't changing. Although we know over ten years, it does something.

Secondly, if you give a drug, you're perturbing the steady state. So you're perturbing the dynamical system. And the simplest model one could make, which is one a freshman taking calculus could do, is we just said you know, the virus is being produced at some rate, and it's being destroyed, being cleared. And let's assume we have a perfect drug—which we knew we didn't— just to get some insights. And we said, well, if the drug cuts off the production of virus, it could just be cleared. Let's just try an exponential decline. And we fit the data to a simple exponential. And over the first week to two weeks, it fell exponentially. And you could measure the rate of decline and get a half-life of the virus. Which totally surprised the HIV community: the half-life turned out to be two days.

And so this was a virus that people thought nothing was happening with over a timescale of years. You come in with an imperfect drug which is blocking some viral production, and it's being cleared with a half-life of two days. And that has to be a minimal estimate, because we know there was some original production.

So conversely, what that says is in a quasi-steady state before you gave the drug, it had to be doubling in people every two days, and the body was simultaneously clearing it. So this was viewed as a real paradigm shift in HIV research, because it changed the view of this being a very slow disease to one where there is rapid kinetics. One could then go and actually compute how much virus was being produced.

So I could actually measure now how much virus was in a patient. We had those measurements. I told you it was just doubling every two days. You just do the calculation, and it turned out, in the minimal estimates, there were about a billion virus particles a day being produced. Well that doesn't sound like very much. It's less than a gram of virus or something.

But if you compute how fast the virus replicates and how error prone it is, it is a virus that on average has about a .3 probability of making an error in replication somewhere in the genome. So that's per virus replication. And now you have ten to the ninth viruses replicating. And what it told us is that every single mutation in this virus was going to be tried everyday, in fact thousands of times. The virus has only 10,000 base pairs. And so every three viruses, one of them is going to have a base change in it.

And so people were developing drugs at this time, where a single base change would lead to resistance. And there was one large drug developed by Merck. It was put into clinical trials, required one base change. And in three weeks, every patient who had gotten the drug essentially was resistant, drug-resistant. So the trial was stopped. That was a $300 million mistake, based on thinking this virus was real slow, and they targeted a reagent that had never been seen to vary.

And then we went on and developed more sophisticated predator-prey models. Virus infects cells. Cells produce virus. Virus kills cells. And it turns out the decline in virus really isn't the simple exponential. And we drove more precise experiments in fact, where people were being sampled every two hours after they were given the drug. And on a very fast timescale, we were then able to identify how rapidly cells are being killed by the virus, and also we estimated how rapidly virus was being cleared. You'd be surprised at a minimal estimate then. We thought the virus was being cleared with a half-life of six hours or less, with the identifiable lifetime of cells that were producing virus of order a day or so.

So now, for the first time, we could estimate how rapidly cells were being killed in people. So I actually had a slide of that. So this was from a paper we published in *Science* in 1996. The data are these viral load measurements actually, the amount of viral RNA in a person's blood. And these are two separate patients. They were being sampled every two hours at the start of the experiment. The solid line is the fit of a very simple couple of ODEs. We started out as three nonlinear ones. And we linearized, because we thought over the timescale of the experiment, we could make another quasi-steady state assumption. It has two eigenvalues. You can identify the two eigenvalues. And it turns out from the biology of the model that one of them is associated with the clearance of the virus and the other for the lifespan of the cells. And as you can see, the theoretical predictions and the data points agree quite well.

Now, you know, this is just theory, very simple stuff. And one really couldn't take this seriously, but the experimentalists did. And we went back and designed other types of experiments to verify whether or not the timescales here were correct, if the virus was really being cleared on a timescale now of six hours or less, and if cells were being killed on a scale of a day or so. And both of those predictions in totally independent experimental observations have been verified.

We went on and then developed more complicated models, saying this is only a perturbation over the timescale. Look here: it's seven days, looking longer. And unfortunately for patients, that rapid fall with the one-day half-life doesn't continue, and the curve is all bent out until what we call a much slower second phase. It varies here between about seven days and a month. Again, we built additional biological models to identify what processes were driving that second phase. The solid lines again are theoretical predictions based on ODE-type models. And we believe this involves other biological processes, either cells that are living much longer periods of time that produce virus was what we had published. And we are actually still working on alternative hypotheses for this second phase.

And so then the bottom line of all of this is because of the repeatability of the viral replication, where the real impact was, it convinced the drug companies and the physicians, after about a year or so of this work, that therapy with single drugs was not the way to go. And I think we had a part in moving into this era of combination therapy where one now uses three drugs or more. And the impact of that, if you can see, starting around '96, when practices changed from one drug therapy to three, the number of cases of death by AIDs—which is this red line here that is hitting a maximum—has declined dramatically. So it's an interesting story where some very simple mathematics really has changed the course of medicine.

It led to some very high-profile papers. Our first one showing that the virus replicates rapidly was published in *Nature*, and it was about the most highly cited paper in '95. This direct virology was three papers in *Science*, in *Nature*, and we had a total of five, with some other more esoteric aspects of the kinetic aspects of this disease.

And it's also engendered a whole change in philosophy in at least the virological community. Where when we started I had trouble talking to any of the groups, trying to get data from people, to a position now where I probably have fifteen collaborators. People are collecting data, publishing it for the use of modelers. Almost every major HIV group is now associated with somebody with some mathematical skills to analyze data and drive model-development. And this has all been in the course of about five years. So it's been quite gratifying.

And work is still continuing. It's moving on into a variety of other viruses. There's work going on now in hepatitis B and C, some other more esoteric viruses, cyclomegalovirus. A whole set of chronic diseases where timescales were thought to be ten, twenty, thirty years, where we and other people now have shown that there really are events on multiple timescales with things happening as fast as hours, even though these diseases can take twenty or thirty years.

So now let me go into something that hopefully is of more relevance to your daily life. I hope we don't have any HIV-infected people here or any hepatitis-infected people. If we do, I'll be glad to talk to you later. But let me move now to an application of the agent-based models of the immune system. And I think like in the field of neural networks, that early in the development, people were building models of neural systems. And they had models, and they ran them on computers, and they did all sorts of things. Well people in immunology have been doing the same thing, including my group.

We know a little bit about immunology. We build simulation models. There are probably five different simulation models that people have done, or people who write ordinary differential equation models of the immune system or nonlinear differential equation models. You can get oscillations. You can get multiple steady states. And you interpret them as memory, disease, you know, all of the standard stuff. But it's just sort of game-playing. I mean, we can all do this. You can all write equations, and you find steady states. You find some dynamics and you associate it with something.

So as in the field of neural networks, which I think only took off when people tried to apply them to real-life problems and show that you could do very serious pattern-recognition problems with simple things like Hopfield nets, what we are trying to do is ask, can we really use immune systems models to say anything of practical interest?

Now what I would like to show you today is an example relevant to trying to develop vaccines for viruses such as HIV, which constantly vary. We've not been very successful in developing vaccines for any type of pathogen that changes, that's evolving on a reasonable lead, rapid timescale. And the one place where we've been trying to do that is in influenza. Influenza varies from year to year. So what do we do? We try and chase it, and we make a new vaccine every year.

And for many of us, at least for me for example, I mean I try and get a flu vaccine every year. My employer gives it to me. My kids get it. The local hospitals give out flu vaccine for free. And we try and constantly do this. It turns out if one looks in the medical literature and asks, is it a good idea to get flu vaccine every year? That's not so clear. And in fact, it's only been recommended for people age 65 and older until this year. And this year, the CDC, the vaccine committee has lowered the age to age 50 and over for people who should get flu vaccine every year, plus people in high-risk groups, if you have asthma or respiratory problems.

Now let me show you data. There are two large clinical trials—one in the '70s and one in the '80s—that compare how well the vaccine does. It's called vaccine efficacy, which is really one minus the fraction of people who come down with the illness, for people given vaccine twice. You got a vaccine last year, and you got one this year. So sort of the idea was whether or not you should get an annual flu vaccine, relative to somebody who just walks in off the street and got a flu vaccine for the first time. The horizontal line at one is the break-even point. Anything greater than one means if you've had multiple vaccines, you've done better than somebody who has had a first-time vaccine.

So the first study is the Hoskins study, which showed that in 1974, people who had gotten the vaccine last year and this year only did as well, sort of 30% as well as people who were first-time vaccinees. So somehow it looked like there was a negative effect of having gotten vaccinated the previous year.

A much larger trial in the '80s ran over five years. And you can see in two of the years, it looks like there is some extra benefit. One year there was some substantial extra benefit from having multiple vaccines. While in the other three years, there wasn't much. And in fact, in one of the years again, there was this very strong detrimental effect to being vaccinated the previous year.

Within the context of flu, there was an interesting nonlinear effect that had been observed in the 1950s in trying to develop vaccines for flu. And that was a population of people who were immunized with a vaccine. And what was noticed is that in the older set of people in that population when they were tested, it was noted that they did not generate antibodies, which are a protective molecule, to the flu vaccine. They generated antibodies, but they were antibodies to an influenza that had moved through that population in a large way a few years earlier.

So rather than responding to the vaccine, for some reason, they responded to something they had seen in the past. It was given this nice name called 'original antigenic sin.' And the people who were younger and hadn't seen this prior infection responded correctly. They've developed antibodies to the vaccine. So somehow there was a historical effect going on. And if you had had a prior infection, sometimes it looked like your immune system got things wrong and responded inappropriately.

So a Ph.D. student Derek Smith, he's a computer science student, University of New Mexico, and Stephanie Forrest of the Santa Fe Institute and one of my long time collaborators in building simulation models, and I tried to look at this process.

And in the interest of time, let me just show you a quick cartoon of how we believe the immune system does pattern recognition. It's a process called clonal selection. And this is sort of the basis of our model. So let me give you one biology slide. You may know that your immune system contains white blood cells, lymphocytes. And there are two types, T and B cells. T cells are the ones infected by HIV. I'll talk today about B lymphocytes. They are the ones that make antibody molecules.

There is a wide diversity of these. They come out of the bone marrow. And on their surface, they have a Y-shaped receptor molecule. And to a large extent, each cell has a different receptor molecule. And I've labeled them here. Cell 17 has receptor 17, cell 23, receptor 23, etc. And we think the diversity in the mouse is about ten to the seventh different receptor types.

In humans it's not known. It may be as high as ten to the tenth on the different lymphocytes in your body. But by and large, you can think of each cell as having a different receptor. And by chance, when something foreign comes in, the hope is that it will find a receptor that matches it. If it does, as in this case of these red hex nuts—which many viruses sort of look like—and by chance it interacts with cells expressing receptor 23, those cells go aha, something foreign is out there. And they get activated. They divide, clonally expand. Some of their progeny differentiate into something called a plasma cell, which is just a fancy lymphocyte that secretes large amounts of antibody molecules. And the antibody is just their cell surface receptor made into a soluble form. So it's the actual molecule that you know binds the red hex nut. And some of them go back into a resting state again. And those have been called memory cells. And they stay around until other encounters go on. So that's sort of a crude picture of what's called the antibody or humoral response.

Well, our student Derek Smith implemented this in an agent-based model. So an agent here was a B lymphocyte. B lymphocytes have receptors on their surface. Receptors in this case were implemented in terms of binary strings. I'll just show you a quick little picture. Strings and simplest representation could be binary. Everything that's foreign, like the molecules on the surface of a virus, would also be a string. You could use a string-matching algorithm to determine when something bound to the receptor. We know that the matches need not be complete. For binary strings, you can just do an X or compute the number of positions that match.

You can make it more sophisticated. You don't have to use binary alphabets. And in fact in simulations that we ran, we used four-letter alphabets on strings of length twenty. But those were for some technical reasons to try and mimic some of the properties of the clonal selection process. And once a cell gets stimulated, it followed those procedures that I showed you. Its colony expands, makes progenies, secretes antibodies, some become plasma cells. So one can make a direct agent-based simulation of this.

Well, how does all of this relate to this original antigenic sin and flu and everything else? Well, let me just give you one other part of the story. The way the immune system works and what we implement in the model is this process of that cartoon clonal selection, where each cell has a random receptor. About twenty years ago, a colleague George Oster at the University of California at Berkeley and I did some calculations to ask, is this a good strategy for the immune system? Can you really recognize large numbers of pathogens just by making cells with receptors at random and hoping that by chance, things will bind?

And we introduced the idea of what we called shape space. And here, a point in this shape space is meant to represent the shape of a receptor molecule. We believe that as in lock and key mechanisms for immunology, that there is a complementary shape that it binds to. So the idea here is there is an antibody or B cell receptor. There is a complement in the center of that ball, which is the exact thing that the receptor interacts with.

And then for the immune system to work, each receptor can't only match its exact complement. If it did, you would need a receptor for every possible molecule one could ever construct, and it's just too many. And in fact, we know experimentally that if you take an antibody, not only does it bind to "the complement," the thing that stimulated it, but nearby molecules, things with similar chemical shapes, may also interact. Immunologists call this cross-reactivity. And so we've represented it here by saying this antibody recognizes not only one thing, but actually a whole ball, a whole region of shapes. And that region, we just said just has some volume in some abstract shape space.

And the idea is if you make antibodies now with random shapes, as long as these recognition regions, recognition volumes, cover the whole shape space, then you can recognize anything. And it turns out, you can do an experiment to figure out what the volume of one of those recognition regions is. And that's by finding a B lymphocyte, as we said. It covers that little volume and throw antigens at it. Throw foreign things at this immune system and ask what's the probability of stimulating that B cell. What's the probability of landing in that ball? Sort of like doing a Monte Carlo evaluation of multi-dimensional integrals. And it turns out that that volume is about ten to the minus fourth, ten to the minus fifth in total shape space, which is sort of the probability of hitting that ball when you do an experiment.

Well, as I told you, even in the mouse, the immune system has about ten to the seventh randomly-made shapes. So it means you have a coverage of this shape space with balls of that size of order a hundred-fold coverage. It's a great recognizer.

Okay, so what goes on now in flu? Well, within the case of shape space, let me say I have now these cells, these antibody-secreting cells. Their shapes are these Xs, sitting in this shape space. It's actually multi-dimensional, not two-dimensional. And I throw in a vaccine, throw in something foreign. And now rather than dealing with complements for the purpose of this diagram, let me say that the B cell that recognizes that vaccine is exactly at the center of the ball.

Now because of this notion of cross-reactivity, in fact, that vaccine won't be recognized by only one B cell, but nearby B cells, and this whole world of ten to the seventh or so B cells. And all of the B cells that recognize it are inside. Now we call it a ball of stimulation. So all of these different B cells might respond to that vaccine. And this is somewhere, you know, a few weeks into the simulation, where the density, rather than being light, as the outside, is now much higher, because these B cells have proliferated the vaccine-made antibody, etc. So that would be the result of this.

There's also a little bit of contraction in there, which is a process called affinity maturation/somatic mutation, which I won't get into. But it's something that in simulations—our organizer Yaneer and I actually had some of his students simulate it a few years ago.

So that's what happens when you get a single vaccine, and everything works fine. Now what happens when you get a second vaccine? You've had your flu shot last year, and now in the context of this model, the year goes by, and you get another vaccine, which we'll call vaccine #2. Well, flu varies from year to year. So the second vaccine most likely or may be different than the first vaccine. It's moved over a little bit in this shape space. Now you'll notice there is an inner section region here. And in that inner section region between these two balls of stimulation, there are cells stimulated by the first vaccine that are in this ball, that have been amplified, really rearing to go. And they recognize or can respond to the second vaccine.

Well, what is the immune system designed to do? Well, after you've been vaccinated once, if it sees the same thing again, it's supposed to eliminate it. So what happens is you give them a second vaccine, and if it's close enough to the first one, all of the cells that have been stimulated to the antibody that they gave, do their job. It eliminates that second vaccine. And the net effect is all of the cells out here, which should have grown up by vaccine #2, never really have a chance to, just because the molecule was eliminated too rapidly. So the immune system is doing exactly what it does. But it's not what the physicians who vaccinated you want it to do. Now that actually depends on the distance. If these two balls are very well separated, you won't have this effect at all.

Now what happens when you come in, or what happens if you're exposed to a real influenza virus? Now, the people who designed vaccine #2 are hoping that that's pretty close to the epidemic strain. And sometimes they are right, and sometimes they are wrong. And these are just some cartoons of possibilities. So here is one possibility, where the epidemic strain is sort of pretty close to vaccine #1 and vaccine #2. In fact, it's encompassed within both of their balls of stimulation. And that will get eliminated very, very rapidly. Those two vaccines were great for you. You're in the ball of recognition of the first vaccine. You have all of these high levels of B cells to antibody. The encounter with the second vaccine actually caused a little bit more of clonal expansion in those cells. The epidemic strain gets eliminated real quickly.

However, the vaccine developers aren't always right. And they may be off a little bit. So maybe the epidemic strain is a little bit further away from vaccine #1 and vaccine #2 than you hoped. And in this situation, we think that one would find real negative interference to be of a sort that I showed you in those clinical experiments. So, there's basically no expansion of B cells that will recognize this epidemic strain. So it's as if you almost have not gotten vaccinated, because none of the cells here are expanded.

Now if you look at someone who did not have vaccine #1, then when he had gotten just the second vaccine, it would have given exactly the same response as over here. I mean everything would have grown up. The vaccine #2 would have just been a first-time exposure. It would have caused memory. And you could have a high density of things here, so the epidemic strain would be eliminated. So these, we think, are the situations that we saw in these clinical experiments, where people who had been given vaccine a year before might not have survived well. And as you can see, it all depends on distances.

And so what we did is we said, well these are just two-dimensional diagrams. So we took our agent-based model, built everything with these repertoires, and then exposed them in this typical experimental situation. So in this simulation, we took a population of individuals. We vaccinated them once. A year later, we gave them a second vaccine. And then two months after the second vaccine, we challenged them with "live virus." And then we looked to see what happened, depending upon the distances between the vaccine #1, vaccine #2, and the epidemic strain in bit string space or shape space.

Now typically the results fell into two categories. One where protection wasn't very good, and the virus would grow to very, very high levels. And we would call that disease. And the other situation, the virus would grow to fairly low levels and be eliminated quickly. And we said that was a case where you didn't come down with symptoms.

To restrict the main, the balls of stimulation here at radius seven. We said the vaccine #2 was pretty close to the epidemic, was distance two away. And then we looked at all possibilities of the influence of vaccine #1, what happened twelve months ago—there were distances between zero and seven—and scaled all possible distances. And then ran 200 simulations of every individual with every distance, and just asked, what would we expect to see? And I don't want you to really look at this, other than sort of each vertical line here. Each box is a different distance. Black vertical lines meant you got disease. The lighter lines didn't. The reasons each individual would behave differently is they would have different immune systems that were generated at random.

So let me just show you a summary of what our recommendation is. And it's a complicated slide, but at least it has a public health recommendation, if one believes computer simulations. So as I said, the distances that we studied went between zero and seven. That was the radius of our ball. And so this is the distance between the vaccine, the first vaccine that you got a year ago, and the epidemic strain.

These were the distances between the two vaccines. And this column is what happens if you only received a vaccine a year ago. If you received a vaccine this flu season—it's down here, V2 only—in this simulation, 55% of the people would come down with flu. Now that number is wrong. Here, every single individual in the simulation is exposed to flu. And the average population, we believe somewhere between 20% and 25% of the population, would be exposed to influenza.

So all of these numbers would have to be scaled by a factor of four or five. That's pretty shocking too, to think that 20% of the U.S. population is exposed to influenza in a given year. But those are the numbers from the CEC.

So if you had a vaccine this year, it's .55. Distance with two, if you had vaccine fourteen months ago, 87% of the people got infected rather than 55%. So it's better to get flu vaccine this year than a year ago. But if you had both vaccines a year ago and this year, you'll see some numbers are bigger than 55%, and some are less. So depending upon the actual distance relations, sort of as in those diagrams I showed you, sometimes you get a better response than if you would have just been vaccinated now, and sometimes you get a poorer response. So you can get either both positive and negative interference.

One thing you can look at though is you look across any horizontal row— so if you look across a row—pick row two. So if you'd only been vaccinated a year ago, you'd have in this simulation 87% chance of getting this disease if you're exposed to it. And now you ask, should you get vaccinated again this year, the current flu season? And what we see is no matter what the distance is, the chances of getting this disease again are always less. And that's true for every single row in here. These numbers to the right are always less.

So if you were vaccinated last year and somebody asks you, should you get vaccine this year, the answer according to these simulations is yes. You will always do better by having this current vaccine. However, if you had a chance to go back and change history and ask, should you have gotten vaccinated last year, the answers are sometimes yes, sometimes no. And so clearly it's something you can't quite change right now. But we think if we understand what's going on, we may be able to use that for designing vaccines.

But last, do we believe these simulations? Does this actually tell us anything about this historical data? For reliability, I will have to show you two more slides. So it all depends on distance. And this is something that we would really like to know. What is the distance between two flus, or the distance between an antibody and its ability to recognize? Well it turns out that in trying to make vaccine decisions, what the CDC does is they immunize experimental animals.

And it turns out ferrets are the experimental animal of choice in the flu business. And that's another story. They immunize a ferret. Get its blood, which contains antibodies. It's called antiserum. And then they test it against the strains that are circulating, to ask, does being vaccinated provide protection against a broad range of things? And in particular they can ask, if I had the vaccine for last year and that anti-serum that's raised for it, does it protect me against the strains that I expect to be circulating now? And if I find a strain that's circulating now that has poor protection against the vaccine last year, that strain is a candidate to be included this year.

And so they do these tests. And these numbers and these tables tell us something about how well the antibody raised in that serum can bind to new strains. And so here you infect the ferrets with different types of flus and ask, how well does it bind, and do direct antigen antibody binding tests.

So these numbers in here, we think, are related to distances. And the way the numbers are generated, it has to do with a dilution, two-fold dilutions. They are called titres. And so all we did is we took log two of these numbers, because everything here was factors of two. And then instead of two-fold difference, or distant change, or one in this log scale corresponds to a distance one, sort of the zero with order thing that you might do. This is a test of antigen antibody reactivity.

So we got the distances for all of the strains of flu that were in those two clinical trials. And then we looked at our big tables from the simulations and said, how would we expect people who'd gotten the vaccine two years, you know, last year and this year, to do compared to those people who'd gotten the last year's just this year, based on the distances from these antigen antibody tables?

And we were really shocked as to how well this simple model of just taking that computational table I told you about, compared to the clinical trial results. So what you see in red are the predictions of this model, just taking our fraction of cases that we would expect to come down with disease based on this simple model, so in red. The blue is what was actually observed and what I had showed you before and if one plots that in sort of a usual regression-type analysis of observed versus predicted over those eight years of outbreaks.

So based on just a single study, it looks like this very crude agent-based model is actually capturing the variability that we see in the efficacy of annual flu vaccines. It's not a perfect simulation. If we actually try and predict the number of cases that come out in any particular year, we are very far off. But when we looked at the ratios over two years, we seem to do extraordinarily well.

So what are the implications for this? Well, one need not only look back historically. First we'd like to test this in a variety of ways. And we're trying to get people to do experiments, say, in this ferret system, where one can directly test the predictions. You could say, if you've given this vaccine now and this next vaccine, what's going to happen? And the model will make predictions. And we'd like to see that tested.

Secondly, what we were very surprised about is we've only looked back one year. Many of us get vaccines not twice, but three times, four times, five times. Does it have an influence? No one has asked that question. I was trying now to find some clinical data looking back over three years. We run simulations now over a three-year time horizon. But also we'd like to go forward so you can say, well, what happens if I've been vaccinated last year, and what should I do this year in making my vaccine? Should my vaccine just be my best guess of what the epidemic strain is going to be? Or maybe should I modify a little bit, knowing that a large fraction of the population had gotten the vaccine last year.

And what we're finding is that sometimes you don't want to just do the obvious thing of making the vaccine identical to your guess of the epidemic strain, especially if you guess the epidemic strain hasn't changed very much from last year. If the vaccine this year is very similar to the vaccine last year, you're sort of maximizing this interference effect where the vaccine is being eliminated.

And we're making predictions. We're talking to people at the CDC right now to see whether or not they think it's reasonable or not. We're presenting some of this material at very large influenza meetings for the first time. We've gotten a number of people very interested in this idea of distance. We've looked at a few of these immunological tests. We've got two major flu groups now going back and finding all of their immunological data to see if we can put them into this distance context. And we're trying to see whether or not we can make a correlation between immunological distance — which is actually very difficult to get in terms of having to do lots of data analysis— and trying to correlate that with DNA sequences for the viruses. So can we somehow go from sequence information to immunological distance? And if we can do that, it would be extraordinarily exciting.

And lastly, we're looking at "the dimension" of shape space. So once you have all of these immunological distances, if you believe them, you can ask, what is the minimal embedding distance, that dimension that you need to sort of reconstruct distances out of the shape space? And a colleague of mine at Los Alamos is using multi-dimensional scaling types of techniques. And at least the preliminary results, looking at all of the data that we have now, suggest that for these influenza experiments, the dimension of the shape space is not very large. It's about four or five dimensions. It's not two-dimensional, but it's also not twenty or a hundred.

And still, one can also start asking the question, can you develop enough vaccines to try and cover the shape space? How much of shape space in the world of influenzas have we actually seen? How much out there is still novel to these? And these are all sorts of questions that we're in the midst of trying to answer.

And again, as I said, the idea is taking a simple agent-based model, simple rules, embedding it into a real-world problem, trying to look at real-world data, making the correlation between them, and seeing if we can actually use the models to answer questions that are very, very difficult for the experimentalist to look at directly. [Applause]

M: You said that if you've got AIDS or HIV, you make a billion viruses a day. How about for other viruses? Is that typical?

Alan Perelson: It's starting to look very typical for chronic infections.

M: How about flu?

Alan Perelson: Don't know. No one has done that calculation yet. We're hoping to. We've looked at hepatitis B and C, which are also chronic infections, and the numbers are even larger there. For hepatitis B, at the peak of infection, our latest estimates are ten to the thirteenth viruses per day. In chronic infection, it's probably more like ten to the eleventh. Hepatitis C, our current estimates are putting it up around ten to the twelfth. And chronic infection, I don't know what it is. There's something called cyclomegalovirus, which causes problems in the retina, which also goes on for years. And there, there have been estimates also ten to the ninth, ten to the tenth that have just been published, based on drug perturbation experiments.

M: Okay. The other question: you said dimension of shape space was four or five. That's just four plus? Is that its overall shape space? And how come you can't cover it so well you can't catch anything?

Alan Perelson: Okay, so that was the admittable embedding distance that we calculated based only for flu data, for these antigen antibody reactions. George Oster and I, when we first did these shape space calculations, actually had predicted that the dimension would be of order five to ten for any type of immunological shape, based on the approximation that we said that shape space might be Euclidean.

And the idea was as follows. I mean, if one of these recognition balls covers say ten to the minus fifth of all of shape space, and then you try to think of what a characteristic of shape is, you know, a length of a binding pocket or something, and you said, how sloppy can that be? And it seemed to us, you know, the wildest imagination is that it covered 10% of all possible distances in one dimension. And so if you thought of something in one particular dimension, if it covers one tenth of it, what's the dimension that you need to get a volume of ten to the minus five? So it's dimension five. To get the dimension to be much higher, you need the recognition to be much, much sloppier. See, it's algorithmic dependence. So it had to be a very small dimension, if one could use a Euclidean type of idea for computing volumes and dimensions. You can sit down and do the little calculation yourself.

M: I could think of other ways you might get a handle on this.

Alan Perelson: Yes.

M: Yes.

Alan Perelson: And it would be very nice to do it. And it's still really an unresolved problem. But so far, both our theoretical arguments in this one data set are suggesting low dimension for the relevant immunological shape space.

M: Fabrice Saffre from Brussels. First of all, I would like to thank you for this most inspiring talk and for making it clear in the beginning of it that it was quite easy to build fancy models, but not always to make accurate and realistic models. But here is my question. I would like to ask you if you would say that by stimulating your immune system with a vaccine, in fact what you do is overspecialize your immune system and maybe reduce its plasticity and its ability to respond to previously unencountered germs?

Alan Perelson: A little bit. But I mean the idea, my idea is that, you know, we believe there's this huge extra coverage of shape space. I told you that because of these shape volumes that we think that at least in the mouse, that there is a hundred-fold excess coverage. In the human it might be quite a bit larger. So my view is that a large part of the immune system can maintain itself being fairly randomly distributed and covering.

And that as you learn, as you expose to vaccines and to real pathogens, what happens is the densities in shape space do enlarge in certain regions. So rather than, say, having ten to the seventh clones distributed at random, let's say I only have ten to the sixth distributed at random, and I have the other nine times ten to the sixth now specialized. Well that still gives me a ten-fold extra coverage.

So I still have the extraordinarily high probability of encountering random things. But I also have put the other clones where experience told me pathogens are. And so I think the immune system has this trade-off. And you're right. If you push it enough, you could say ultimately you will lose the ability to recognize new things. And that's a characteristic of older people. That it does turn out that as we age, that we do lose a little bit of the ability to recognize new things. Whether or not this is a reason, I don't know. But it's at least consistent with that.

M: Regarding the possibility of multi-year correlations, what's known about the timescales over which the population of the memory cells decays?

Alan Perelson: In general it's not well-characterized. For influenza, what is measured is the level of antibody in the serum. And that does have a substantial decay over a year. And it's known that your level of protection does decay. It looks like protection is probably very good for about three to four years. And depending upon the severity of the illness, and by the time you're out somewhere six, seven years, it seems like it's totally gone. So if you have influenza A—there's multiple strains of influenza floating around—but say you have a bad case of influenza of a particular type, it's very, very unlikely that that same strain of influenza would infect you again until the six or seven years that your natural protection should last, unless the virus varies a lot.

M: So you really expect when you look at the data that you will see these kinds of correlations and anti-correlations up to about five or six years?

Alan Perelson: We think so, yes. But it does decay. Whether it decays exponentially, I don't know. But it is a correlation length that's fairly long. So potentially it's a problem.

M: You mentioned prion diseases and viral diseases in the same breath. Obviously, they are similar in the timescale of symptom onset and very, very different in the mechanism of replication.

Alan Perelson: Sure.

M: Can you say anything about whether there are similarities in, for example, scavenging by the body or anything else that would make it possible to have one modeling or possibly even treatment method for both types of diseases?

Alan Perelson: I haven't thought about it. I don't want to get into prion disease. But there are other viruses that are not prions that are also fairly slow types of infections.

M: This is a question, a real outsider's question from the world of *E. coli*. There are things called defective phage in bacteria, and as you know, they act as parasites on the actual phage, so that they interfere with the replication of the phage. You can imagine that if we were giant bacteria—which is a vision

I have, perhaps not yours—but if we were giant bacteria without an immune system, we could engineer viruses, which would only be able to replicate in the presence of the replicating wild type and would interfere with the replication of that wild type in which they'd be parasitic. Now I don't know whether there're perhaps people who are doing that, but if they are, is there scope for modelers to play in that world?

Alan Perelson: Yes. And in the case of human viruses, there are defective interfering viruses that have been discovered. They work in the same way as the macrophage system—David Baltimore of retroviral fame and infamy. I've done a lot of work with his wife Alice Wong about ten, fifteen years ago on defective interfering human viruses.

M: No, I was going to say in principle one could construct an HIV, which would interfere with the original HIV. But maybe there is a good medical reason for not doing that. But it would be fun to model it if it were possible.

Alan Perelson: Right. I agree totally. [Applause]

Jim Collins
Dynamics in Multiscale Biology

Yasha Kresh: Well, thank you very much for coming back. We're going to continue. And our next speaker is Jim Collins, who is the Core Director for the Center of BioDynamics at Boston University. And Jim is going to speak about dynamics in multi-scale biology.

Jim Collins: Okay. Well I'd like to thank Yaneer for the opportunity to speak to you all. My research group at Boston University is in the Biomedical Engineering Department and also part of the Center for BioDynamics. And what we focus on in my group is taking techniques and concepts from nonlinear dynamics and statistical physics, and using those techniques and concepts to develop devices and techniques for characterizing, improving or mimicking biological and physiological function.

And the nature of the title of my talk comes from the fact that we work on many different levels in my group. We work at the level of the whole body, where we are attempting to develop techniques for characterizing and enhancing human balance, so how you maintain this erect stance. We are also working at the level of organs, for instance. We're developing control techniques for eliminating arrhythmias in the heart.

And what I'm going to talk about is actually lower level in terms of spatial scale, recent work that we've doing where we are developing techniques for enhancing sensory function, where I will talk about our operation at the level of sensory cells. And very recently we've moved into the level of genes, where we are developing and constructing artificial gene regulatory networks. So I'll basically talk about two things that are more or less unrelated, but to give you a flavor of what we're all about.

So the first topic is this idea of enhancing sensory function with noise. It's a counter-intuitive notion. Counter intuitive, because anybody who has been through an engineering curriculum knows that we spend an awful lot of time teaching people how to get rid of noise. So in your hotel room here, you would basically tune out the static in your radio. For those of you with your cell phones, you basically will pay a premium price in order to get a cell phone that has very little static. So the idea that you would want noise to improve signal detection information transmission is counter-intuitive.

But going back now twenty years ago, groups of physicists began playing with this idea called stochastic resonance. And the idea of stochastic resonance is that you can have a nonlinear system, and if you input noise, it's possible to enhance the response of that nonlinear system to some weak periodic input. This was originally proposed in the context of some climate modeling by two separate groups in the very early eighties. And what they did is they considered a very simple model to try to account for the periodic recurrences of the Earth's ice ages.

So the very simple model they considered was just a bi-stable well potential that had some barrier use of (not). They modeled the system dynamics as just say a marble in one well. And what they did is they considered for different geophysical dynamics, an input periodic signal that was itself too weak to cause transitions between the two wells, and then considered what happens if you super impose noise. And then they looked at the effects of what happens as you change the intensity of the noise.

So the way to intuitively understand this phenomena -- which is really quite simple -- and understand this model is imagine that I'm holding in my hand a two carton egg carton. And in one of the cartons, one of the wells, I have a marble. And let's say I'm going to move my hands back and forth periodically. You can't see my hands. All that you can see are the jumps made by the marble from one well to the other. So the only way you get information about the movement of my hands is from these jumps.

So let's say I'm rocking my hands back and forth periodically at an amplitude that's too low to cause transitions. Well, you're not going to get any information about what my hands are doing. But let's say Yaneer actually has a vibrating platform that he can place me on that's going to randomly vibrate. So now if you place me on this platform and introduce a small amplitude of random vibration at the bottom of my feet, what will happen is that the noise gets transmitted through my skeleton. And if I'm -- given I study motor control, let's assume I can still maintain my periodic movement. That every once in awhile, the random vibration will combine with the movement of my hands to actually cause a jump from one well to the other, and likewise back again repeatedly.

Let's say Yaneer has a very fancy platform that can actually introduce a huge amount of noise. So now I'll stand on the platform. I've got a massive amount of noise being transmitted through my skeleton. Well now what will happen is a very high noise, you're going to get an awful lot of jumps between the two wells.

So what you'll have is a lot of jumps. Some of the information about my hands will be contained in that noisy time series, but not so much, because it will be swamped. And likewise with the small amount of noise, what will happen is you will get a little bit of information about the movement of my hands, but not so much, because you just don't have that much information contained in that very sparse time series.

Well, if you now plot say some input/output coherence measure, which we won't define. Let's just say it's some generic measure that says, how similar is your output in this case, the jumps made by the marble, to your input of interest, which is just the clean sign wave? What happens is that if you plot it as a function of noise intensity is that say at low noise -- as we talked about -- you get very little information in the output about your input. And at high noise, you also get very little information. But there is some intermediary Goldilocks level of noise where in fact now the noise being transmitted through my skeleton will optimally combine with the movement of my hands, so that you will get a peak in the information that you are actually getting transmitted or detected through the system.

And this is the phenomena of stochastic resonance. It's not a resonance phenomenon. They call it a resonance because you get a resonance like curve where you are varying a parameter, and you're seeing some measure of the system go through at a peak and come back down. This has been demonstrated in many different systems for the last twenty years, experimental and theoretical.

In the very early nineties, Frank Moss from University of Missouri at St. Louis began to demonstrate this in different biological systems, in particular, the tails of crayfish, did it experimentally and theoretically. I met Frank at a meeting in Montreal on February '94. And Frank pulled me aside, knew I was in a biomedical engineering department, and said, you know, Jim, you really need to think about medical applications for stochastic resonance. Nobody's got one yet. If you come up with a good one, you'll never have to write a grant again, and I'll just drive a truck up with money, up to B.U., and just dump the money into your lap.

Well, I -- like all of you know how distasteful it is to write grants -- I thought this was suitable motivation to actually think about, could we come up with an application? As you'll see, I came up with an application. The truck has not yet arrived at B.U., but I remain hopeful. But flying back on the plane, I was thinking, instead of worrying about say a crayfish and its natural environment and the background noise and/or its noise intrinsic to its nervous systems, realize that we as engineers -- biomedical engineers -- can actually introduce noise artificially into systems. That's one.

Two, is that all of our sensory systems are threshold based. So if I have you close your eyes and I'm going to press on your fingertip with some load, I need to press on your fingertip with some particular load before you actually sense that I'm there. So if you're young and healthy, it might be say half a gram a load. If I'm below that, you don't know that I'm actually pressing on your fingertip. But what happens as you get older -- say you hit seventy or so -- you might now, instead of having half a gram a load needed to have your neurons fire, it might be now five grams a load. If you develop diabetes and get peripheral neuropathy due to diabetes where you're going to have deadening to the nerves, it now might be ten grams a load.

So with age and different disease states, your thresholds get elevated. This poses significant promise or interaction with your environment. So older people have difficulty sensing where they are in space. They tend to fall. Likewise with diabetics. Diabetics have difficulty say sensing pebbles in their shoes. It can lead to a cut. They don't know there's a cut. It can get infected and actually end up with very serious consequences such as amputation. And in this country alone there are 50,000 amputations resulting from diabetic neuropathy each year.

So we thought well maybe you could actually blow noise into a system -- in the human systems -- and lower detection thresholds. Well our first step that we needed to go after was that Mrs. McGillicuddy, our seventy year old, doesn't only interact periodically with the environment. She actually interacts aperiodiocally. So it was necessary to consider, could we generalize stochastic resonance beyond just a periodic sign wave? Could you do it to any sort of input signal to enhance it?

And our first starting point was on the modeling side. So we considered a very simple model of a neuron, a FitzHugh-Nagumo model. It's just a two dimensional system where you've got a fast variable V that represents the membrane potential and a slow variable W. Typically, S of T, that's our broadband input signal, (A to FT) is the noise. So ignore that for the time being.

With this system, A can also be considered a current input. If A is suitable large, your system behaves as a limit cycle oscillator. This is just a 2-D simplification of the Hodgkin-Huxley model. However, if A is sufficiently small, you have a system that's basically dead. Your computer neuron is not firing. It instead has a stable fixed point. And it actually has (re-sodable) dynamics. That is, you can kick the system. It will then have its system state point zip around the (no-clines) and come back to its original point after firing, like an action potential. So it has nice dynamics, stochastic resonance type studies.

So we basically set it up so that it was in the sub-threshold regime. We added a sub-threshold, broadband, aperiodic signal SFT, and then we added noise of different intensities. And what we found is that when you look at different input/output coherence measures -- in this case, just cross correlation and co-variance type measures -- there is a function of noise intensity, D here. We found stochastic resonance type behavior.

So the solid symbols are basically simulation results. And the line is actually analytical treatment. And to take away from this is now with this broadband non-periodic input signal that was previously undetected by the model neuron, we blow in noise of different intensities. So as you jack up the noise intensity, the input/output coherence increased to peak and then decays, increase to peak and then decays, indicating you could generalize stochastic resonance type effects to arbitrary inputs. Okay, very nice, it's a model. You can only goes so far with models. And you can only convince so many people when you do model data.

So we actually formed a collaboration with Peter Grigg at UMass Medical Center, who is in the physiology department. And what Peter works on our rat neurons in skin. So touch sensation, in other words, is sensitive to both stretch and compression. And briefly what we did is we took a patch of skin where we could actually make direct recordings from a bundle of nerves leading from a sensory (afron) of interest. We then took a computer controlled indenter arm that would apply a sub-threshold, mechanical, broadband input signal. And we then used the same indenter to actually superimpose noise of different intensities. When we looked at these neurons -- we actually looked at a dozen neurons - in eleven of the twelve neurons, we actually found stochastic resonance type effects.

So we're showing results here from three different neurons where I'm looking at in input/output coherence measure versus input noise intensity. And to what you take away from this is as we jacked up the noise intensity in these neurons, the input/output coherence of this broadband input signal previously undetected of the neuron, goes through basically, as you jack up the noise intensity, you increase to some peak and then decay, increase to peak and then decay, increase to peak and then decay. Indicating you can use noise now in this biological prep to enhance its response to a previously undetected signal.

Now as much as I like to do these types of experiments, one of my career goals is not to enhance sensory function in rats. But I do have interest in enhancing such a function in humans. So we moved up the food chain in one of our next studies and actually set up a series of psychophysical studies with B.U. graduate students and undergraduates who we knew were at least physically healthy, brought them into my lab, had them sit in front of a computer screen, used our indenter arm from the rat experiment, and now used the indenter arm to introduce a mechanical signal that was sub-threshold. So they couldn't detect it. And we used it also to input mechanical noise.

We had it so that you basically -- it was a discreet mechanical signal. And the task was you had to have the individual identify when he or she actually felt the presentation of the stimulus. And they would be basically told of the upcoming presentation from the computer. You present a stimulus or you don't present a stimulus. They don't know what's coming. You randomize it. And the task is, can you actually detect it?

So you bring it sub-threshold. You expect basically if you're looking at your input/output coherence measure, say your percent correct that with say sub-threshold measures, you're about 50%. That what we found is that as we ramped up the noise in nine of these ten healthy young subjects -- sorry this transparency machine is not properly sized for these transparencies -- but for input noise, that we found in nine of the ten subjects, stochastic resonance types effects. So as we jacked up input noise, that their ability to detect a previously undetected signal -- in this case, a mechanical stimulus -- increased to peak and then decayed, increased significantly to peak and then decayed, increased significantly to peak and then decayed. So what we're using here is mechanical vibration to enhance the ability to detect mechanical signals.

Now from the standpoint of feet say for motor control, we actually are now working on developing devices that would be say the equivalent of a vibrating Dr. Scholl's insole that you could put in your shoe. From the standpoint of gloves, it's unclear. What would you actually do for mechanical vibration in the glove? And you could think of (pneumatic), where you would have this huge device that perhaps would be too slow. You'd have air blowing on it, but it wasn't clear what you could do.

Well it turns out that you can actually electrically stimulate quite easily. There are actually devices that are available where you can electrically stimulate various parts of your body.

So what we wanted to do was see, could you actually do cross modality SR? So we went back to our set up. And now we use the indenter arm only to deliver the mechanical stimulus, again, the previously undetected mechanical stimulus. But we also use it now as an electrode. So sorry, this laser pointer doesn't also work too well. But we're basically set up where now we generated a noise signal, sent to that as an electrical signal, passing through a current isolator, and now had electrical noise being presented as current at their fingertip.

Well, what we found is that now in nine of eleven subjects, we could actually get cross modality SR. So as we introduced the electrical noise and ramped up its intensity, that the ability of the subjects now to detect a previously undetected mechanical stimulus could be significantly increased. That is, it would basically increase significant to peak and then decay, increase significantly to peak and then decay, increase significantly to peak and then decay.

Now we found that both with mechanical noise and electrical noise, that the effect was repeatable. So you could bring subjects back in two weeks after the initial session and you'd get the same type of effect.

Well -- question?

M: Yes. Was the stimulus a tap or a vibration?

Jim Collins: Both. So we did both a tap and a vibration separately.

M: So for high frequency stimulation, so how would you interpret it?

Jim Collins: I would -- yes, a tap would be basically a high frequency initial ramp, and then a hold, and then we'd drop. So we basically do ramps and we've also done vibration. So in fact, the next set that we did was, where do we go from here?

Well, we decided, could we actually help Mrs. McGillicuddy and her buddy. Who he, while Mrs. McGillicuddy with her boombox seems to enjoy the noise, her buddy is not too happy with the noise. But we actually set up a collaboration with Lou Lipsids at Harvard Medical School, brought in a group of older subjects. Brought in a dozen older subjects, aged sixty-four to eighty-four.

And now instead of doing a stochastic resonance type protocol where you've got to ramp through wide range of noise intensities, we said, well let's actually just take one noise value and compare noise versus no noise. For reasons doing that is the psychophysical protocols are much more psycho than physical. They are really difficult to do. You have to worry about attention and the people getting bored. You're basically trying to feel something that they can't feel. It's very frustrating and they take a long time when you do these SR sweeps. So we decided, let's just try basically 90% of the detection threshold of the mechanical noise as a source. And I should say that in these previous studies I showed you, that the noise actually was not detectable at the levels where you've got the peak for SR. So it was actually sub-sensory noise.

Well, what we also then did for changing the protocol now, instead of doing percent correct, we basically did kind of a standard psychophysical protocol for threshold determination, where you basically start out with a signal that's super threshold. Then you're bringing it down to try to basically estimate the thresholds where you say, okay, can you feel it? Yes. You reduce the size. Can you feel it? Yes. You reduce the size. Once they can't feel it, you increase it so they can feel it, then you reduce. And you basically kind of zoom in on the threshold measure, shown here for one set of subjects, one subject where you will do fifty presentations. In this case, twenty-five with noise, twenty-five without noise.

Okay, well what did we find? We did nine such trials on these twelve subjects. We found that in every trial for every subject, that the noise trial reduced the threshold relative to the no noise threshold. And we found that it was significantly different on every one of the elderly subjects. Again, they are now sixty-four to eighty-four. We compared them with twenty-one to thirty year olds. We found that now in the end, that these are the group means, that there were significant differences between the young when you had no noise, compared with the elderly with no noise.

And there was on the difference, on the order, about 95% greater for the elderly. When you now compare the elderly with the noise, there were now no longer a significant difference between that value and the young value. And in fact now the difference is only on the order of about 30%. And bear in mind, we're not actually optimizing the effect. We just picked one noise value and said, okay, can we do something with that?

Well, what we then did with Lou's group over at Harvard Med is we actually decided to see, could we do it in a patient population? So we brought in diabetics and actually ran the same protocol 0-- the work with the elderly was done on their fingertip. And with the diabetics we did the same sort of thing with their fingertip. Again, got significant reductions for each patient. And in fact, lowered their values below that of the young subjects. So the diabetics were aged about fifty to seventy-four. And we'd actually now bring their detection thresholds lower than that of our twenty-one to thirty year olds. And we in general got about a 30% reduction in detection threshold for the diabetics.

We did a similar type thing at the foot, and also got about a 30% drop in diabetics. We haven't done actually foot trials on young people yet, so we don't know what the comparison is. We then went also to a kind of surprising population.

That's how thresholds will also get elevated. Stroke -- as most of you know, if not all -- it's a central phenomena. You basically get a part of your brain wiped out from having a clot. Blood doesn't go there, nerves die. And yet as a result of that, your thresholds at the periphery go up. And whether or not it's a central event, it's unclear. But we brought in stroke subjects, did the same type of protocol on their fingertip, and in fact, found significant differences in four of the five stroke patients, where we could actually lower their detection thresholds by 15%.

And what we're now doing is we're extending these studies, as well as actually developing some of these noise type devices that can be used for more functional type tests, such as seeing, if we in fact lower Mrs. McGillicuddy's detection threshold down to the level of the young person, do we also correspondingly improve things such as balance and locomotion, each of which actually tends to reduce with age.

ARY: Jim, does that persist after you turn the noise off?

Jim Collins: No, you have to have the noise there. What we've found is -- that's a good by Ary. Ary asked, does it persist when you shut off the noise? No, you need to have the noise there. What we found is that there is no adaptation to the noise. Now we don't know that -- we don't know that if Mrs. McGillicuddy has a vibrating Dr. Scholl's insole on for forty-eight hours, what happens. We haven't done that. But over the course of say two hours of testing, the effect is the same at the beginning and end. And that's the value of the noise.

You know, you can ask, well, why use noise? Why not just use a DC offset? Why not just preload the individual? Well what happens is our sensory neurons are really good at getting rid of constant signals, let alone regular signals. So for instance, we typically don't remember or realize that we're wearing clothes -- which can be a constant load to your skin -- until I say, oh, you don't remember you're wearing clothes. And you say, well yes, this shirt really doesn't fit me too well; it's kind of scratchy. So in general, we're pretty good at getting rid of that. We're not good at getting rid of noise at the periphery for the effect. And we're hopeful that the lack of adaptation will continue.

Okay. So that ends the story of the first part, which is at the cell level that kind of goes up to the brain, so at least organ level. And what I want to tell you now briefly about is our new area, we've done now down to the level of genes. And I started out really as a systems level guy, so I worked on whole body. And up until just two years ago, if I put up a slide like that, in my talk, that would mean does not apply.

But what happened was going back now three/four years ago, Charles Canter -- who was one of the leading scientists in the human genome project and was Chair of my department -- called me into his office. And I was not yet part of regular teaching faculty; I was research faculty. He said that our department at B.U. had made it to the final round for a Whittaker Foundation development grant -- which is this multimillion dollar deal -- and they were going to be site visiting. And he says, well our theme was to go from genes to cells to organs to whole body. And he said, Jim, I know you do control work at kind of the high level. How about you try playing with some of these ideas down at the gene level. I said, okay, maybe we'll try. If you have some models, that's really a good starting point. He says okay. So as I was leaving, he said, oh, but Jim, by the way, you have five days to do something. The site visit will be here then. I says, well, I'll see what I can do.

Well, we actually didn't do anything in those five days that was of any use. But we really began to explore this space and saw that there was an awful lot of activity going on. So you heard from Stu Kaufman earlier in the week. And Stu and many others have been doing some really nice work on the theory side, where they are developing different theories on gene regulation. It's becoming hot again. There are more and more people getting into it.

But what we found, even though that many of the people doing models were taking information about data, there really wasn't a link back to experiments. So as all of us know, there is a huge amount of work being done in the genomic area for experimental work. That's really the big science in biology right now. And they're moving more and more on to the regulation side. But there isn't that much talking going on between these two groups.

And so what is really needed -- and this is really starting to happen now -- is that there's a need for an integrated approach, where you actually have people doing models that are actually coupled with experimentalists, where the models are making predictions. So you're not only trying to validate models -- which we still need to do in a lot of these areas -- but you're making predictions that are influencing experiments and having a feedback.

Well having said that, we also looked and saw these natural gene networks as being tremendously complicated, and perhaps too complicated for us to kind of make a stab. So being ingenious, we thought, well -- even though this is a complexity conference -- I thought well, why not tell you about stuff we're doing more on the simpler end, which is we thought, let's actually instead of reverse engineer dynamics, let's actually try to forward engineer dynamics. And by that I mean that we actually thought, well why don't we actually try to go ahead and actually construct artificial gene regulatory networks that had specific dynamics in mind, that could at some level be used to validate models, but really to use to demonstrate interesting dynamics.

And really, the first low hanging fruit that we thought of that was an interesting dynamic to consider was what we call the genetic toggle switch. So this system consists of two genes and two promoters. So what you have here is promoter two turns on repressor one/gene one. And you've got promoter one that turns on gene two.

And what we did is we set up this system so that the protein product of gene two would try to shut down promoter two. And the protein product of gene one would try to shut down promoter one. Okay? So what you have now, or basically these genes, two genes trying to shut each other off. Gene one wants to shut down gene two. Gene two wants to shut down gene one.

So what you can have now is basically a bi-stable system in principle, where you'll have in stage one, gene one is on and gene two is off. In stage two, gene tow and gene one is off. And so perhaps you could fight between these two, and you get these two stable states.

Well, what you can do with this system in principle is flip it between the states by using an inducer, a temporary inducer. Okay? One of the key things first, say for people on the engineering side and physics side as we move more and more on to the genomic world is language and just jargon. I mean anybody who has tried to open up a microbiology text -- if you have my background, which comes from physics -- you know, you're in paragraph two and you're already in alphabet soup, and you're trying to figure out what's going on. Well, there's also just in terms of things that are different for dynamicists.

So for instance, in this case, I learned we had to use inducers. But in fact inducers served to shut off stuff instead of inducing things. So what you would do now in this case to actually flip this state is we had to put an inducer two -- let's say if gene two is on -- you throw inducer two that will shut down gene two, that will now allow promoter two to turn on to activate gene one. Now what happens is you now have just a transient presentation of this inducer that would allow gene one to come on sufficiently to now shut down gene two, so that you don't have to have your inducer present after that, and also vice/versa. Okay?

Well, very nice idea. And what we did is Tim Gardner was the Ph.D. student working with me. We started where we're most comfortable, which is just on the modeling side. So we did very simple modeling, where here -- very quickly -- U is just say a concentration of protein from gene one. V is a concentration of the protein from gene two. You've got alpha one is a synthesis, rate of synthesis for gene one. Alpha two is a synthesis for the product of gene two. You've got beta. Here is just a co-optivity of repression for -- in this case -- promoter. It's linked with gene two, and likewise gamma, the same. It's just more or less a competing species model. You've got nice degradation terms. You've got synthesis and competition terms here. Okay? A very simple type of analysis.

Well, what you can do is you can do very simple phase-plane analysis and actually show that if you look at now the (no-klines) in say the UV space, is that if you actually have now your promoters of more or less equal strength -- so they're kind of battling; they've got kind of equal abilities -- that you can actually get bi-stability.

So what happens with bi-stability -- provided you've got nonlinearities in your beta and gamma terms where you're basically looking at multimerization factors -- is that you can have now intersection of your (no-klines) at three points, where you've got an unstable fixed point and two stable fixed points, where they correspond to say state one/gene one on/gene two off, state two/gene two on/gene one off.

But if you say have imbalanced promoter strengths, you would get a mono-stable system. Basically, one gene wins. So say in this case, state two/gene two on/gene one off, if in fact the promoter for gene two is much stronger than that for gene one, than in fact gene two wins, and you get the single state.

Well, we can do -- we did various analyses. We published a paper in January in <u>Nature</u>. I have some copies if you want. And again, we were very excited. We went around to several biologists and said, what do you think? Can we actually build this thing? And every one of them said, no way, you're not going to be able to build this thing in a real living cell. Cells are too complicated. You guys are crazy. Nice idea. Stick with your modeling.

Well we went to Charles Kanter, who is a real cowboy in this area. And he says, yes, I think you can do it. And he actually let us into his lab, gave us some resources to do this. And what we did is we constructed a number of different plasmids to actually try to create these toggle switches.

So here's in general kind of what we're looking at. We've got a promoter -- in this case, a trick promoter -- that's linked with the (C-its) gene. We then had GFP as our marker. So this would basically be used. So what we're going to measure to see, is this gene on or not? And then we've got the PL promoter linked with the laci. Well, it turns out that the protein from the laci shuts down (P-trick). And the protein from CIT shuts down PL. Okay? So it's a nice system. In this case, there's actually a chemical temperature, inducible toggle switch.

So what did we do? Well, Tim Gardner made these things, got them into bacteria, E. coli. And we did a first demonstration that you've got by stability, is he took a culture of these guys' colony, basically divided it in half. And so what we want to think of here is, say with low state for GFP, assume that state one/gene one on/gene two off. When it's high, assume it's state two/gene two on and gene one off. And what he did is he divided the culture in half. One he left alone. So control, he didn't give it any inducer. And you can see, it stays in its original state one state. The other he blasted with IPTG for about six hours. And he actually can flip it into the high state and it stays in state two in this case.

He then made a series of plasmids where he basically was just varying the promoter strengths. And what I'm showing here are results from two general types of toggles that we constructed. On the top plot it's a temperature sensitive. I'm sorry, chemical/temperature toggle. On the bottom, it's a chemical/chemical toggle.

So IPTG is just going to form a sugar. In this case, we stick the system initially in state one, so we've got four different plasmids just varying in their promoter strengths. You blast it with IPTG. These guys go up into state two. You remove the stimulus; they stay up in state two. You then hit it with temperature to now flip it -- that's the inducer -- to go to the other state. And you knock them all back down to state one, and they stay there.

For the chemical/chemical, we've got out (iPdG) one chemical, and then we've got tetracycline, another chemical. We basically zap this guy with (iPdG). It gets up into the state. Remove the stimulus; it stays there. Then you whack it with the chemical; it goes back down. This guy you see, the (pike 105) -- in fact, we zapped it with (IPG), got it up. But when you remove the inducer, it in fact went back to state one. And in this case, the reason was that in fact we found that the promoter strengths were imbalanced. So in fact, you got pulled back down to state one, where gene one is on and gene two is off. And so in fact, it was a mono-stable system, not bi-stable.

You could actually look at the swishing threshold. And in this case, it's for one of the top guys, the PTAK, now as a function of IPTG concentration. So as you ramp up the concentration, you go from state one, then very clearly up to state two. And in fact, we can look at the dynamics going on in this transition by looking at points two, three and four shown down here as histograms. So here you have your cell basically in state one. Then when you get to point three here, you see that you have basically bimodality. So you're kind of going through your transition, until in state four, you're actually flipped over.

In terms of switching time -- now this is again the chemical temperature system -- we've got a system, say in state one, you blast it with IPTG. After about six hours, it makes the flip. So you see here, you started state one. At three hours, you're still in state one. At about four hours, you start getting your bimodal distribution. Until six hours, you have complete switching. Whereas with the temperature side, it's actually must faster, because you're very rapidly destabilizing your repressor protein, so that you're on the order of about thirty minutes.

I presented this stuff for the first time at a DARPA meeting on amorphous computing back in September, so there is a lot of talk about biocomputing. So we waved our paper and said, oh, we've actually done a toggle switch, a flip flop; it could form the basis. And they're all excited. And I said, oh, but one little thing: it actually has a flipping time of six hours. And there was just, shhh shhh, air was just let out of the room, like, oh no. So there are some things that need to be considered there.

To finish off, kind of tell you where we're at and what we're thinking about in terms of applications. Tim Gardner, who just finished his Ph.D., coined the term genetic applets for these synthetic gene regulatory networks, the idea if you're going to construct these little synthetic networks with different applications in mind. We've already done the toggle switch. We're about a month away from doing it in yeast, so cells of the nucleus. So we call it (u-car-iats). And we're going to be launching mammalian testing soon.

We have also just about finished in bacteria a genetic sensor. So a genetic sensor is a slight modification of the toggle, where now instead of having it activated by transient presentation of chemicals, they can actually activate on the basis of the level of a chemical. So say as a chemical rises, once it flips some level, the gene will be turned on. Once it drops below a level, the gene will be turned off. And I will talk about a medical application in a second.

We also have design for a genetic oscillator. We were all excited. It turned out after we got our paper, or as we're getting excited, had it accepted in Nature, we found out that Stan Leibler's group at Princeton with Michael Elowitz had actually designed and built an oscillator in E. coli that was very similar to what we had in mind. They got it working. And in fact, it's in the same issue in nature. If you're interested, they did very, very nice work coupling also modeling with their experiments.

And we're also playing around with multi-gene, multi-stay networks. What happens if you have three? What if you do four genes? Can you actually flip it between different states?

Briefly on applications -- and this is it; I don't have any other transparencies, so I don't have the Chair of the session get too worried -- for gene therapy, you know, there are some very obvious applications. There are gene switch technologies available. All of them require the continual presence of some drug to actually have your therapy to gene go between states. So let's say you if you have your transfected gene in the off-state, to get it into the on-state, you need to have a drug -- say tetracycline -- there to basically bind to the promoter, to either go from the on to the off.

For the toggle switch you don't need that. You basically just have transient presentation of a chemical to flip it between your states. You can also use it -- and we're now also, I have some people in my gene jock team back at B.U. working on designing a kill switch. So can you link it, so just with a trans-delivery of chemical, you can flip it into a kill switch state.

Biomedicine on the sensor side, there's obvious applications for things such as diabetes. Can you design a sensor that's sensitive to glucose? That as glucose levels go up, you flip a switch that will produce insulin or release insulin. That once the glucose levels drop as a result of the insulin change of the permeability of the cells to glucose, you would basically have the system flip off, so you could create a genetic applet based artificial pancreas.

Biotech, there's a lot of interest in DuPont and Dow -- these chemical giants -- for using cells as factories; that's what they're doing. It's a decades old industry. There's interest in trying to get tighter control over gene regulation such systems.

And then of course biocomputing. I think that the idea here on biocomputing is not so much, are you going to have a jar of E. coli on your desk doing your tax calculations for you? Now ruling that out, you still have to figure out what are going to be the so-called copper wires, as Mark Spano told me the other night. There's a lot of things you have to worry about. But I think you can think of artificial gene regulatory networks as systems that have computational ability, certainly programmable ability. I mean with your toggle, you now have a memory element, the simplest memory element. So you could in principle program cells to do different things.

And I would say on the technology, the nanotechnology side, there is a lot of interest in micro-electromechanical systems, designing them. We think of them now, these genetic apps as just kind of nanotechnology in a wetter form. That is in living cells where you can design these circuits with specific functions in mind to program cells to go do things.

So I'll end there. And thanks for the floor and the entertaining ?? . [Applause]

M: Yes, two questions about each half of your talk. On the stochastic resonance for the TAP response, do you have a model for that mechanism or just phenomological observation?

Jim Collins: We're just kind of fine-tuning a model. Chris Richardson, a Ph.D. student in my group has been focusing on that. It turns out there aren't very good models for (mec-an-er) receptors in the literature, and particularly models that would account for say skin properties.

M: Is it the high frequency of the TAP that you're then amplifying with the stochastic resonance?

Jim Collins: Well it depends on the neuron that you're getting at. So with stochastic resonance frequence, it's not an amplification phenomenon. So that's where there has been some misperception and people have thought it could be used to enhance and cause problems when you use your cell phones, this sort of thing from background noise. It's really just a boost. So it will basically just shift your signal up more so than it normally was. So when I say it depends on the neuron, it will do that for your high frequency neurons. And then for your low frequency response, it will basically shift up the --

M: That too?

Jim Collins: Yes, absolutely.

M: Okay, that's a mechanism I also have in mine. We should talk later. But the second question having to do with the DNA, the concentration of ten to the minus four (molar) I believe, how many copies of the molecule per cell does that correspond to?

Jim Collins: Oh, good question. You know, we don't know. We actually didn't characterize that. I mean there were certainly cells that had some, that had multiple copies of the plasmid.

M: Well, I meant of the inducer.

Jim Collins: Oh, in the inducer. Oh, I don't know.

M: You said you blasted the cells.

Jim Collins: That we blasted the cell. Oh, I mean, that that's -- don't associate quantitative with my term blast. That's just color, colorful. But I don't know. I don't know the answer on how many.

M: But the interesting feature is in natural biological switching networks, regulatory networks. The number of inducing molecules can be very small. And so there's very interesting issues there, going away from a statistical or a thermodynamic description of the switching to dynamics really.

Jim Collins: Yes. On that topic, there are defense people who have been talking to us. They are very keen on using this for bio-warfare detection. And specifically on their case, their interest is to get it for small molecule detection. And we have not engineered these things specifically for that, but it's possible that you can tweak systems to get down at the very low --

M: Low concentration.

Jim Collins: -- low concentration, small molecule number, small number of molecules, sure.

M: I think this stochastic resonance principle -- I think -- is very exciting. And twenty-five years ago I proposed a model of enzyme and catalysis, which can be explained in terms of stochastic resonance, because enzymes need thermo fluctuations for it to work. So that thermal background could be regarded as the noise input. If you cool down the enzyme, the noise goes down and the enzymes don't work. You have to raise the temperature so that you have enough noise for it to work. So I think your stochastic resonance can apply to enzymic level. And I think it can be applied to the brain level. And there is a way to test this.

My stepdaughter Michelle always studies better with her TV on. And apparently if the TV is off, there is no noise, and her mind doesn't work as well. And I wonder if you can test this.

M: What channel does she watch? [Audience laughter]

M: It doesn't matter what channel; something has to be on. And it would be very nice to gather a few groups of teenagers and test their cognitive efficiency in the presence or absence of TV noise.

Jim Collins: Yes, a couple of things. Firstly, I would be interested in getting a copy of your paper on the enzyme thing. And there have been a few people who have been doing some modeling on that end. On the brain side, Mark Spano, who I came up with, has actually done a piece that was published in PRL, where they looked at a brain slice. So it's separated from a student studying. [Audience laughter] But a brain slice where they were actually able to demonstrate stochastic resonance type affects in brain neurons.

Now on the experiment, I don't know of anybody who has actually done say EEG measurements, where you are looking at what actually is happening to the brain with the noise background. I am actually intrigued by the opposite phenomenon, which is could you actually use the noise to help shut down processes, the calm processes. And it is self interested, because I have an eleven month old at home who, you know, at times, anybody who has had a baby, it is very difficult at times to get her asleep. And we actually would put white noise background --

M: I think it would depend on the noise intensity. You have a bell shaped curve. So the rate of efficiency would be bell shaped with the respect to noise. So you can have increased efficiency or decreased.

Jim Collins: Yes, I think that that's very true. And you know, say people do give these little things you can buy on airlines to kind of block out background schemes, where the unpredictable nature and broadband nature of the white noise tends to be a benefit. But perhaps, I know there are people in the audience who are much better experts than me in the psychologic/psychology area, and perhaps they can point us to studies that have been done that have looked at this in a quantitative fashion.

M: Like turning a fan on to fall asleep in the evening.

M: Yes. So you guys have managed to make logic circuits very impressive, and memory, flip-flops. And now you guys are creating sensors. The next question is: can you create – can you program intercellular communication?

Jim Collins: Yes. I think you can. Obviously, that's one of the next steps to do. Tom Knight at MIT in the AI lab is looking at schemes of doing that via light and actually fluorescence. I was just talking with my group about that as one of the next big things we all need to think about, particularly if you ever do want to get into computation.

I had some members of my group who were considering some pretty clever ideas where you could actually get molecules that could somehow get through the membrane of the cell and get to the nucleus and actually have impact. But both of them were recruited away by biotech industries before they could actually do anything. But I think that you could program that.

I mean, the theory of mine is that this is a wide-open space. And I think that it offers a lot of opportunities for people with modeling backgrounds and interesting complex systems. And it's a matter of getting in there and actually learning enough that you can do things. And there's just tremendous opportunities for designing this stuff right now. But it's -- that's one of the next big areas people got to get at.

M: My question is back to the first part of your talk about using noise to increase sensitivity. So I'm coming from a practical point of view; I bike to work. And after five or ten minutes of biking, my hands start getting numb from the rolling of the bike. And I use silicone gel gloves to reduce that. What about the longer or medium term effect of noise on the sensitivity, especially if you're using it with patients, but whatever?

Jim Collins: Yes. I don't know the answer to that. I know over at least the two hours, that we're getting same effects at the end and the beginning. And the critical difference between say your bike experience and what we're doing is that the noise is subsensory. So that is with just the noise of the subject, you can't feel the noise at the level we're presenting it. So that we're not actually with the noise, generating responses from the neurons. Where say you are doing that say with your mountain bike just cruising along. You're stimulating the neurons so that you can actually feel it from basically the first time you hit the pavement. And what we're doing is we're not actually eliciting action potentials, o at least action potentials that are registering at the brain.

M: But you can do that with a different kind of a bike. Let's say I am using a mountain type bike, so the tires are very knobby and all of that. But you can get a tire that's very think and very soft and all of that, and you go on smooth pavement, so you don't -- and you still, hands will get numb. And not just from the vibrations that they're so audible, but just from being on the bike and --

Jim Collins: Well, so if it does become a problem -- so there actually are some engineers that are not with my group, device guys, looking at this. And one thing they are considering is that you may not need to have the Dr. Scholl's in-sole vibrating all of the time for Miss McGillicuddy. So say she's sitting here being bored by my talk, she doesn't need to have it on when she's sitting in the chair.

M: Oh, okay.

Jim Collins: That's one. Two is you may be able to actually sweep through different frequencies to actually change up the presentation. Because there's not only one band of frequencies that works well, there's a lot of bands; that's one. Second is that you can also just sweep through in terms of just on/off. And that's a problem that we will I think have to face down the line, if in fact this gets to device end.

The critical next step that in that area is, can you get functional benefit? So this is all nice and we have very well controlled clinical type studies, but does it matter? Is it something beyond just a lab curiosity? Will it help Miss McGillicuddy to function out there?

M: Thank you.

Jim Collins: Thanks. Thanks for the question.

M: Thanks. [Applause]

Tim Buchman
Multiorgan Failure

Thomas Deisboeck: The next speaker is Dr. Tim Buchman. Tim is a trauma surgeon in ?? , St. Louis. He's going to talk about multiorgan failure. Tim, please.

Tim Buchman: Sounds good. Well, the question is: how does a trauma surgeon get interested in this particular topic? This patient from a few weeks ago -- actually now a couple of months ago -- they have something called all-terrain vehicles out in Missouri, a lot of rural areas. These are small vehicles that cruise in the countryside. And he does down an embankment.

We get a call from the medical folks on the scene saying extrication is prolonged. Which is hard to imagine, since he's out in the middle of no where riding an ATV, which doesn't have a hood on it or anything like that. A helicopter on route, say there's a little bit of facial trauma, and they helicoptered the guy in directly from the scene.

And the problem is that Spike here has a little problem.

What you're looking at is a piece of reinforced steel bar, otherwise known as re-bar. It's what goes into steel that reinforces concrete, which is going through his neck. It actually went through one of his carotid arteries and one of his jugular veins. And the point is that he is some where far distant from his normal physiologic attractor. And he is now in my hospital and I have to do something to get him back from very deranged physiologic space, into yet a different physiologic space, and get him back to some level of normal.

Well, what I'm going to do for the next few minutes is talk to you a little bit about multiple organ failure, what it is, why do I care. What one of the operating hypotheses is about the origin of multiple organ failure, talk about (Carston Schaeffer's) description of normals, and then get into a discussion of uncoupling with a few papers, and finally tell you a little bit about where things are going.

You need to understand that multiple organ failure is not an old disease. It's not like streptococcal infection, which has been around for a few million years. In fact, multiple organ failure descriptions didn't appear until the mid-seventies. And I remind you that the last Apollo mission was in 1975, just about the time the first descriptions of multiple organ failure were appearing.

If you look at that little photograph on the right, in dead center there's an arrow. And at the tip of the arrow there is a little baby. All of the machinery there is to try to keep the babies physiologic systems somehow normal. I borrowed this slide from a colleague of mine, Jim Fackler at Hopkins, only to give you some idea of the context in which we physicians operate. The machines are all there to lock the patient's physiologic system at some arbitrary value that we physicians think is a good idea.

The story begins somewhere around 1973 with this paper by Nick Tilney, who was looking at not pediatric patients but older folks whose aortas had ruptured. And what Nick observed in his work up in Boston was that some of the patients did well. But by and large, most of them had sequential failure of their multiple organ systems.

This is a figure from that paper. And what Tilney showed was that this was not a random series of events. In other words, the organs failed in a relatively predictable fashion. The names have changed a little bit since this initial report, but in general, the lungs always fail early. The liver and the GI system fail next. And the last thing to quit are the kidneys.

Interestingly -- and we'll come back to this as an idea -- this is the exact reverse of the sequence in which the orders come on-line during embryologic development. The kidneys mature first, then the gut and the liver, finally the lungs and the infamous born. But it is a sequential system failure and this description has been unchanged for about twenty-five years.

This also is a diagram from his original paper. The details of the diagram are unimportant. But it should be evident from that mass in the middle that he's not dealing with a simple hierarchical system, but he is dealing with some kind of interconnected network.

Well, people looked around and said, you know, this is not only happening after aneurysms; it's actually happening after a lot of other diseases as well, mostly commonly major trauma. Art Baue, who at the time was in St. Louis, wrote this editorial which he entitled, "Multiple, Progressive, or Sequential Systems Failure: A Syndrome of the 1970s." And what Art pointed out in this paper is that this disease of modern society didn't exist until we had multiple organ support.

Kidney dialysis is only a few decades old. Ventilators outside the operating room, only a few days old. And it wasn't until the Vietnam conflict that we were able to bring all of these different types of organ supports to bear on individual patients. In the old days, if you had a single organ failure, we didn't have a support; you simply died. Certainly if you had multiple organ failure, you just died. Now we have patients who are maintained with multiple organ support.

I've mentioned the characteristic sequence: lung, then gut, liver, then kidney. The point is, is that individual organ support we discovered was necessary, but insufficient to predict recovery. In fact, about two-thirds of the patients that we try to support this way go on to die. Not because of our therapy -- we hope -- but in spite of it.

Now I want to take you forward a little bit to this relationship. Hiram Polk, living, he's Chairman of Surgery at the University of Louisville. And he and Don Fry in the early 1980s said, you know, it's not just trauma. It's not just ruptured aneurysms. It's infection. And he associated infection as the most frequent predecessor of multiple organ failure. And in fact, these are the datas, we have them today.

Despite the development of huge batteries of antibiotics, (guerilla) mice, (encepha-kill-a-mol), budget buster (cillin). I mean we get all sorts of antibiotics each year. In fact, more and more people are dying of sepsis. Now the problem is the precipitant is sepsis, but the cause of death is multiple organ failure. These patients are dying not of the infection, but as a consequence of something that the infection is doing.

We talk about causality, sepsis and death. We get into the story of inflammation. This is Kevin Tracey. He is a neurosurgeon affiliated with Cornell. He works at Long Island North Shore Hospital. But when he was a resident, he worked with Anthony Cerami at Rockefeller. And he got interested in the problem of sepsis and used injection of endotoxin or LPS as a surrogate for authentic human sepsis. He showed not only that when you caused a sepsis type syndrome in animals, they would die, but you could block it by putting in an anti-inflammatory molecule, an anti-TNF alpha in this particular case.

And in the mid-eighties, this set off an enormous amount of enthusiasm and interest. And people said, wow, when we talk about sepsis and multiple organ failure, the real problem has got to be the explosive release of this one molecule that disturbs a network. And my gosh, the culprit has to be TNF, because in animal models, this TNF level grows rapidly. The inhibition of TNF in animal models prevented death. And in human plasma levels of TNF correlated with mortality. So it was real simple: block TNF alpha and save lives.

Well, as with all simple ideas, it didn't turn out quite to be that way. This is work done by Charlie Fisher. Gary talked about it last night in his session. But basically this is representative of about two dozen clinical trials that were performed at a cost of about ten billion -- that's "B" as in billion dollars -- over the last fifteen years. The idea here was to put in a soluble receptor to soak up the TNF. And in fact, compared with the patients who got placebo, there wasn't even a helpful change. It didn't even stay level. In fact, the therapy was worse than the disease alone. We were killing people with this stuff.

Well, neither prevention of infection, nor eradication of infection appeared to modulate this multiple organ failure or multiple organ dysfunction syndrome. Up to half the multiple organ failure actually have no evidence of infection. Well, if infection doesn't cause MODS, what actually is going on?

I need not go over the importance of this book to this group, the notion that physiology is a dynamical process. And certainly Ary was the first among many to point out that health is fundamentally not a physiologically stable, but rather a physiologically variable state. And I'm going to share with you some work that we did originally that led to a hypothesis that sustained uncoupling of multiple organ systems, gave rise to physiologic regularity.

For this group I probably don't have to go over the fact that you can use non-invasive data to get a window on physiologic variability or regularity. This is an EKG tracing. And the series of intervals defined by the arrows there, the so called R-R intervals -- the intervals between heartbeats -- is more or less regular. And I saw more or less because sometimes it's more and sometimes it's less.

The experiment we did now about six or seven years ago was to take human volunteers, medical students and fellows -- I suppose you can call them normal -- and inject them with low doses of endotoxin. And we recorded their R-R intervals for about eight hours. There was a randomized crossover study with placebos, so that we were able to do internal comparisons. And the important outcome of this study was that their was a reversible loss of variability associated with this clinically relevant, very mild stress. The metric we used was approximate entropy. We changed the sign here.

But quite simply, if you subtract out the approximate entropy of every hour following the injection from the baseline level, you can see that at about three hours, significant loss of variability, increase in regularity of the R-R intervals after endotoxin injection. And this goes away after about twelve hours. But very reproducible, very predictable phenomenon.

Now Steve Pincus is a mathematician at Yale, had suggested in a paper in Mathematical Biosciences that loss of variability at any node corresponded to isolation or loss of interconnectedness in complex systems. And we suggested that uncoupling of one organ system for another might actually be a disease mechanism. But we actually needed a little bit evidence that uncoupling and recouping occurred as part of normal physiology. Normals have this characteristic behavior. They will, if you look hard enough, abruptly shift their heart rates. And you don't have to be a cardiologist to figure this one out.

The reason for this wasn't really clear until this paper came out. This was a paper by Carsten Schafer from the Potsdam Group, published in Nature in March 17th, 1998. The first couple of sentences of the paper, "It is widely accepted that cardiac and respiratory rhythms in humans are unsynchronized. However, a newly developed data analysis technique allows any interaction that does occur in even weakly coupled complex systems to be observed.

And what Carsten did was to look at the relationship between heart rate and respiratory rate. And I emphasize that this work was done in healthy athletes at rest. He used a Hilbert Transform, because it's very robust to noise, a couple of other good reasons. But this is the critical figure that he published in his paper. What you're looking at is what he called a cardio-respiratory synchrogram. The top portion of the picture has to do with phase.

And basically what it does is it addresses the question, in normals, what's the relationship between the respiratory cycle and the cardiac cycle? And if you look carefully at the pictures, you will figure out that the coupling between the cardiac and respiratory cycle changes from five to two, through a transition zone of six to two, a cycle here. The particular numbers are irrelevant. People do this with a variety of different coupling frequencies.

But the point is it happens over physiologically meaningful intervals of about twenty minutes or so.

So Carsten was able to show that in normals, not only does coupling exist; but coupling, uncoupling and recoupling is part of normal physiology, at least within these two systems. And it's suggested that pathology could reside either in an accelerated uncoupling or diminished recoupling or both.

A little data from our own shop. And then I'll go on and talk about other folks' data. We looked simply at loss of heart rate variability in patients who were undergoing aortic surgery, 109 abdominal aortic surgery patients immediately post-op. Two-thirds of them were aneurysm. A third of them were occlusive disease. There were three deaths. But basically, we put Holter Monitors on them, looked at them in the first post-op day. We basically said, you know, if they're going to be in the hospital more than seven days, we're going to call it long. Less than seven days, we're going to call it short. It's a simple median split. And we compared the two groups.

We looked at twenty-four hour time and frequency domain, uh, measure of heart rate variability and alpha, and looked at spectral plots, FFTs of the (R-R) time series. Alphas are the reciprocal of frequency. It's basically a slope of the representation in the power spectrum. And the timescale we're talking about are changes between a minute, an hour, in terms of the frequency. These are typical tracings from our patients, just to give you an idea of what the power spectra actually looked like. And here are the data.

If we look at folks who had length of stay as less than seven days, they had a significantly smaller alpha than the folks who were going to be out there for greater than seven days. And if we did the multivariate analysis, the significance of the difference in a power spectra are really quite substantial.

So basically from this study, early failure to recouple post-op suggests that this patient is going to have a long stay in the ICU. What that means is, is there's a substantial fraction of our increased length of stay is dealing with organ insufficiency, but finding it before the organ actually has clinical evidence of failure.

Now it turns out that you can find similar data from many groups. This is from Winchell and Dave Hoyt's group in San Diego, published in The Journal of Surgical Research some four years ago. They were looking at spectral analysis of heart rate variability in the ICU. And they were looking at ratios of high and low frequencies, basically a relatively low sympathetic tone and low total power. And they found that this really did correspond to poor outcomes.

Here are the data. And if the patient had low heart rate variability and high HF to LF ratios, you can see that the mortality versus the control patients, whether the underlying disease was a cardiac disease or not, mortality was increased eight fold. So there actually is hidden in the heart rate variability data significant information about how the patient is going to do long before the outcome occurs.

They had a little bit -- these are data from Yien, published in Critical Care Medicine in 1997. And basically they were looking at relationships between systemic arterial pressure and the heart rate -- two measures in the cardiovascular system -- as a prognostic tool for predicting patient outcome in the intensive care unit. This is probably the key figure from their paper. And the point is, is that survivors, people that who were going to survive their stay in the intensive care unit tended to recover their variability very early.

The top two panels are arterial pressure. The bottom two panels are heart rate. Over on the left are very low frequency. On the right are simply low frequency data. And these are simply data from two patients, actually from four patients, two survivors and two who died. You can see the significant difference between the open circles and the closed circles. Basically, you needed variability to survive. Progressive increases in both low frequency and very low frequency. Both heart rate and blood pressure were strong predictors of a good outcome.

Well these are independent measures. So we get in the issues of coupling. I wanted to present some work done by colleague Brahm Goldstein, published a couple of years ago in The American Journal of Physiology. He basically said, let's take a look at the autonomic and cardiovascular systems in acute brain injury. And what Brahm did was the following. He suggested the acute brain injury resulted in decreased heartbeat oscillations and baroreflex sensitivity indicative of uncoupling in the autonomic and cardiovascular systems. What he did was look at twenty-four patients with a diagnosis of acute brain injury.

Here are the data. Across the top I've listed GCS. GCS stands for Glasgow Coma Score. It's a measure of how well the brain is functioning. Everybody in this room has a Glasgow Coma Score of fifteen. If you're brain dead, you have a Glasgow Coma Score of three. Glasgow Coma Score of nine is the middle column. And that represents a moderate degree of coma. Patients who have a Glasgow Coma Score of nine or below are usually intubated, on a ventilator, and really very poorly responsive.

What he has plotted here are heart rate variability, power spectral density, systolic blood pressure, power spectral density, and then the transfer function magnitude down there at the bottom. Basically, what you have here is an indication that the interaction between these two systems -- the heart rate and the blood pressure -- is changing significantly with depths of brain injury.

I point out to you that there is a significant change in the scale in the patient who is brain dead, over their Glasgow Coma Score of three. When you look at the power spectral density of both the heart rate and the systolic blood pressure, those scales are one one-hundredth of their adjacent panels. So there is a significant change with near-zero levels for all variables during brain death and in fact, a pretty good predictor of survival versus non-survival.

As you can see over there on the right, the transfer function, a significant difference in how much coupling there is between the survivors who are pretty well coupled and the non-survivors who are even more poorly coupled, and the gradient between the non-brain dead and the brain dead, as you can see in the bottom panel.

So what Brahm concluded was that as a severity of acute brain injury increases, there is proportionately greater degree of physiological uncoupling among the autonomic and cardiovascular systems. Now he has previously shown that neurological recovery after acute brain injury results in restoration of heart rate variability toward healthy levels. And it's at least consistent with what Paul and I proposed in terms of functional relationships. That is, that you have to be able to recouple if you're going to get better.

These are new data which Brahm let me give you a view on. They're in press; they're not out yet. But again, data on kids, now looking at sepsis. A little snapshot of data from thirty PICU patients. He's looking at very brief recordings, only about 128 seconds, once a day. And basically what he's seeing is a significant decrease in variability as the patient goes into sepsis and septic shock. As the patients do recover, however, they do recover their variability and the changes can be seen significantly in the low frequency power, as you see down there on the lower panels.

What you're seeing is the patient who is -- the patient's heart rate over on the left. The blood pressure over on the right. The point is, is that the patterns in both of these cardiovascular parameters not only echo one another, but they actually talk to one another.

So I would suggest to you there is evidence that healthy humans normally are coupled, and they go through uncoupling and recoupling as part of normal physiology. And there are experimental and clinical data that support at least the idea that critical illness can be associated with uncoupling and more importantly, failure to recouple.

There are some barriers to further work in this area. There is first, no shortage of patients. There is also no shortage of analytic tools. And I don't think there is a shortage of interest of scientists and clinicians. What there is -- at least from the clinician's standpoint -- is the failure of the scientific community to agree on the "metric." I need not mention to this audience the tremendous variety of metrics that have been put out to try to measure what's happening in time signals. And the problem is clinicians don't want twenty metrics, they want one. They want a number that they can put in a box and have it mean something, go up -- if it goes up and down.

The second problem in this area is that there really has been a failure to design prospective studies that test explicit hypotheses. Most of the data that have been generated basically has to do with a retrospective look at time series data that have been collected off-line and simply analyzed at a later convenient time. That tends not to persuade patients. Rather, it tends not to persuade granting agencies that you're actually looking at something useful. They look at it as data dredging, not as useful science.

I put this up only to mention that Spike actually got better. He was one of the lucky ones. He was in our ICU for a couple of weeks, going through a variety of infection and organ failures. But this is a picture taken about six weeks after his injury, with his wife there holding the (re-bar), and one of my ICU nurses on the side.

I would be happy to answer any questions that I can. Thank you so much for your attention. [Applause]

M: The order of failure of different organs, is that determined by the order of micro-circulatory failure?

Tim Buchman: The question was, the characteristic order of failure of the organs -- the lung, liver, gut, then kidneys -- is that based in micro-circulatory failure? We don't think so. There is good evidence that the micro-circulation is as good in any of these organs at the time they are functionally failing as the micro-circulation is in anyone seated in this room. Although there is some evidence of micro-vascular thrombosis in all of the organs, it appears that it's not significant enough to change at least gross parameters of profusion.

GARY: I enjoyed your talk very much, finally to hear it after reading your papers. I would like to hear your opinion on whether or not you think that the media of the coupling are physiologic parameters at the organ level, or whether or not the coupling is an observed emergent phenomenon from underlying molecular and cellular processes.

Tim Buchman: The answer to your question Gary is yes. It probably is both. And the reason we can make that kind of statement, there are strong data from the transplant literature that when you take an isolated organ -- and let's face it, if the heart is sitting the bucket, it's uncoupled from its recipient or the kidney or whatever organ it happens to be -- there is a functional improvement above and beyond the simple reperfusion injury that it has to overcome over time.

So at least part of it is simply an emergent phenomenon, if you will, because the heart is physically still denervated from the rest of the system. It's seeing the hormonal milieu; it's seeing lots of other things, but it's got to be both of those pieces. I'm not sure the two are separate.

M: What are your ideas about the normal function of coupling in the first instance and preserving the homeostatic state or responding to stress, beyond the mere correlation of the two, the cardiopulmonary events?

Tim Buchman: I was struck by the comment about stochastic resonance that we heard earlier today, because I think the importance is that we are not talking about simple feedback systems. That in order to respond to an ever-changing environment, we have to be able to move within at least the fairly confined physiologic parameter space from one to another. I suspect that coupling is essential to do it. But I also expect that there has to be a little bit of random fluctuation in the system to permit us to move, to respond, if you will, to the outside influences.

What we end up doing in the intensive care unit is what we think is right. By God, that PH is going to be 7.42. The potassium is going to 4. The glucose is going to be exactly 120. And if the residents haven't made those numbers exact the next day, they'll hear about it. So we induce a behavior that says we have to lock the patient at certain outcome measures. It's not clear that that's the rational thing to do. And in fact, most patients seem to do better when you relax the constraints just a little tiny bit.

The significance is that as Canon articulated homeostasis, he was working off Bernard's ideas. Bernard never said there's a feedback mechanism that drives a particular number towards a particular value. That was meant to be a first approximation. Unfortunately, Canon got a hold of the ideas and bastardized them just enough to give the entire physiologic community for the next hundred years the idea that the number had to be fixed. In fact, I think Lawrence Henderson -- who was a contemporary of Canon's -- had a far better idea, that it needed to be approached in the way we're doing it today as a systems approach, not as a fixed number.

Yasha Kresh: Someone was asking and you were answering about recoupling. And we've studied patients who have given up one of their hearts and got another one. I can tell you that the recoupling, because physiology is really a function order, there is no recoupling as you will splice to wires. You don't return back to where you started from.

So even patients who get a new heart or a different heart -- and if you believe that they renervate, they don't renervate in any meaningful way that looks physiological. I mean the heart reorganizes in some way and finds a new -- sort of if you want to think of it -- a tractor to operate at. But it's not anything that was before. They can't play tennis as they played before and they can't really run as fast as they ran before. But you reconstitute something, but you don't reverse the process when you reconnect the heart.

Now we're lucky that when we get into your unit, you do dial in 7.4 and you sort of reconstitute us, because you rely on the rest of the system that is quite capable of repairing itself to reconstitute something that looks like physiology. Again, because we are normal and his dagger was out of his neck and he walked out.

We are breaking for lunch.

Yaneer Bar-Yam: Briefly.

Yasha Kresh: Yes.

Yaneer Bar-Yam: Can you explain the process?

Yasha Kresh: Is the process to bring lunch there?

Yaneer Bar-Yam: Yes.

Yasha Kresh: Here, you can explain it best.

Yaneer Bar-Yam: Lunch is just sitting right outside the door. So just go out, take your lunch and come back in, because we'll just continue over lunch. There are three choices of different sandwiches. You can't pick which one you want.

[Break]

Ary Goldberger
Fractal Mechanisms and Complex Dynamics in Health, Aging, and Disease

Thomas Deisboeck: We continue with Ary Goldberger. Ary heads the lab for the analysis of nonlinear dynamics and cardiac arrhythmia over at Harvard Beth Israel. Ary, please.

Ary Goldberger: Thanks. It's a pleasure to be here. And Tim made my life a lot easier by defining some of the terms and the concepts that are going to apply here. This is going to be a very general talk. And I would like to address things with respect to your personal physiologic health and thinking about what makes you as individuals or people in your family healthy or diseased in the dynamical context, and then talk a bit at the end about a new research resource that will hopefully generalize our ability to carry on the type of research that Dr. Buchman addressed; namely, how does one take the information at hand and make it actually practically available and test out ideas, because that's a huge challenge.

Well, so I'm going to say just a bit about nonlinear dynamics in general in cardiac physiology. And it's interesting, if you go to the medical school where I am or pretty much, I don't know what it's like where you are or what it's like on the other campuses of Harvard, like the MGH. But if you go around asking people to define nonlinear dynamics or fractals or chaos theory, my suspicion is that perhaps more among students, but very little among faculty will you get much in the way of an informed response. Is that accurate?

M: Deer in the headlights.

Ary Goldberger: Deer in the headlights was the -- well, just -- there are two quotes that I came across, that these are nonlinear musings. So just as a quiz for the audience here -- this is a very well informed audience -- just to pose to you. There are two quotes. The first is, "I see the world in very fluid, contradictory, emerging, interconnected terms, coupling.

And with that kind of a circuitry, I just don't feel the need to say what is going to happen or what will not happen." And then the second one, "As a net is made up of a series of ties, so everything in this world is connected by a series of ties. Anyone who thinks that the mesh of a net is an independent isolated thing is mistaken." Also getting at Dr. Buchman's point about coupling and connectedness and nonlinear coupling in particular.

Does anyone know who said the top one here? Does anyone know who said the bottom one? I think we'll give out a prize if there's anyone -- hmm?

M: Al Gore.

Ary Goldberger: Al Gore for the top one is actually close politically. Jerry Brown is the top one. And the Buddha is the bottom one. So there's also a connection I believe between those two as well.

So part of what we do or think we do, and part of what we try to interest our colleagues in is the idea of finding hidden information. The idea that if you look at a system, look at the output of a system, that people who have a background in dynamics may be able to extract information, which is interesting conceptually about the mechanism that's driving the system, or mechanisms. And also, the information may be clinically important. It may be of importance diagnostically and prognostically.

One of the outputs that we look at -- not the only one -- are heart rate fluctuations, which Tim showed you. And why do we look at that? Well, they're accessible, which is nice. But more important, they give us a way of quantifying healthy dynamics, because they're the output of an integrated neuroautonomic control system, as we'll see in a few minutes. The heart really is at the maelstrom. It's controlled by a very complex set of feedback systems that involves the autonomic, the involuntary nervous system, as well as other systems that are coupled to that.

So in a way, when you look at heart rate fluctuations, that really is far beyond the province of a traditional cardiologist. That's really looking at integrated neuroautonomic function, so that a variety of perturbations will affect that. And that may be associated with advanced pathology, as in the case of multiple organ system failure. Or on the other side, subtle perturbations that may be pre-clinical might change the way the heart rate regulatory system is acting and might give you an early warning.

That would be very exciting, if one could look at fluctuations in some accessible output -- say heart rate with or without other physiologic parameters -- and get some information about your health that might be relevant to what's going to happen in the future before something catastrophic supervenes.

So now there's some big problems here. I don't know, can people see these graphs from where they are sitting? That would be another big problem. We dim the lights. Is it possible to -- is that a -- because if you can't see the data.

So here are the -- here are some of the big problems in looking at this type of data. This is heart rate, which is the inverse of the inter-beat interval. This is in beats per minute. This is normal. This is from a patient with sleep apnea. And what we try to tell our trainees and remind ourselves is the first thing one does looking at data is to simply look at the data; look at the time series, which is a very different type of approach to data than one generally sees in a physiology laboratory. The idea of looking at this particular output, of looking at the fluctuations is remarkably foreign to most of the people whom I deal with.

But here, this is the normal physiology. And what one sees here -- and once again, in keeping with the themes of this conference, which have to do with noisy variability -- is first of all, the time series. And this is a healthy person who may be at rest. So what I'm telling you here is not an epi-phenomenon of activity.

So in this normal subject, if one looks at this time series, it presents two problems that make it basically entractable with respect to conventional statistical analysis. The first problem is that it's highly non-stationary, that the statistical moments change as one goes down the data set, that the mean obviously changes; the variances changes, and higher moments of the time series may change as well.

And the second property that one can show embedded in this time series is that it is nonlinear. And you get the visual kind of intuition about it. It's nonlinearity; it has to do with a patchiness. If you look at a time series and it looks patchy, heterogeneous, that's a pretty good clue, if it's asymmetric, that probably the process that's generating it is nonlinear. And it turns out that that is in fact the case for this normal time series.

In fact, it turns out rather remarkably. And this is -- I pose this as a challenge here. I'll make a statement that may sound wily over blown. But it's hard to keep people awake after lunch, so you have to say something controversial. I believe that this signal here, which in a sense should be the most innocuous, benign, uninteresting signal in physiology.

The normal heartbeat, according to Walter B. Canon -- who was quoted before for his perversion I guess of Bernard -- would say that the normal heartbeat of a person at rest should be ultimately homeostatic, regular like a clock, metronomically predictable. In fact, if this is the data -- and this is in fact the data in a young healthy person -- the statistical approach to this, the mathematical approach to this turns out to be at the frontiers of complexity. Because the understanding of this requires not just fractal analysis -- which we'll talk about briefly -- but it turns out this system is actually, appears to have features of so-called multi-fractality, which makes it one of a class of some of the most complex signals in nature, including turbulent like flows, where one requires not just a single scaling exponent but multiple scaling exponents. And the system is nonlinear and fractal and then some.

So this signal turns out -- for those of you who are physicists in the room -- to be as challenging as anything you'll see I believe in the real inanimate world, the natural world, the physical world. But it's kind of like turbulence plus. It's that complicated, but the system is alive, so that's a big challenge.

So when they do all of this stuff with the theory of everything, you know, and you get down to super strings and stuff, they're still not going to understand, this is going to be there the next day. This is -- and there isn't any single metric that you can use to encompass the complexity of this healthy variability, this person sitting there at rest, generating the ultimately type of complex behavior. It would be naive to think that you could summarize that type of complexity with any single metric.

So the people who were asking the question that Dr. Buchman posed, which is the clinician saying give me one number, well, that's not going to work. You require a matrix of numbers. And that will, won't work, anymore than you could write the one novel about human existence and that would suffice.

The other interesting thing that turns out -- so this may be as complicated as anything in nature. And what we need to do to analyze that then becomes a deep challenge to understand the control system. But in pathology, we see changes that become of interest as well. And this is an example of a patient who has sleep apnea, a condition where people periodically stop breathing at night and then start breathing again. It's an important condition because it's associated with daytime sombilance, traffic accidents, increased risk of sudden death, high blood pressure, and a variety of other problems.

But if you look at this time series compared to this one, does anyone want to comment? Is there any pattern that emerges?

(Dr. Meyer): Lower dimensional.

Ary Goldberger: Lower dimensional. That's a quick calculation that Dr. Meyer (crested). He is very good at that, but he seems to have calculated a lower dimension. And that -- how did you do that?

(Dr. Meyer): Just looking at it.

Ary Goldberger: Just looking at it. That's good, because that's actually -- the eye is a very good signal processor here and pattern detector. And what you see here is the emergence of something that's much more periodic, much more predictable, lower approximate entropy, if you were to make that measure here. But still not stationary, still this is a complicated signal that would challenge most of the conventional signal processing techniques. But this periodic oscillation of the heart rate actually correlates with periodic changes in ventilation where the person stops breathing and starts breathing abruptly again. But those changes are not random. They are highly structured.

So one of the themes and indeed the theme of what I wanted to say is that one has, under healthy conditions, this highly complex, what turns out to be multi-fractal type of variability, fractality plus a certain type of nonlinearity. And that with a wide class of diseases, there is a breakdown of this type of complexity, not any type of complexity.

And one of the modes of breakdown has to do with the emergence of more organization of a pathologic type and more predictability, which is different in different diseases, but gives you a fingerprint or a profile that you may then able to use diagnostically. It turns out that the data is even more complicated because you can get artifacts, you can get extra heartbeats. And you need to analyze not just one signal, but you would like to have multiple signals simultaneously. So this is a very challenging program.

M: What was the horizontal scale?

Ary Goldberger: The horizontal scale here is in minutes. I'm sorry, it's thirty minutes of data.

M: Are both those graphs from sleeping subjects?

Ary Goldberger: I don't want to -- it would be convenient that the top one is, but that's what people look like during sleep. I'm not sure in this case. But the complexity of the healthy heartbeat is of that ilk, whether the person, for people who are not doing anything.

This, on the other hand, is the way that I, if I were giving a clinical talk or if I was listening, forced to listen to one of the type that one is forced to listen to in the venues that I hang around in, this is the way people show data, right? Yasha is here and he looks at data like this. Not himself personally, but what people will do -- this is the way physiologists look at data.

They would -- the question is we have health and pathology, but you would also like to know about health in young people and older people. So a question is what happens as we get older? Well here is some real -- it's real data. To me it's very interesting data, so it's really interesting data. And it's a young person and an older person and it's about fifteen minutes of heart rate data. And it turns out that these two subjects have identical mean heart rates and the error bar is the same, so the variance is the same.

So that's the secret to aging. If you wanted to know what happens to your heart rate dynamics with age, this is what would be done in the conventional analysis. People would -- and then they would compute a -- you don't have to compute any tests here. There's no significant difference here.

So that's -- and it's hard to publish these sorts of data. To me this was like some of the most interesting data we had seen. And the reason is and part of what -- see, it's important to -- on average, this is the Mona Lisa, right? So part of what I spend my day doing -- I want to share this with you because I hope to unburden myself, so you can take part in this challenge -- is that in talking to people in different communities, one of the lessons -- and this may seem like a trivial one to the people in this room -- is to convince physiologists that there's more information than is just simply in the mean value, the first and the second moments of the variance.

And just as there is some anomalous behavior of this Mona Lisa -- although on average, it is the Mona Lisa -- that looking at time series, looking at fluctuations, looking at the hidden information in the dynamics is of extraordinary importance and is a defining feature of the mechanisms. And that's a very important discussion, which you need to get into with your colleagues.

This is, in fact, the young subject on top, the older subject on bottom. For which is the dimensionality higher here, (Godfried)?

(Godfried): The top.

Ary Goldberger: The top has a higher dimensionality. Well, they have the same average, same variance. The approximate entropy, which Tim mentioned on the top is 1.09, on the bottom is .48, which is a highly significant difference. The lower signal is more predictable, less complex by that definition, lower dimensional. So there is a change with aging. There is a loss of complexity. The mechanism of that is something that needs to be worked out and is probably multi-factorial and may in fact have to do with uncoupling of different parts of the control system.

The other thing which would be of great interest, if we can now quantify cardiovascular health. And in the past, we really didn't have any way of talking about what it was to be healthy. But if we can quantify that and quantify what it is to become older, than if you have some intervention that you think is going to make you go from bottom to top, which would be what I would call the Ponce DeLeon Effect, this thing, the elixir of youth that most people in this room have some interest in, that would be of interest. But you'd want a way of testing that. You'd want a way of knowing whether in fact what you just took or what you just did made things better or worse. And this is a way of getting at that. I'll talk about one potential intervention, namely meditation at the end.

So the idea then is that the healthy system has this sort of fractal property or properties. And from a -- once again, in talking to my colleagues, I can talk to them about fractal structures, because they can see self similarity, the idea that if you look on different scales that there is some property or properties that have this internal look alike feature.

And the different terms of scale and variance, self similarity, power loss, scaling, long range correlations are things that one can convey, at least for the architectures inside the body. So the idea that the complex anatomies that are within us, the blood vessels, the lung tubes, the brain folds, the bowel folds are all fractal like is something that one can communicate relatively readily.

The more difficult problem is with respect to the time series, the idea that heartbeat control, the control of human gate. And also of interest to us, at least, is the outputs of the human imagination, what we call the biology of imagination may have certain aspects of this type of fractality, classical music being one. But what's common to both is the idea that there are many different scales of length -- or in the latter case -- of time. And there is no characteristic or preferred size or frequency, that nature doesn't play favorites with scales of time or length with respect to, at least to physiology.

There are bounds, obviously. But within those bounds all of the different frequencies have some ecologic relevance and equivalence. This seems, as I said, to apply not just to structures within us, but to the structures that we both seem to enjoy being around like trees and like this cathedral, which is where?

M: (Inaudible audience comment).

Ary Goldberger: (Inaudible response to audience comment).

I won't talk today about the notion of fractality and some of our creative outputs. And I usually don't even admit to my colleagues that I published first of all, an article about fractals in <u>The Birth of Gothic</u>. But worse than that to most of the people I work with is the fact that I published it in a journal called <u>Molecular Psychiatry</u>.

So that clearly puts me in a rather awkward position vis-à-vis the usual cardiologic community, which likes to look at this structure, namely the fractal cardiovascular system and in particular, the cardial-arterial tree, a rich redundant system of blood vessels. The electrical wiring of the heart that maintains the normal dispersion of electrical information is the so-called (hisperkingee) system, which once again is a tree like, fractal like network. So nature likes this type of system, at least structurally.

If you look and you dissect out the heart - in this case, of a dog -- you find that the coastline, the internal architecture of the heart is not smooth and regular. It looks like this -- this is called a pectinate muscle. It actually has this tree-like -- it's like a tree embedded within the heart, which is quite interesting. You can also look at the breakdown of fractal organization in the vascular system. This is the normal tree-like arterial system from a normal kidney arterial tree injected with a radiographic dye. On your right is the same arterial system, but from a patient with a disease called sclera derma, where there's an obvious change. What's the dimensionality on the right, lower or higher?

M: Lower.

Ary Goldberger: Lower. Radiologists refer to this as pruning of the arterial system. And there's been some interest in recent years. And actually what you'd like to do is to put a number on how much pruning there is. And the conventional numbers that people put on radiographs don't allow for calculation of fractal dimensions. But that actually would be one way some measure of fractality or lacunarity would be one way of getting at this.

There is interest now in actually making those measurements. So that would raise the question of whether in fact there is topological disease, or whether topological metrics could be used to quantify disease, which is an interesting proposition. I went to the literature. There wasn't much in this. The only reference I could find to that was actually from this cartoon, which says, "I'm going to be very frank with you Mrs. Boswell. You've got cubism." Well, that -- I guess cubism would be the opposite of being gothic, which is a multi-scale state. So that is a pathology.

And not only can you get cubism, excessive structure in this context, in the structural context. But you can also have a breakdown of fractality in a dynamical context. And once again, this is the normal heartbeat shown below, as it's represented and as it looks and as it feels, if you feel your pulse, this apparently metronomically regular sequence of pulses.

This is the cardiologist's view of the heart. The actual view of the heart is a little bit different. If you look at it from the dynamicists or the neuroanatomicists point of view, the heart no longer fills up. It's like a New Yorker's view of the world. Remember those maps, you know? Well it's not like that really.

The heart is really embedded in the center of this highly complex feedback system where it's being buffeted by the autonomic nervous system, this sympathetic part that tells it to speed up, the parasympathetic that or vagal part that tells it to slow down. There's input from up above in the cortex, from down below. This is an enormously complicated couple feedback system.

And that's what's really driving the heart. That's what makes it such a powerful indicator of overall physiologic health or disease.

And once again, this is the heart rate time series of a healthy individual. And this could be someone at rest or asleep. Still highly complicated, highly non-stationary. And the idea in terms of the gist of the idea that this is a fractal like output can be gleaned from this type of non-quantitative representation where the idea that you can have self similar architecture in space, and on the right side, the idea that you could have self similarity in time. That if you look at a time series over hours or minutes or less, that there are certain statistical features that are built in on those different timescales. And if you cover up the absyssa, you don't tell people what the timescale is and renormalize to the same variability, that you can't tell them apart, that there is something that's going on over a period of hours or minutes or seconds that's statistically self similar. And that's the gist of the idea of this type of fractality.

But as I mentioned, this is -- the healthy heartbeat, it can't be characterized by simply one scaling exponent, that it requires multiple scaling exponents. And in fact, it is multi-fractal, which is the topic of a paper that the first author of which is (Plomin) Evanoff, a colleague at Boston University in Gene Stanley's group. And that was published in Nature back in June of last year, if you're interested in the idea of multi-fractality. But once again, just in a simpler way, the time series of the healthy heartbeat, its broadband spectrum on a log-log plot, the one over F or inverse power law scaling.

So the question is -- we saw that you could have a breakdown of fractality in a geometric or anatomic sense in the pruning of those arterial vessels with sclera derma -- what happens in a temporal sense to this idea of healthy variability? Well, this actually sorts out in a rather straightforward way.

Because if you start with the idea that under healthy conditions that healthy dynamics are multiscale and have this long range order, these long range correlations -- the correlation property where things that go on now have an effect statistically of these, not just one or two beats ahead, but hundreds or even thousands of beats into the future, like one would see in a critical phenomenon -- that there are only two major ways that this type of dynamic can breakdown.

One would be if you lost all correlations. And then you would have on the lower right there uncorrelated randomness. It would look like white noise, for example, static. And in fact, that's the heart rate pattern that you see in a cardiac arrhythmia called atrial fibrillation. Over short timescales it is in fact essentially random and the person describes a highly irregular pulse. So not everything irregular is fractal.

Of more interest for this discussion is that the other way you can break this system down is something that has multiple timescales, can collapse in a pathologic sense to something that has a single dominant timescale. That would be a very anti-fractal place to be, to go from having multiple scales of time to having a single dominant characteristic frequency. And that's what you see on the left there, where the heart rate rides up and down in not quite a (sign) way, but close to that. And so there's something going on that's repeating itself every minute or so, in this case.

And this person actually turns out to have severe heart disease, severe congestive heart failure. And patients with severe congestive heart failure are like those with sleep apnea, start to get the emergence of a characteristic timescale. And that's always a marker of pathology. When you go from having multiple timescales -- remember they said that Gerry Ford couldn't chew gum and walk at the same time? Remember that little political joke?

Well, the idea is that you should be able to do more than one thing at the same time. Your physiology is infinitely complex. If you believe it's multi-fractal, it has infinite singularities. So that's what you would like to get back to. But on the left, lower left there is the complete loss of complexity with one characteristic timescale. He's either walking or chewing gum, but not both.

So the underlying concept here that we've tried to develop is that of the decomplexification of disease, which is the output of many systems actually becomes more regular and more predictable with pathology. And that I can tell you as a clinician that clinical medicine is not feasible without such stereotypic periodic behaviors.

The clinicians, if you go to see a doctor, they're looking not to assess your multi-fractality -- because otherwise you'd get out of the office quickly -- they're looking for patterns. They're looking for the emergence of some pathologic patterns, whereas the healthy function once again, which has a broadband spectrum. And whether it's determinist to chaos or includes that or not is not something that we need to talk about now, but it's certainly fractal like in its variability.

The other side, the loss of fractal complexity resolves this clinical paradox, which is that patients with a wide range of disorders often display strikingly predictable or ordered dynamics, despite the connotations of the term disorder. They're not disorders, they're reorders.

So Parkinson's Disease, which has been in the news with Michael J. Fox, the actor who's tragically developed that syndrome. But what that's characterized by is -- and if you saw the last TV show that he did, he had his hands in his pocket a lot of the time, because he has an uncontrollable tremor like Janet Reno. But it's a highly periodic dynamic. It's uncontrollable periodicity.

On the other side, if you think about various neuropsychiatric diseases and if you think about personalities that you may not want to be around, people who are obsessive compulsive drive you crazy because they keep repeating themselves, they keep repeating themselves, they keep repeating themselves. And it drives them crazy too. And a whole long list of periodic syndromes. Those are not exceptions, those are really more the rule. And if you're in a position where you're watching your cardiac monitor, if it starts to look like a sign wave, that's an extremely bad sign and you should do something to change that quickly, because your doctors probably will be jumping on your chest shortly.

If you look -- this is not simply cardiac -- but if you look for example at the electroencephalogram; the brain waves of a healthy adult in deep sleep -- and Dr. (Meyercrest) was one of the first people to look at EEGs from a dynamical point of view. You don't need him to tell you that the dimensionality of the top tracing, this is stage three/stage four sleep -- which may characterize many of the people here -- is a lot more complex than the EEG during hepatic coma, which once again starts to smooth out, that visual smoothing out. The loss of complexity there is visually apparent in this case.

So if you -- the warning here is that excessive regularity is bad for your health. And the examples are not just more subtle things where one needs dynamical probes. But there was an experiment that was done inadvertently in Japan a few years ago where there was a Nintendo like display. I guess is was a Pokemon figure that was flashing on and off, but it was periodic. It was -- so that's called photic or stroboscopic stimulation. And the kids came pouring into the emergency room with seizures and they were throwing up. So the brain does not like excessive stimulation. That's not good.

No part of your body likes that. If you sit at a keyboard all day and you're being stressed and you're excessively being excessively periodic, you know, you get carpal tunnel syndrome. If you bang your head literally or figuratively against the computer screen, that's what daily life is all about, right? And then you get migraines, you get stressed out. That's excessive predictability, excessive periodicity.

So there was this nice ad in The New York Times last week. It said, "Fight Saneness." That's the great battle of modern life. Bu it's also, that sums up this dynamical battle. You're fighting excessive regularity. You're trying to restore complexity. And the way to do it is with art. Monet is -- that's the way to do it. That's the answer. That's the cure.

I wanted to summarize for some colleagues where I thought chaos complexity were going in biomedicine. So I picked out a bunch of areas, including going beyond the human genome project to integrated molecular bio-dynamics into ideas about medical practice, how we disseminate information, how we look at drugs, what we heard from Dr. Perelson about immuno-dynamics. I just wanted to highlight one or two areas.

One of the ways that we represent on a signaling basis, whether it's in the immune system or whether it's in any system, if you pick up -- this is from Science a few years ago -- if you pick up any description of how signals are communicated within the body, usually you see this type of diagram. The idea -- and this is a very linear construct that we inflict on medical students and ourselves -- the idea that there are arrows. This one doesn't even have any feedback in it, much less multi-scale variability or nonlinear coupling, all of which are missing.

So how should we think about -- how are signals really being transmitted? How has the coupling that Dr. Buchman talked about really going on? Well,

I have no idea how one would actually represent that. The closest I could come in a cartoon-like way was to take this, borrow this from Leonardo DaVinci's picture of turbulent flow. And the idea is that this -- if you're going to represent something, it's probably better to use something that at least captures complexity and that kills it off.

So if I were speaking to medical students, I would suggest that what's going on here may involve what you could call spin cycle signaling and the idea that physiologic control in some way incorporates coupled cascades of feedback loops, a very different notion that's on the previous slide, and that those operate far from equilibrium. Once again, very different from the homeostatic set point, engineering notion of the way people think about or represent physiology. How this, one would actually go about modeling this is a bit of a challenge. The analysis we did of the multi-fractality of the human heartbeat would once again put the output of that system well within this type of turbulent-like cascade.

More examples of chaos/complexity: smarter monitors for diagnosis and prognosis; in pulmonary medicine, better ventilators, the idea that you might want to incorporate noise; as (Belesukian) colleagues proposed, the detection of sleep apnea; new ways of monitoring gate; new ways of enhancing complexity through exercise and potentially through meditation. Some interesting work going on at Los Alamos on robo-chaotics: Can nonlinear machines be made more adaptive and creative? We talked about anti-aging interventions. Is there a Ponce de Leon effect?

And I want to mention finally, a new NIH research resource for complex physiologic signals, which may change the way we do the type of business that I think we want to do. Before I close with that, let me just pose a question to you, which is during meditation -- which would seem to counter what we were just talking about -- the idea is that if you meditate, you're kind of smoothing out variability and making yourself more placid, more like a silent lake, rather than some cascading turbulent-like flow.

But it turns out, interestingly enough -- and we started to look at this systematically and actually published one paper on this -- when we looked at heart rate dynamics during two forms of meditation that were being performed by young to middle aged individuals -- one was traditional Chinese chi meditation where there is a lot of visualization of like a lotus in the abdomen and people are actually breathing slowly, and then a second form of meditation that seems unrelated, yoga, kundalini yoga meditation, which is a bit more active in some ways -- but if you look at the heart rate dynamics during the meditative phases here, there's no smoothing out. Things are not becoming more linear or more calm. In fact, you actually get more variability during the time that people have started this meditative protocol. And the amount of variability here is dramatic.

These young people -- if you can see the scale on the left -- during the time that they are meditating -- and that's associated with slow deep breathing, which we can also demonstrate -- three or four times a minute is what their respiration goes down to, which is quite remarkable. Their heart rate is going up and down with a variance of nearly forty or fifty beat per minute in some cases. That's a tremendous range of variability.

So the idea that we've been exploring is that perhaps meditation, far from making things in a sense more placid dynamically is actually -- it's more like a Nordic Track for your autonomic nervous system. You're taking your autonomic nervous system and making it more plastic, making it more variable, restoring variability. And similar things may go on during sleep as well. So you're kind of re-injecting or allowing the system to go back to this plastic, highly variable, highly adaptive state from which it can recruit a whole bunch of different responses.

And when you're in this zone -- and that's what athletes call it, it's being in the zone -- you're actually the most nonlinear, the most complex. Although it feels in a sense linear; it feels kind of smooth and kind of frictionless. But that's not the physiology, is extraordinarily complex.

So this, if you wanted to give a drug, it should look like this. If you give a drug and it smoothes things out, it reduces variability, that's a drug you should avoid. And the problem with it, when the pharmacology people in the pharmaceutical industry assay for the effects of drugs, they have not been doing dynamical assays. They are targeting a single drug for a single receptor. And then when they put the drug out there, they find it's killing people, like the drugs that were used to treat arrhythmias back in the last decade. All of those drugs that turned out to be highly toxic and were associated with several fold increase in mortality when were looked at in retrospect, they killed off heart rate variability and eventually killed off people.

So the final thing I want do here, which is basically to let you know about a new resource that's funded by the NIH, by the National Center for Research Resources. It consists of three components: PhysioNet, PhysioBank, and PhysioToolkit. This is a multi-center resource that I direct. It involves ourselves at The Beth Israel Deaconess at Harvard. It involves colleagues Roger Mark and George Moody at MIT. It involves Gene Stanley's group at B.U. and Leon Glass at McGill.

And the idea here is basically to do unto physiology what had been done unto molecular biology via the gene bank, where by having data, in which case, in genetic sequences. A huge amount of work was catalyzed because there was a common database. In physiology now, if you try to get the sorts of data that Dr. Buchman was talking about or if you want to look at heart rate variability, if you want to test or re-check to make sure that we did the analysis correctly on any of the data that I've shown you, you have been unable to really obtain those data sets.

So we thought and in the past we've been involved in kind of, on a small level, in making data sets available. But what you really want is to be able to test hypotheses and develop new tools in training sets, and then go out and test them in large scale test sets is the data. And the data has to be high quality data. And it has to be available to everyone.

Open source data is extraordinarily important. It's the only way this field is going to move forward. Otherwise, you have a bunch of people with their own proprietary data. They won't share the data. They won't tell you what algorithms they're using, because that's proprietary. And no one is going anywhere with that.

So our solution to this is to basically -- we want to give away the shop. We want people to have all of the data we can possibly put out there. We also want people to contribute their data to this, under the PhysioBank component. But this is more challenging than the genome project, because you need not just the data, the annotated data. You also need the new tools to handle non-stationary, nonlinear data sets. Because the statistical books fall apart there. They say if your data is nonlinear, non-stationary, we can't help you.

So the PhysioToolkit part is to provide the algorithms and to discuss their uses and limitations that can be used to handle these sorts of data. So that's the challenge that we, as a first -- just to show some of the power of this, what we've put up here in collaboration with a group in Marburg, Germany, Thomas Penzell and colleagues, they have a large amount of data on patients with sleep apnea. And one of the challenges is how do you recognize sleep apnea without doing a whole type of polysomnographic nocturnal recording where you have to hook people up to EEG and so forth? But you recognize the changes that the -- for example, those that I showed based on changes just in heart rate. And just put a heart monitor on people, recognize a syndrome, treat it, and see if you've affected it.

So we have as a competition, a time series competition that's made available through the resource that challenges the community to see if they can diagnose obstructive sleep apnea just from the electrocardiogram. And the results of that competition will be part of the Computers in Cardiology 2000 meeting coming up in the fall. And if it turns out -- if it already turns out that people actually have very good algorithms to do this, those can then become part of actual devices that will enhance diagnosis and prognosis.

But without the data, it was impossible in the past to even pose that question. Now the data is there, test set and training set. People can go to that, they can go back to it. You can go to that and you can take a look at that. You can download data here. Your students can download real live time series.

And one of the other features here is that we strongly believe that anyone who publishes an article, particularly with tax payers support, having to do with physiologic signals -- it doesn't have to be heartbeat; it could be electroencephalogram; it could be any signal -- that there really should be the expectation that those people will make available the raw data, the signals on which their conclusions are made.

And if that if you won't make those data available, then your article basically shouldn't be published. And if you're funded by the NIH, that that should be looked at in a negative way. The people who -- the good guys in the community who make their data available and those data then go on to be used by others, should be credited with that. That's a huge contribution. That those who are unwilling to make their data available, that should be something that has negative reinforcement. I think we all benefit by that and medical diagnosis will also benefit. But the challenge is on the community to make those data sets available.

And the other payoff, why should people do that? Why give away stuff? Well the answer is because if you give away stuff and other people do as well, then you have access to much larger data sets than you would have initially. It provides and breeds the best combination, we think, of competition and cooperation. Having the data out there means that there is more data. But there's also competition to get there first, to get the right answer, the better algorithm. Because the person who does that is going to be a winner.

But there's also the control, which means that if mistakes are made inadvertent or otherwise -- for example, the Wolf algorithm for (Le-off-on-off) exponents had a mistake in it, in inadvertent mistake, that one of my colleagues Joe (Mitus) discovered. We let Wolf know about it. But the community at large, even if we published that in a journal, it would come out six months, two years later in a letter to the Editor. PhysioNet allow you to correct mistakes and also to discuss possible discrepancies immediately, tomorrow, today. So we're very enthusiastic about this. We invite you to participate. And hopefully, this will move this whole field forward with a challenge that Dr. Buchman described. Thank you. [Applause] Yes sir?

M: Can you speculate, or are there actual research results, on what mechanism in the body is responsible for the extra regularity in the case of disease?

Ary Goldberger: The question has to do with what causes the loss of variability, the extra regularity with disease. And there, as far as I know, what we think is that it has to do in part with the mechanisms that Dr. Buchman talked about, which have to do with loss of coupling. As you pull out different components of the system, they give you a loss of certain types of variability.

I would also raise the possibility that it's not just any type of coupling, that excessive coupling of a certain type may also -- if things get over coupled, too tightly coupled, and they stay mode locked, that's also pathologic. So if all of the components are coupled together, which is what you see in heart failure, there is excessive coupling of certain components. And there may be a combination of that, plus a breakdown of other types of coupling. So that's a deeply complicated question.

But the challenge once again is now that the data are there, you can go and, if you have a model, a punitive model, you can test that model out on real world data. The physiome project was discussed before, the idea of modeling different systems. Well to do that, you need to have the data. So this is a way of providing data against which to test exactly the type of model that you were referring to. Yes sir?

M: I think you answered my question really. That's a similar question I was going to ask you, and you answered beautifully. Now to test to your answer, can you do the following? Is there any quantitative relationship between the number of times scales needed to describe your time series, normal heartbeat and the number of organs to which the heart is coupled? How many timescales do you need to describe normal heartbeat?

Ary Goldberger: The question is, how many timescales do you need? Well, if it's a fractal system, in a sense, there are over a range, there are multiple, if not infinite timescales. The question is, does that correlate in some way with the different parts of the feedback system? You know, there's a given time -- does a given frequency correlate with a given anatomic or neurologic locus?

In the case -- and the answer is yes and no. If you breath in and out at a frequency of .2 hertz, say twelve times per minute, that will show up as a spike in your heart rate spectrum. But the heart rate over a range of hours or longer, has a broadband type of spectrum, which is in a sense, scale-free. And what we think is that the feedback across and among components somehow generates this complex variability that doesn't -- you can't point to, in a sense that it will, a linear, you know, you point here. This particular frequency corresponds to this input, or -- and I don't think it works that way.

But once again, the other thing that we can -- the other point to make is that not all one over F spectra are equivalent. You can in fact generate a linear, a system that is linear where you have the super position of a whole bunch of different frequencies. You can sum up, if you pick the frequencies right and the right amplitude, you can get a one over F spectrum. How do we know that the heartbeat is not just that type or kind of, what I would call a trivial or a linear super position of different frequencies that happen to add up? Because that's the way the body is. You know, you've got different components.

Well, the test of that is in the multi-fractal analysis. That if that were the case and it were a linear system, if you took the time series and randomized the phase, got rid of whatever nonlinearity was there, it wouldn't change the scaling. And for the conventional Fourier analysis, in fact, you can tell, because we don't look at phase information in the convention FFT.

But in the multi-fractal analysis, the multi-fractal spectrum, which shows a broad range of scaling exponents, that property is killed off by shuffling and randomizing the phase. So in the real heartbeat time series, there is something more than simply the information that's encoded in the scaling of the amplitudes. The one over F scaling is not the same for the heartbeat as it would be for a contrived, one over F spectrum, where you simply added up different amplitudes at the appropriate frequency. So that's the proof that this is something much more than that super position phenomenon.

Yasha Kresh: Ary, it's also a low dimensional noise, a signal. In some ways, it's very hard to find it.

Ary Goldberger: Yes. The other debate is: to what extent is there deterministic chaos, which through a relatively small number of components, could generate an extraordinary amount of complexity? And that's, once again, testing that hypothesis requires the data. That's a big debate now about to what extent there is low dimensionality versus higher dimensionality.

But the problem with that debate has been it's the old lime and the elephant parable. Is that people were looking at different data, at different data sets, different elephants, in fact, with different algorithms, different sensory modes. And then they were coming to meetings with conclusions that no one could agree upon. We think it would be a lot more exciting if you could come to the same meeting, if everyone came here looking at the same data set with algorithms that they would make available.

So any disagreements, we could presumably, at some point, converge on. We would find that we disagreed because someone was misapplying an algorithm, or it was a different scaling regime, or when you tested an algorithm on surrogate data, it didn't come up with the answer you knew was already in the data. So I think that these are now questions that we can begin to address.

Timothy Buchman: Ary, you might mention what we do clinically, manipulating the autonomic nurses system, and what's the concern both in terms of partly, variability, what we clinically have here, like in the instance with beta blocking agents and what that seems to do to patients tolerance of stress.

Ary Goldberger: Essentially and pharmacologically, the best way to kill of variability is if you give an agent that block the parasympathetic, the vagal side of the nervous system, like atropine, that almost looks like a heart transplant patient. It's a very flat type of time series.

Yasha Kresh: Not my heart transplants.

Ary Goldberger: Your heart transplants?

Yasha Kresh: They're not flat.

Ary Goldberger: They're not flat?

Yasha Kresh: No.

Ary Goldberger: Even initially?

Yasha Kresh: They're not flat, because I think you were kind of touching upon it. The heart has its own nervous system. So even in a heart that's out of the body in a petri dish --

Ary Goldberger: It still has some residual -- but actually, if you --

Yasha Kresh: But maybe in Boston, they have -- (Laughter)

Ary Goldberger: Everything in Boston is flat now. (Laughter) If you give beta-blockers, you can actually -- it depends on the dose and the setting -- but you can actually enhance variability. Which is interesting, because there are very few drugs that have been shown to increase mortality in cardiac patients. And beta-blockers were the first ones that were shown.

And it's a paradox, because you think that by blocking the sympathetic system you would get rid of variability. But if people have too much sympathetic tone, that's a bad thing. It's like being stressed out. That will reduce your variability. By giving an appropriate dose of a beta-blocker, you may, in fact, allow the system to kind of relax back. It's maybe like a pharmacologic meditation. It allows the system to re-expand, to unclench. And in fact, the data are very good that beta-blockers enhance mortality, improve mortality after myocardial infarction, and they also increase variability.

So anything that increases your heart rate variability in a physiologic way is good for you, generically. Anything that reduces it, you should keep away from, whether it's your in-laws or whether it's some drug.

Yes?

M: Your meditation data is really interesting. I want to know if you could put it up again. Because it seems to me that they really are very different. And I think that for the Kundalini meditation, there is an increase of variability. But for the chi meditation, it seems to be really homogeneous. And if you take into account, the breathing part of it, I'm not sure if I can agree with the loss of dimensionality. And in that sense, it wouldn't really go along with the idea that something that is particularly healthful would be more variable than something that is not.

Ary Goldberger: Yes. What you basically heard was a difference of opinion in the way I described the data. And without -- the answer I will give you is this. The article and the data are available on PhysioNet. You can download the data. You can reanalyze them. We really didn't make any claims about dimensionality, so I don't think you will be offended. But I don't think we said it. But if you want to go on now and do a further study -- which I think -- and the hypothesis you raise, I think is a very interesting one -- those data are available, and they are free. And you should do that.

M: All right, I will check it out.

Ary Goldberger: And we would be very interested to hear from you. So those data are up there. There are some flyers that describe PhysioNet, that I've placed at various places as well.

M: Yes, I have a very quick question.

Ary Goldberger: Yes?

M: Can you comment on your statement that excessive regularity is a bad thing, or can be a bad thing? It seems like a very general statement, and you would have to talk about timescale and perhaps a particular system that you're talking about.

Ary Goldberger: Yes, the idea, when I said excessive regularity is bad for your health, how literal should you take that? And what I showed before that, I guess was the idea that if there was a compression and a compassion of timescale, so that you lost information that was previously disseminated across many timescales, and if something is perfectly regular, by definition, there is only one timescale. And that is in fact a bad place to be physiologically.

So if you're looking at the physiologic output, if it becomes metronomically predictable, that in all of the circumstances that you've seen, is an extraordinarily bad place to be. And it usually correlates with brain death, or death of your personality, or something. It's not -- you just -- it's an un-physiologic place.

M: I understand it in terms of the regularity of a physiologic signal. But you used repetitive stress disorders as an example of an environmental input that's regular that can be problematic.

Ary Goldberger: I was, yes, I took a bit of a stretch there by suggesting that if the environment that you're forced to encounter is excessively regular, that that might be bad for your physiology, if it had to cope with that over a sustained period, as opposed to having a more syncopated, jazzier environment. And I believe, that's a testable hypothesis, in the sense that you could take a bunch of people who were all working on a computer keyboard, and see if those who became afflicted with repetitive stress injury, Carpal Tunnel Syndrome, in fact had a more periodic type of typing motion. You could test that, you know, the old water torture technique was not to make people face Niagara Falls, but it was that very periodic water dripping, which you know, the sound that drive you crazy, literally. So I think that those -- but that's a testable.

So I think the other side, which I think is a more pleasant thing to think about is things, if that is true, than things that we would do to make our lives more healthy and not just more pleasant in some aesthetic or cosmetic way, but actually more healthy, better longevity, better immune system function, and so forth, would be things that would enhance variability, enhance -- and so anything that you're doing, if you want to reconfigure your day, anything you're doing that excessively periodic, think about it, and figure out a substitute, and see if you live longer.

M: Actually, there is --

Ary Goldberger: A survivist talks, almost.

Yasha Kresh: There is a conspiracy between excessive regularity and clinicians. Because as you can imagine, a clinician couldn't do a bloody thing unless there was regularity to the pathology. So I think there is a conspiracy there.

Stephen Small
Medical Errors

Yasha Kresh: Our last speaker is Mark Smith. I'm sorry, yes. [Applause] I'm sorry, Stephen Small is our last speaker. And the topic is Medical Errors. And he is here.

Stephen Small: I haven't pitched this talk to an audience of this size before. I'm a physician at Mass. General. I used to make house calls and other prosaic things. And now I'm an anesthesiologist. Hopefully, I won't put you to sleep, the way I do with folks everyday for a living. Laser pointer? Is there a laser pointer?

I don't have data to show you. But I'd like to just give you a qualitative presentation on what thinking about complexity has meant for me in terms of what I've been devoting a lot of my time to in the last six to eight years, which is to try to develop a more humane, rationale system of care in healthcare. The talk today, perhaps you can think of it as an index case of undesirable things, events in complex socio-technical systems. Socio-technical meaning, a place where people touch dangerous technology, and people touch complex things all of the time, where things don't just happen with automation.

Is there another device? For those of you, it's hard to avoid this, since last November, even if you don't read the newspaper or watch TV. The National Institute of Medicine came out in November and said that 100,000 Americans die every year due to preventable adverse events in healthcare. You go into the hospital expecting to have a knee operation, and you come out with brain damage or no kidney or something else. And for those of us who have been working in healthcare for a long time, this is nothing new. And in the context of the way things emerge in complex systems, this problem, if you will, the tipping point in Massachusetts was the Betsy Lehman case in 1994, where a Globe reporter died of a preventable drug overdose at Dana Farber Hospital.

We've known about this for decades. And the National Institute of Medicine report essentially was triggered by many, many things: multiple deaths across the country being picked up by the media, and managed care demanding that healthcare be more accountable and more efficient, etc. And so what the report came out on the web and said was, there were six or eight chapters. And these are the chapter headings, demanding a comprehensive approach to fixing this complex system that was healthcare. Complexity theory, that we've been talking about in the last few days, was not represented in this document. You can download it from the web for free.

They did, however, talk a lot about things that Don Norman writes about, for example, as to how to fix complex systems by making them less opaque, simplifying them, reducing variability and these things. But people have not yet applied complexity theory to this issue. And there really is no (literance) that I'm aware of as far as errors in healthcare.

Can someone help me with this? There we go. You've got to be pointed right at it

President Clinton next demanded that all of the federal agencies that report to him having to do with healthcare address the policy directives from the National Institute of Medicine report. So you had about twenty or thirty agencies, of which I've listed five: HCFA, which pays for the care for 70 million Americans; Medicare and Medicaid; The Department of Defense; The Veterans Administration; The Federal Drug Administration; The Center for Disease Control. And a few months after the NIOM report came out, these agencies said that they would follow the National Institute of Medicine recommendations, which will really have a dramatic impact on the way healthcare is paid for and the way accountability is addressed in our system.

And these are all of the different things that came out of that directive. The Leapfrog Group is a bunch of purchasers, large companies like General Electric, who have billion dollar healthcare budgets, who are saying, we need to find ways to teach healthcare how to control their system dynamics the way we have done in production of industry.

If you just sort of conceptualize what the nature of adverse events in complex systems is, over the course of the last century and a half or century since the Industry Age, as things have become more complex, they've actually indeed probably become safer. Right about here, we stopped having aircraft engines falling off jet airplanes due to metal fatigue; that problem was solved. Lots of other things that used to happen no longer occur.

But what we're now facing as we cross the threshold into the 21st century, is that we've now become victims of our own success. We can essentially operate on you before you're born and operate on you after you die. We operate on babies before they're born and do intra-uterine surgery, which is amazing, if you think about it. And we also, some of the sickest patients I've taken care of have been organ donors.

If you get hit by a car and your brain is squashed, then your blood pressure drops out to the floor, your body gets very cold. And when I support an organ donor whose organs are being removed for someone else to use, those people are very very, very sick. And yet, we can warm them; We can pump them up with blood and fluids and drugs, and get their heart going, and essentially, keep them going indefinitely until their organs are harvested. And then we can keep them going after their organs are harvested.

So these are the kinds of problems that we are facing. We are not dealing with third world problems, such as recycling IV fluid, and trying to sterilize, use old instruments that should be cast away that are one time use instruments. We've solved a lot of those issues. Richer is indeed safer. And so the problems that we're facing in a complex, technologically driven society are issues of complexity and not being able to understand, predict, or control the many interactions that we have when we take care of sick patients.

You can be the best practitioner you can be; you can know everything. But unless you're working in harmony with the rest of the systems and the technologies that you are using, we will have this almost ineradicable level of preventable adverse events.

And if you think about what the numbers mean: four billion prescriptions, thirty-five million anesthetics, one tenth of 1% of that translates into 100,000 dead people a year. And I personally think it's a lot higher, having been a part of some of the research teams that have done the outcome studies. I'll get to this later in my talk. It's extremely difficult to get at the kind of sensitive nuanced data as to things that are happening to people, because of the nature of the way the data is walled off, if you will, in organizations, due to medical/legal reasons and other issues.

I don't want to go into this too much, except to say for me, it's an interesting refection point intellectually as a clinician, quality is something everyone understands. But safety is something that is poorly understood, at least in healthcare. It's being talked about as a dimension of quality. But in contemporary terms, you can't really talk about quality without talking about cost, value and cost. If you have infinite resources, then your quality is really a determinant of just what you're willing to pay.

But in reality, we all live in an era in which cost is being rationed. And it's being rationed implicitly as well as explicitly. And from my point of view, safety is really a code word for respect, how much resources you're willing to allocate. Safety, we can determine what is safe and what is not safe. And then we can deliver that to the quality people and say, you make a decision. If you want to make a decision for less safety, that's an ethical choice, as far as a human life is concerned.

The construct of safety, whether we're talking about task analysis, cognitive task analysis, or heuristics, or user rate of design for new technology, is language that is not used by quality people, who are mostly clinicians and managers who are looking at better diabetic control, or reduction in mortality from heart attacks.

There are a number of theories -- are there any organizational psychologists out here? There are a number of reigning theories of adverse events in complex systems. One of the most popular models has been Jim Reason's model, where he has described the sharp and the blunt end of organizations. The sharp end being the actual nurses, pharmacists, ambulance drivers and physicians that are laying hands on people.

And the blunt end meaning the decision makers, the managers, the resourcers, the CEOs, who come in and out of systems over different timeframes and whose decisions about purchasing and about the types of technology that is being used, staffing, resourcing, policies and procedures, often have downstream effects that maybe five, ten, or fifteen years later. For example, the Challenger disaster, Bhopal. If you look at organizational disasters on that level, the policy makers are often very distal from when the events are actually occurring.

And then Reason's model, he talks about latent pathogens, which is something clinicians understand. Latent pathogens are, for example, something very simplistic, a desk bolted to a floor in front of a fire escape would be a latent pathogen, if there ever were a fire. And you can take that to its natural extreme.

If you walk through a hospital or ambulatory care clinic, there are dozens of these latent pathogens which, in the presence of an active trigger, such as one person calls in sick, another person has to leave because a family member passed away, all of a sudden you're short staffed, coupled with the fact that there is an epidemic of Strep Throat, and blah, blah, blah. You then have a cascade in your system, and then your latent pathogens are able to interact with whatever contingencies are occurring on the face of your system that day, and you can have an unpredictable adverse event.

Perrow popularized a model called normal accidents, in which he said that there are some technologies that are just too risky for humans to control. He was writing, I believe, this was in 1984. I think Three Mile Island was '79. And his model, based on Three Mile Island, was that there were two drivers, one was interactive complexity, and the other was the degree to which these things were coupled in organizations.

So that if you had a technological process that people were controlling, in which there were many, many interactions below the surface, that you then, if you entered a tightly coupled phase of operations, then things would then quickly cascade into a disaster that was unpreventable, because it was completely -- the people did not have a mental model for how things were occurring in real time, and there was no way that they could solve this, in order to prevent the catastrophe.

I think that complexity theory, at least in my analysis of the many adverse events that I've studied in healthcare, has a lot to offer. There are a lot of fresh thoughts in complexity theory that I think are helpful to modeling adverse events in healthcare. I have found that the initial starting conditions, for example -- and again, it depends on what your stop rules are for what you're looking at.

If you're looking at an operation in the operating room, the initial starting conditions -- just like Lorenz's experiments with the weather -- I think that the team dynamics, the status of the technology that you are using, the health of the patient, the quality of the information that you are using, have a lot to do with how this case is going to wobble in and out of the risk space that you're going to be working with for the rest of the day with that patient.

And certainly, nonlinearity, as far as tiny things having enormous effects, it's clear that one little tiny piece of information that you may communicate to a surgeon or a nurse that's incorrect, just like a tin can conversation, these types of communications can wobble through the system and have tremendous effect several days later in a patient's care. And clearly, the way we study adverse events now and try to learn about them for preventability is we study them in isolation. We don't study the connectivity of these types of things. We're just beginning to start to do that.

Higher Reliability Organizational theory is another model that I have found very helpful in trying to design solutions to some of these problems, about how do you understand complex systems and get more control over the output. There are four ingredients in the HRO model that have found to dominate. And we're talking about nuclear powered aircraft carriers, power plants, other types of high hazard industries that have been studied from an ethnographic point of view by organizational theorists.

And the four things that organizations who should have more than their share of disasters and don't, do, is that they have leadership that's really committed to safety as a core value, and they resource that in a sustainable way. They also develop something interesting called the culture of safety. And for those of you who look at the NECSI list server, there has been an interesting discussion in the past few months on culture and cases, culture and complex adaptive systems, which I've found interesting.

A culture of safety is something that's hard to put your finger on. I have a few more slides about it. But I think it is probably the most important component, in that you develop a net of attitudes, beliefs and values throughout the organization, so that everyone is operating with the same decision premises as to how to react flexibly and adaptively to contingencies that occur. And of course, systems designed, redundancy.

For those of you who have been patients in hospital or had loved ones in hospitals, and encountered some of the foolishness in hospitals, it's really astounding that we really are working more in the fifties with nineties technology, as to how poorly some of these things are designed to work together.

The other thing that HROs do is that they're constantly in alerting mode. And I don't want to go into this too much. Some of you may be experts in learning organizations or managerial theory. But a nuclear powered aircraft carrier, for example, is always in a learning mode, both structurally, as far as the way they cross train, and the way they constantly move people in the positions in which they are uncomfortable with what they are doing, so that superiors constantly have to ask juniors what they are doing or how to solve a problem.

And they are constantly using simulation. They are constantly in a training mode. And they are constantly weary that they are going to get bit by something they haven't seen before, that they have not seen the world of experience that they need to do, in order to be error free, that there is always something out there that they haven't experience yet. And they choose to learn from everything that they possibly can in their environments.

Most of the literature in my specialty now in anesthesia and in critical care, has focused on individuals. How can we enhance people's memory? How can we enhance the way people work with a computer interface? How can we create a context so people will be less liable to be distracted? And this is a very limited approach, I think, to getting at what I was showing you before, is that asymptote of adverse events in complex systems.

You can only go so far by taking incredibly bright, motivated, well trained people, even if they have been up two days in a row, sort of the sore thumb of medicine, the way we take call. Even if they are exhausted, even if you had to sleep for twelve hours a day, trying to deal with the limitations of individuals I think is really the short answer to the problem. There are some positive attributes to human cognition, such as our ability for (pana-) recognition and to adapt to new situations.

But Dietrich Dorner and many others have shown that people are typically object oriented. At the point of contact, at the point of care, with hands-on tasks, they tend to not be abstract and not be systems thinkers. And much of our -- many of our complex tasks are routinized. And so, in logic of failure, Dorner discusses simulation experiments that he did where people are allowed to work in these complex cognitive simulations for several days, solving complex problems.

And they usually get bitten by the logarithmic change, and they get bitten by things that happen. For example, if you're running a Sub-Saharan African economy, or if you're a manager in a company that exports say candles to the Far East or something, it's not something that you normally do in your daily work. But given the context for the simulation and given information, it's essentially a fair game for anybody who has the capacity to deal with the problem.

People tend not to anticipate change well. People tend not to anticipate contingency well, maybe become overwhelmed, and they don't anticipate feedback well. And so, and Dorner actually began to show that you can actually use, with the proper curricula and hands-on experiential teaching, you can train people to be better systems thinkers.

Work, in this context, has also evolved over the last hundred/hundred and fifty years. For many hundreds of years, individuals have been perceived as operational units in work. We've abused the human element in work. And only in this century have people been -- mechanistic models of people as cogs in machines began to change in the thirties and forties. And Gibson and other ecological psychologists began to develop ecological models. Can you back it up one more? One more. Thanks.

Began to develop ecological models of work where the task and the person and the tools were looked at as the ingredients for design, as opposed to seeing the person as another tool to be designed around. The character of work has changed considerably to where information is a primacy. For those of you who, again, there may be experts in the audience who do process control, many, many engineers who understand this issue, that people have moved back from the assembly line to become more managers of information.

In healthcare, physicians are really very much hands-on, very much object oriented, very much committed to one-on-one interactions with people. But even in healthcare now, physicians and nurses are becoming more managers for information. This has been a real paradigm shift. And there are no curricula for people in how to deal with this. I think that what we really need is we need a lot of these complex systems ideas to be melded into some new curricula.

The ACGME, the American Council that accredits all residents in graduate medical education in this country, has developed six new competencies that they are going to be actually measuring young physicians on when they graduate from training. And one of them is going to be around this issue of systems thinking and managing information, not just being good physicians from a clinical perspective.

This is a slide from (Yen Raspisin). And I think in this room, it's pretty hard to see. But I think the idea that it conceptualizes is powerful. And what Yen's is depicting is, is that over the last thirty years, there has been a paradigm shift in control theory. Here, we see that there were normative prescriptive theories and models, essentially saying that this is how people screw up, and if people don't follow the rules, you punish them. And through design theories, diagnostic decision making, decision research, organizational theories, if this is the nineties, and this is the sixties, seventies, and eighties, and we're moving in, all of these different fields have been shifting from normative prescriptive models to descriptive models, in terms of deviation from norms.

And now, systems are now being thought in terms of risk spaces or constraints and affordances and opportunities for action, to where the user has become the valuable person in the system, as we realize that these complex socio-technical systems are really complex adaptive systems in which the user is your best defense against an adverse event.

We run into a problem where we often don't even know what is a preventable adverse event, what is an error, and what is a failure of expertise. I mean typically, in the newspaper, when a plane crashes, you always see, pilot error is the first thing. The pilot is always the first one at the scene of the accident. And it's the same in healthcare. But if Mark McGuire hits a single instead of a homerun, is that an error, or is that a failure of expertise? If he strikes out, is that -- I mean, if he hits seventy homeruns, is that good or bad? What does that compare to, what is the norm?

And because we don't make patients and we don't know how they work, and because it's very, very difficult to get the nuanced kind of information that we need about performance, because of medical/legal issues in healthcare, it's very, very, very difficult, if we were try to do a log/log plot of little tiny events, bigger events, and major events, is that we can't find these things. We can't put a seismograph in the hospital like we can out in the desert and count the number of earthquakes, or put a camera and look at the traffic patterns. It's very, very, very difficult to get this kind of data.

The concept that accidents evolve has been around for about ten/fifteen years. And as I made this slide, it struck me, I hadn't even considered this whole idea about evolution in complex adaptive systems. I mean, again, I'm not a student of that field. But evolution means different things to different people. And in a very simplistic, very, very simplistic sense, evolution meaning just change in progress, organizational theorists and clinicians have understood that accidents evolve, but they don't just happen.

There's a lot more thinking, a lot more conceptual thinking that can be applied to this. It's also thought that accidents cascade in the same way that blood clots. So that when blood clots, you have initially a stimulus, you have multiple proteins interacting. And then there are multiple feedback loops and the process really speeds up.

I think the reigning model, the iceberg theory of accidents in complex systems leaves a lot to be desired, because it's very linear. The idea that there are this whole host of inconsequential things, white noise is happening here; I forget something, but it has no impact. Your drug is delivered to you in the hospital an hour or two late, but has no impact, because it's within the degrees of freedom that you can give that medication. And then you have a near miss, which is where I get his medication, because our names are similar. And because the medication isn't toxic, it doesn't hurt me. I just got his medication, but it was a near miss. Had it been something I was allergic to, I might have died. And at the tip of the iceberg, you have very, very few cases where people actually die, or ships go down, or planes crash.

And I think that we need -- again, many of you could probably help us with this -- I think we need much better models than the iceberg model. Although I think there's something conceptually appealing about it. Because when you go out and count these things, there are certainly many, many more inconsequential things than there are accidents. And it's valuable to look at near misses, because there aren't legal implications associated with those, and you can learn more. I think that the beef is probably going to be in some of these nonlinear interactions and the interconnectivity between these things, as to whether or not they break the surface or not.

So as we reframe the error issue, we live in a very judgmental society. We live in a culture of blame, which prevents us from learning. And we're in a "see change" in the way that we're now thinking about this. There's almost a forgiving mode, if you will. We're trying to get away from this concept of human error and think about system competency. And this is a huge thing for policy makers. It's a huge thing for elected officials. We're always looking for the person to blame at the sharp end of the problem. And it's very difficult to take a step back and look at the system as a whole.

We have tons of research on the performance limits of individuals, as far as vigilance, as far as short-term memory and distraction. But we have very little data on system vigilance and very little data on system competency, at least in healthcare. And we're trying to reframe the discussion from one of human error. If I screw up, usually nothing bad happens, because there is a lot of redundancy in the system. And the system is very complex and it adapts to that.

But the death of the patient is really more of an organizational accident. There are things that are happening beyond my control in the system that are setting the context for this to occur. And there are usually multiple breaches of the defenses in the system, and multiple interconnectivity and nonlinear interactions that are bringing these injuries above the surface of the water.

The most powerful thing is probably humans redesigning systems to enhance the performance of both, as opposed to what we've been doing, which is "systematically" dealing with the human error component and isolation.

Well what have other industries done, these HROs, over the last twenty/thirty years, to reach some of these levels, levels of safety? What are some of the things that they've done that we can use in healthcare? And one of the things that we've been looking at carefully in the last few years has been their information gathering mechanisms.

In aviation, there was a, I think it was a USAir flight that almost crashed in Pittsburgh in 1975. And then there was one about three weeks later on a different airline that actually did crash for the same reason: the confusion with the air traffic control; the language problems; the different mental models about the approach; the latent factors in the environment with the terrain; and the confusion in the procedures, etc.

The information was in the system; but the information was not transparent to everybody using that system. And that one incident created -- just like the Betsy Lehman case -- it created a tipping point at the right time. And Congress mandated a confidential near-miss reporting system in aviation. When the FAA tried to start it, nothing happened; nobody subscribed to it, because they didn't want to report to the state police that they were speeding at seventy miles an hour on the way home yesterday. And nobody in their right mind would do that.

So the system was offloaded to NASA as a trusty, squeaky clean, third party that didn't have an ax to grind. And NASA created the aviation safety reporting system for near misses in aviation, which is actually still autonomous, although it's confidential for ten days. For ten days after you file your report, they can get back to you and debrief you and get more narrative data about what actually happened.

In the last thirty years, however, these systems have moved into second and third generation systems where they are no longer autonomous. They are confidential, such as an American Airlines, who has a very, very advanced safety reporting system, which is much too complex to go into. But I was testifying in Washington about three/four weeks ago, about what anesthesia has done to increase system safety.

And Scott Griffith, who is the Chief Safety Pilot at American Airlines, was talking about the ASAP program at American Airlines. And the Committee on Blood Safety and the Surgeon General grilled him for about an hour and a half, because it's just so complex, that they didn't quite get it. But it's really a fascinating way -- it's a fascinating way that a complex socio-technical system has organized itself around learning and control, to be able to get the information out of this black market of people at the sharp end, so that people whose hands are on the technology are actually able to tell, you know, what's happening with the technology, without getting blamed in retrospect for dealing with highly nuanced and fluid situations at the sharp end. And hopefully we'll be seeing some of these systems piloted in medicine, if the lawyers will give us a buy for a few years.

The other two characteristics of systems that are changing are, is that they are focusing more on near misses than actual adverse events. We are always looking for the bodies in the body bag to blame somebody and put it in the front pages. Near misses are pretty boring. Safety is about what hasn't happened. But there are so many near misses, if you look for them, that you can trend them, you can quantitate them, and you can learn a lot more from them.

These are the characteristics of these successful reporting systems. At the end of the day, all reporting systems are voluntary. Massachusetts has a mandatory reporting system. If you look at the state law, you have to report any incident in which a patient was seriously injured or they could have been seriously injured. And I believe this law has been in effect since 1987 or '94, I can't remember. But I think they've got about 800 events in their databank, mostly from complaints from consumers and the media.

And if you look at the Harvard Medical Injury Study that published in '91, the extrapolation being that there are 100,000 deaths per year in the country. Or the study that we did at the Adverse Drug Events Study, that at Mass. General, in six months, there were 1,000 serious preventable adverse drug events. That translates to tens of thousands of events in the state of Massachusetts. And yet they have 800 events over ten years reported. Essentially, nobody reports anything. It's a non-functional statute, and it serves to repress learning, if you will. It's an example of how not to try to manage a complex adaptive system.

And it's useful to think about these elements as far as, if you just sort of configure the users of this system, and the people reporting the information as agents in some sort of a computational model, this is really a very democratic type of approach to how a complex socio-technical system should run and serve its users. Another characteristic of this is, is that patients will never be safe unless the caregivers feel safe. It's an ecology. The safety is a characteristic of the system, if you will. And just like in a family, if there's one member of the family that doesn't feel safe in that family, then the whole family, at some level, is not going to be safe.

This is sort of a visual capture. It's a little bit out of focus here, if I could focus that. What the Brits have done, and what we're starting to do in this country, is to have this nested model for reporting, where at the sharp end here, highly, highly sensitive data that people do not want to report is trapped in a human factors reporting system that is actually like a cell within the organization, to where managers don't even know who is reporting what to whom, but things are being stimulated and fixed locally in sort of a proceduralized way, where there are multiple protections for this data, and people and trust is developed over several years, so that people realize that they can discuss these things without fear.

And then within the organization, this data is surfaced at a higher level for trending and cost and efficiency and etcetera. And then mandated at a national level, there is a mandated system to satisfy the accreditors, where this information loses a lot of its character, and it's highly filtered, so that it can be gathered from multiple airlines and multiple sites, so that very rare events, for example, can be studied, and not in isolation.

So as we move into developing reporting systems in healthcare, it's astonishing how compartmentalized the system is, and how many disincentives there are for the system to learn for itself. The general rules of evidence are incredibly complex. For example, the attorney/client privilege, the attorney/work product protection.

There are even rules such that if I improve my work process tomorrow, because I've stumbled and fell today, there is even a law that says that because I improved my process, that doesn't mean I was negligent when I tripped and fell. In other words, the queue that I improved something, if someone finds out that I improved something, then they can go back next week and say, well he improved it, because he screwed up here. We live in such a blaming culture, that this actually has to be written into legal protection, so that you can't sue somebody for making an improvement.

In the state of Massachusetts, there is also a law on the book that says that if you apologize to someone, if you express to someone that you're sorry -- I'm sorry that your wife died last week -- that that is not an admission of guilt. So there is a law on the books in Massachusetts that says that expressing condolences and empathy is not an admission of guilt. And that's in our legal framework. It's an interesting culture.

So in respect to the fact that I've got a few minutes, I'll try to wrap up and show you a few slides about what I've been doing in simulation for six to eight years. If you frame the simulation, most of you probably work with electrons and computers, the kinds of simulation I do is more with people in immersive environments. It's not virtual reality, but it's a combination of using real tools, computer generated signals, and role-play.

And this is our simulation lab, where we have a six foot long computerized mannequin in the operating room that talks and has all sorts of signals that are generated on these monitors in the operating room that the controller can view and change. This can be automated or run manually. There are headsets in which role players in the environment can use to stimulate certain things to happen in the group that is interacting in the environment. The whole thing is videotaped, and we can go over it later and facilitate a discussion as to what happened.

This is an Israeli soldier that was in training, in tank training in Israel. He died during training. And it was another tipping point, if you will. The paramedics used him to practice invasive medical procedures such as tracheostomy, just after he died. And the family found out about it. And this became a big ethical tipping point as to, what exactly are we doing? How are we learning on people?

We don't have the tools we need to learn. And so we're using indigent patients, unconscious patients, retarded patients, patients who are sedated, dead soldiers, etcetera, and cadavers. And so the push to simulation and to develop simulation tools has accelerated as a result of some of these social events.

This is a simulated emergency room that we developed in the same room you just saw as an operating room. We just put more mannequins in, mixed them up with live patients, and more simulated tools. We now have simulated ultrasound machines. This is in Switzerland at Basel, where they actually have surgeons operating on a simulated mannequin with animal organs in it, and an anesthesia simulator at the top that had been combined. And they can do laproscopic surgery in a simulator environment.

None of them are teaching complexity, per se. But in an informal way -- since I do a lot of the scenario design -- this is a debriefing with senior level clinicians -- I have found that the simulator is actually a neat way to teach complexity theory. Because if you design the scenarios in such a way that all you're teaching is algorithms or proceduralization and rules and how to deal with rare events, that's what you're going to get.

But if you set up the scenarios in such a way that they unfold and you allow people to do whatever they want, when you play the tapes back, you can use in language that is consonant with their domain and where they live and what they understand, you can get them to understand that initial starting conditions are critical, that small things can have huge effects, and big things may have no effect. And they can learn the pedagogy of cause and effect. I think I just have a few more slides.

We also teach them that in complex technology where people have to touch things all of the time and people have to talk to each other all of the time, and the problems are contingent, there is really no way of developing a proceduralized language for everything. And that a lot of our problems come through communication and language, as people begin to define the problem differently, and then it resonates out of control, and people start working on different problems at the same time.

This is a slide that I just had to take from one of the little slots that you get when you fly in a 727, you know, of a model who is clutching the seat and showing you when you put your seatbelt on, you know, how you use the seat as a flotation device. And you know, if you've ever seen the debris on a ocean after a plane crashes, it just struck me that this kind of exemplifies the way that we train people in healthcare today. You know, the hair's not wet, the makeup is on, they're in a swimming pool.

And it has absolutely no relevance to the kind of actual situation that you would have to respond to. In other words, our training for physicians is completely ad hoc. It's not systematic. It doesn't reflect the complex world that they have to function in. And our training doesn't have the requisite variety, and they don't get the skills to deal with the requisite variety in their environment.

This is more of the type of thing that we do in the simulator. Well, thank God we all made it out in time. Of course, now we're equally screwed. So I'll end there. Thank you for your attention. [Applause]

M: Yes, I want to thank you for a most interesting, sensitive, and informative presentation of problems that I think are very serious and complex. The question I have is, the one thing you didn't address is the impact in the last ten years of the shift of power in your industry towards the insurance industry and away from control and decision making by practitioners and providers. And I know that's had a major impact. And I thought being a --

Stephen Small: I think it's had a positive impact and a negative impact. And in many ways, it's had an impact that we don't even understand. But the positive impact is that it's actually forced people to think in terms of systems. It's transformed it from a cottage industry into a system.

The negative impact is that they're trying to institute controls on processes that we don't understand. And I think you may also be referring to the fact that we're mixing up the separation of powers, and that the people, you know, you shouldn't have your hands in the -- if I'm taking care of patients, well, I shouldn't have my hands in their pockets.

And it's pretty clear that in any kind of behavioral theory, your incentives are all screwed up. If I've got a terrible conflict of interest at the level of taking care of an individual patient, that can only cause an ethical problem down the line. I would agree with you there.

But I think that more from the point of view of the context of complexity, I think that the managed care and the systemization of medicine has gone too far down the wrong track of control and not stepping back and saying, we've now gotten so complex that it's no longer controllable, if you will.

One of the examples I think I heard in Seattle at the complexity conference was someone said, try to imagine designing the food supply for Seattle today. You have to buy every frankfurter, every french fry, every institutional dinner, every french bread. You have to plan the food for the city today. And of course, if you try to do that, you get Russia. You get long lines and no food. You try to centralize the control over something that complex, you have to sort of let it happen on its own. And it sounds ridiculous that healthcare should be let to have happen on its own.

But I actually saw last week, there's a new company that's created a website that's allowing patients to find their own healthcare. And it was sort of scary to see that. But what their goal is, is that they want to just go right around the middle managers and managed care and say, okay, you just send in a few dollars a month, and you hit the website, and you can just go out there and find anybody you want that's in our system. And you can look up their accreditation, you can look up their degrees, you can look up how many times that they've been sued, you can look at their websites and see what they know and how they represent themselves. But they're allowing consumers to make their own choices in a free marketplace, if you will.

I mean, I'm oversimplifying it, but I just couldn't -- that analogy between creating the food supply in Seattle and the analogy between trying to design a healthcare system for the United States is, to me, there's a fruitful analogy there.

Yasha Kresh: Thank you very much.

Stephen Small: Thanks. [Applause]

Thomas Deisboeck: Thank you very much, a very interesting perspective. It's difficult to summarize the whole session. It's certainly an extremely interesting study from theoretical modeling, all the way up to very interesting clinical examples. And just now, clearly emphasizing the need for interdisciplinary teams and team education in medicine.

At the end, I think it's very important to state that complex system science will have a tremendous impact in medicine, no question. It still needs conventional science. It's very important for everybody who is scared about the need for interdisciplinary collaborations. As you heard Ary Goldberger and Alan Perelson emphasizing over and over and over again, we need more data, more data, more data. Data are very important for all of these models. It will be a search for multi-quality data analyses and modeling tools.

The next and second big challenge will be the combination of the various models, which will be very difficult, and the creation of new experimental models. We have to try to get as many data from the same experimental setting, if that's a complex systems.

So there's a lot to do for conventional scientists. And there's a need for interdisciplinary collaborations, so the (tone) of that conference anyhow. So thank you very much for all of the speakers. Thanks for making a tremendous session, Yaneer. [Applause]

Yaneer Bar-Yam: So, it's as we opened, so we close. I'm very glad that we were all here. And with this, I will close the 3rd International Conference on Complex Systems. Stay in touch, we'll see you soon. [Applause]